Lecture Notes in Electrical Engineering 57

For further volumes:
http://www.springer.com/series/7818

Dejan Milutinović and Jacob Rosen (Eds.)

Redundancy in Robot Manipulators and Multi-Robot Systems

Springer

Editors
Dr. Dejan Milutinović
Applied Mathematics and Statistics Dept
Baskin School of Engineering
University of California
Santa Cruz, CA
USA

Dr. Jacob Rosen
Computer Engineering Dept
Baskin School of Engineering
University of California
Santa Cruz, CA
USA

ISSN 1876-1100 e-ISSN 1876-1119
ISBN 978-3-642-44139-4 e-ISBN 978-3-642-33971-4
DOI 10.1007/978-3-642-33971-4
Springer Heidelberg New York Dordrecht London

Printed on acid-free paper

Springer is part of Springer Science+Business Media (www.springer.com)

Preface

Over the course of the 50-year history of robotics, we have witnessed the evolution of robotic systems with an increasing trend in the number of degrees of freedom. This trend is visible both in the design of a robot manipulator and in the shift of focus from single- to multi-robot systems. Following the principles of evolution in nature, one may infer that adding degrees of freedom to robot systems design is beneficial. However, since nature did not select snake-like bodies for all creatures, it is reasonable to expect the presence of a certain selection pressure on the number of degrees of freedom. Thus, understanding costs and benefits of multiple degrees of freedom, especially those that create redundancy, is a fundamental problem in the field of robotics and the main motivation for this volume.

Multiple degrees of freedom and redundancy are characteristics of both modern robot manipulators and multi-robot systems. While the reason for introducing additional degrees of freedom and redundancy may be application specific, the presence of redundancy in these systems forms the basis for common research questions. Inspired by this, we organized the workshop "Redundancy in Robot Manipulators and Multi-Robot Systems" at the IEEE/RSJ International Conference on Intelligent Robots and Systems - IROS 2011. This volume is mostly based on the works presented at the workshop and serves as its permanent record. The workshop was envisioned as a dialog between researchers from two separate, but obviously related fields of robotics: one that deals with systems having multiple degrees of freedom, including redundant robot manipulators (serial and parallel), and the other that deals with multi-robot systems.

The volume consists of twelve chapters, each representing one of the two fields. However, to keep the spirit of the dialog between the two fields alive, the chapters are intentionally presented in a mixed order as follows: the first four chapters illustrate benefits and necessary trade offs with respect to the number of degrees of freedom, the number of robots and their heterogeneity; the next two chapters are on hyper redundant robot systems and associated control problems, followed by four chapters about redundant cable robots and the control of multi-robot systems; and the two concluding chapters are about robotic exoskeleton systems.

We conclude this preface expressing our great appreciation to all authors for their valuable contributions to this volume and special thanks to Dragana Milutinović for all her help in preparing this volume.

Santa Cruz, CA

Dejan Milutinović
Jacob Rosen

Contents

Performance of Serial Underactuated Mechanisms: Number of Degrees of Freedom and Actuators

Ravi Balasubramanian and Aaron M. Dollar

Abstract. While underactuated mechanisms have become popular in robot-hand designs because of their passive adaptability, existing systems utilize only one actuator to produce motion in the multiple degrees of freedom in the serial chain of each finger. In this paper, we explore how the performance of an underactuated serial link chain changes as more actuators are added. The fundamental question of what extra capability an additional actuator provides to an underactuated system and how best to implement it has not yet been quantified in the literature. Using a simple linear underactuated mechanism, we show that the performance of a single-actuator system (measured as the average number of contacts made with the environment) quickly plateaus as the number of degrees of freedom of the mechanism is increased. Also, we show that as the number of actuators is increased, the system's passive adaptability improves as the mechanism implementation spreads the actuators across the joints.

1 Introduction

There are two primary approaches in current robotic design to specifying how a robot is actuated. The traditional approach has been to use an actuator for each degree of freedom to produce a fully-actuated system. However, this approach often results in bulky designs requiring complex control algorithms and elaborate sensing modalities for each control input. Recently, underactuated mechanisms that have fewer actuators than degrees of freedom have become popular in robot hand

Ravi Balasubramanian
School of Mechanical, Industrial, and Manufacturing Engineering, Oregon State University, Corvallis, OR
e-mail: ravi.balasubramanian@oregonstate.edu

Aaron M. Dollar
Department of Mechanical Engineering, Yale University, New Haven, CT
e-mail: aaron.dollar@yale.edu

D. Milutinović & J. Rosen (Eds.): Redundancy in Robot Manipulators, LNEE 57, pp. 1–13.
DOI: 10.1007/978-3-642-33971-4_1

research, since the underactuated mechanism's unconstrained freedoms allow it to adapt to environmental constraints without any sensing [4,7,8,10,13,14]. However, most current underactuated hand designs use just one actuator to actuate the many degrees of freedom in a robotic hand. This paper explores how the performance of an underactuated serial link chain changes as more actuators are utilized. We also explore how the routing mechanism used to transmit actuation to the degrees of freedom influences the system's adaptability.

A key performance goal of underactuated robotic hands is to produce power grasps and passively make contact at multiple links, thereby providing the robot the ability to apply forces on the object from multiple contact points and potentially improving grasp stability. This is achieved by routing the force from the single actuator to two [8, 12], three [4–6, 11], and even eleven [10] degrees of freedom in a finger through cable-driven mechanisms or linkage mechanisms (see Fig. 1). These mechanisms permit the distal link to move even after the proximal link makes contact with an object, a property commonly referred to as adaptability [2, 4, 8]. Note that as these fingers are incorporated into a robot hand, these same adaptive mechanisms are used to actuate multiple fingers in parallel as well to provide adaptability between fingers [8, 9, 11]. However, this paper focuses only on the adaptability of a serial underactuated chain.

The limited number of actuators and the uncertainty in object location and shape can lead to undesirable situations in which not all links of the chain make contact with the object. Specifically, a decreased number of contacts results in reduced grasp strength and a reduced ability to resist disturbance forces. Furthermore, unconstrained degrees of freedom (that is, links without contact constraints) permit the hand to reconfigure in response to an external disturbance or internal actuation, which may result in a weakened grasp [1].

In this paper, we explore two specific problems relating to underactuated mechanisms: 1) Given a single-actuator system, how does the system's adaptability vary as the number of degrees of freedom increases? 2) Given a n degree-of-freedom system, how does the performance vary with an increasing number of actuators, and how should those actuators be best routed to the degrees of freedom? Motivated by the robotic grasping problem, for simplicity we will use the number of contacts the system makes with the object on completing the grasping process as the primary performance metric (even though other metrics such as force application capability are important as well). Since the grasping problem in unstructured environments has tremendous uncertainty, we use a probabilistic analysis to quantify system performance.

The majority of underactuated mechanisms described in the robot hand literature utilize revolute degrees of freedom. However, the non-linearity of their kinematics and the variety of contact modes (such as sliding and rolling contact) reduce the generality of the analysis and substantially increase the already large parameter space, which includes joint stiffnesses, transmission mechanisms, and object shapes. To minimize the influence of these factors, our analysis will use a linear underactuated mechanism that we first introduced in [2] (see Fig. 2). The simple geometry of the linear underactuated mechanism's degrees of freedom and contact modes helps retain focus on how the number of actuators influences system performance.

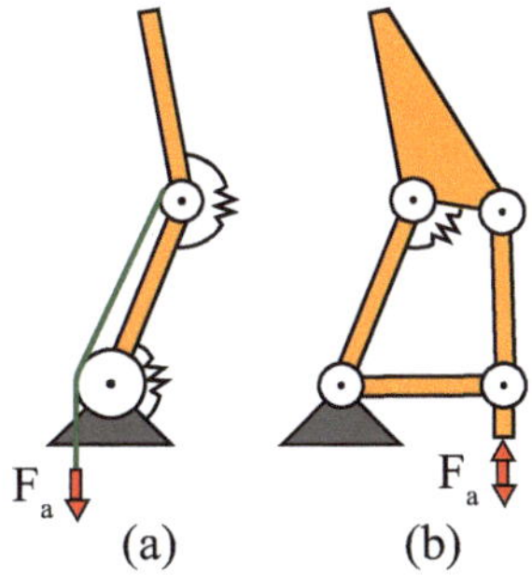

Fig. 1 A schematic of the common underactuated mechanisms used in robot hand designs: a) a cable-driven mechanism (note that the proximal joint is a free-spinning pulley) and b) a linkage-driven mechanism

2 The Linear Underactuated Mechanism

The linear underactuated mechanism studied in this paper operates in a single dimension. Fig. 2 shows a $n = 3$ degree of freedom system, where each degree of freedom is a compliant prismatic joint with unit joint travel. For this system to be underactuated, the $n = 3$ degrees of freedom can be actuated by up to $n-1$ actuators (that is, the number of actuators $m \in \{1,2\}$). Each actuator has some bounded force capability, and the force from the actuator(s) can be transmitted to the degrees of freedom through a variety of transmission mechanisms, which will be explored in section 2.1. Note that the force provided by the single actuator to the distal joint in the mechanism in Fig. 2a causes the proximal and middle joints also to compress in the absence of any contact since the mechanism is linear. Thus, the net displacement of the distal link in free space is three times the displacement of the proximal link.

Each degree of freedom has a "hook" $H_i, i = 1,\ldots,n$, through which it can make contact with the environment, which also has hooks $W_i, i = 1,\ldots,n$. We assume that there exists a hook W_i for each degree of freedom, but its location is not known a priori (Fig. 2c shows the uncertainty in prong location using the thick dotted arrows). Such contacts place constraints on the mechanism's motion. Specifically, if the distal link of the mechanism in Fig. 2a made contact, then the system is locked since the actuator cannot apply any forces on the proximal and middle joints.

With more than one actuator, there exists significant choice in how the actuators can be utilized. For example, the system in Fig. 2b has two actuators, and the system is not locked when the distal link makes contact, since the second actuator can still produce motion in the proximal and middle joints. Note also that the second actuator only causes the distal and middle joints to translate (and not compress) when the mechanism is actuated in free space. Thus, in contrast to the effect of the first actuator, the net displacement of the distal link in free space due to the second actuator is equal to the displacement of the proximal link. Such differences in the capability of actuators inserted at different points in the serial chain have implications

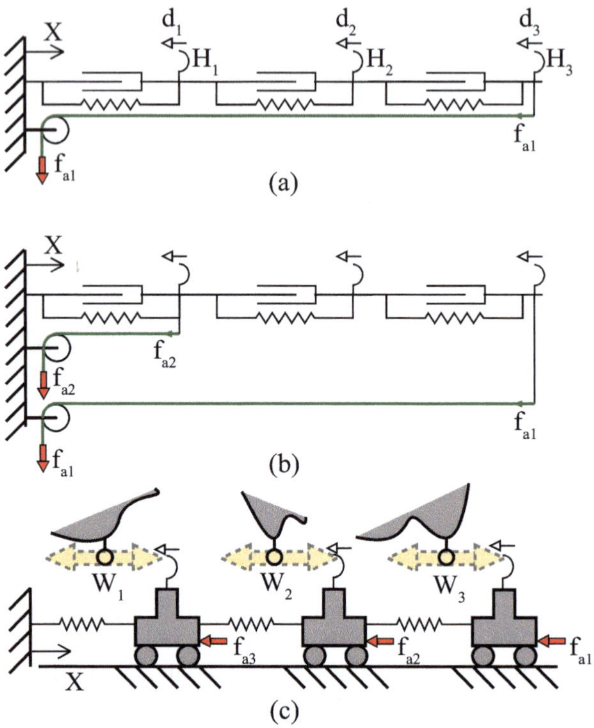

Fig. 2 A three degree-of-freedom linear underactuated system driven by a cable-driven mechanism which routes forces from a) a single actuator (inserted at distal joint) and b) two actuators (inserted at proximal and distal joint). c) A schematic drawing of a linear underactuated system with generalized joint actuation interacting with the environment. The white-headed arrows represent joint travel, the solid (red) arrows forces, and the thick dotted arrows uncertainty in object prong location.

in the grasping process, since joint travel is critical for making contact with the environment. These effects will be discussed in the following sections. Also, the various control policies available to a multi-actuator system will be explored in section 2.2.

2.1 *Transmission Mechanisms*

The transmission mechanism determines the magnitude of forces the actuator can apply at a specific degree of freedom. While the examples shown in Fig. 2 use cable-driven mechanisms, our analysis applies to the use of any actuating mechanism such as linkages or pneumatics. From here on, a joint force will be represented by a force proportional to the actuator force $f_{aj}, j = 1 \ldots m$, which is applied at the joint without explicitly specifying how that force was created. We assume for simplicity that the actuators can only pull, and not push.

The transmission mechanism can be represented as an actuator Jacobian $J_a \in \mathbb{R}^{m\times n}$ such that the joint forces τ can be computed as $\tau = J_a^T f_a$, where $f_a = \begin{bmatrix} f_{a1} & \dots & f_{am} \end{bmatrix}^T \in \mathbb{R}^m$. For the single-actuator mechanism in Fig. 2a, $J_a = \begin{bmatrix} 1 & 1 & 1 \end{bmatrix}$. For simplicity, we assume that the first actuator is always inserted at the distal joint.

For the mechanism in Fig. 2b, $J_a = \begin{bmatrix} 1 & 1 & 1 \\ 1 & 0 & 0 \end{bmatrix}$, where the second row corresponds to the second actuator which is inserted at the proximal joint. Note that the second actuator could have been inserted at a more distal degree of freedom as well (up until the second degree of freedom) resulting in a different actuator Jacobian. Thus, with more than one actuator, there are several transmission mechanisms to choose from. We assume that the two actuators are not routed to the same joint and the higher number actuators are always inserted more proximally.

Also, one can design transmission mechanisms where a particular actuator is routed multiple times to various joints, producing actuator Jacobians that have elements larger than unity (see [2] for examples). In this paper, we will not consider such mechanisms for simplicity. Thus, the number of possible transmission mechanisms with m actuators and n degrees of freedom is equal to the binomial coefficient $\binom{n-1}{m-1}$ (see Table 1).

Table 1 Number of Possible Transmission mechanisms and Control Policies With Six Degrees of Freedom

Number of actuators (m)	Possible transmission mechanisms	Possible control policies*
1	1	1
2	5	3
3	10	13
4	10	75
5	5	541

*for each mechanism.

2.2 Control Policies

With more than one actuator, there are numerous ways in which the actuators can be utilized in the grasping process, each method called a control policy. Each control policy potentially leads to different performance.

One control policy is to use all the actuators simultaneously to produce motion, that is $cp_1 := \{(f_{a1}, f_{a2}, \dots, f_{am})\}$, where the brackets indicate that all the actuators are used simultaneously. If a particular actuator does not produce any more motion because the joint it is inserted at has made contact, then the other actuators continue to produce motion until all actuators do not produce any motion.

Another control policy $cp_2 := \{(f_{a1}), (f_{a2}), \dots (f_{am})\}$ is to use each actuator individually starting from the most distal actuator and then moving to the most proximal actuator. Here the brackets indicate that only one actuator is used at a time.

The switch from one actuator to another occurs only when an actuator produces no further motion (when the joint the actuator is inserted at has made contact) or if that actuator has no remaining force capability to produce motion (the actuator's force being consumed by the joint stiffnesses). Note that when a joint, whose motion is controlled by an actuator, makes contact with the object, only the remaining actuation force is available to produce motion at the other joints that the actuator controls. For example, if the mechanism in Fig. 2a utilized 25% of its actuation force to make contact with the proximal link, only 75% of its actuation force remains to produce motion in the other joints. An alternate control policy $cp_2 := \{(f_{am}), \ldots, (f_{a1})\}$ is to the reverse the order of actuator use and start from an actuator that is inserted most proximally and move sequentially to more distal actuators.

Furthermore, the control policies cp_1, cp_2, and cp_3 can be used recursively with each subset of actuators. Specifically, a control policy $cp_4 := \{(f_{a1}), (f_{a2}, \cdots f_{am})\}$ can use the first actuator individually and then use the remaining actuators simultaneously. Indeed, the number $T_{cp}(m)$ of possible control policies with m actuators grows exponentially as

$$T_{cp}(m) = 1 + \sum_{i=1}^{m-1} \binom{m}{i} T_{cp}(m-i), \tag{1}$$

with each control policy producing potentially different behaviors and $T_{cp}(1) = 1$ (see Table 1).

2.3 Grasping Process

Given an underactuated system with a defined transmission mechanism and a control policy, the grasping process involves executing the control policy completely until all joints can no longer move. With more degrees of freedom, actuators, and a step-by-step control policy such as cp_2 (see section 2.2), the grasping process can occur in multiple stages. Specifically, with cp_2 and the mechanism shown in Fig. 2b, actuator 1 will cause all the joints to compress. If the distal joint first makes contact, then actuator 1 produces no more motion. Then the next actuator in the control policy, actuator 2, is utilized. This causes the proximal joint to compress, produce translation of joint 2 and extension of the springs in joints 2 and 3 until the next contact. If joint 1 makes contact, then the grasping process is complete since there are no more actuators to produce motion. If on the other hand, joint 2 had made contact, then the mechanism can still reconfigure and the grasping process continues until all the joints are locked. Such a grasp is called a power grasp and is a key goal of grasping with underactuated mechanisms.

2.4 Evaluating Grasping Performance

There are several heuristics used in the robotic grasping literature to measure grasp quality [3, 15], but in general the goal of the grasping process is to maximize the number of contacts between the fingers and the object. Each contact would enable the finger to apply an additional force to the object and resist external disturbances.

To model the uncertainty in the grasping process, where the object's exact location and shape (modeled by prong location) are unknown, we simulated many object possibilities in order to quantify a system's average performance on a generic object. Thus, the object prong locations were randomized over the unit joint travel distance (indicated by the thick dotted arrows in Fig. 2c), and each underactuated system attempted to make contact with all the world prongs using the grasping process delineated in section 2.3. The number of contacts that the system made with each randomized object was recorded. For each underactuated system, we compute the probability of occurrence of each contact mode after the grasping process for each randomized object. Then, for each underactuated system, we can compute the average number of contacts expected for a generic object. A system with a greater average number of contacts is rated higher.

2.5 Static Analysis

The linear underactuated mechanism's joint-travel and force-application capabilities during the grasping process can be computed using a statics analysis at each joint. The static balance at each contact state is given by

$$J_a^T f_a + K d_i + J_c^T f_i = 0, \tag{2}$$

where $K \in R^{3\times3}$ represents mechanism's stiffness, $J_c \in R^{3\times3}$ the contact-constraint Jacobian, f_i the contact force at joint i, and d_i the joint travel. In this paper, we assume that the mechanism's stiffness K is an identity matrix even though it is a tunable parameter.

Note that the static equations (2) need to be solved in conjunction with the relevant contact constraints:

Proximal contact $\quad d_1 = 0,$
Middle contact: $\quad d_1 + d_2 = 0,$
Distal contact: $\quad d_1 + d_2 + d_3 = 0.$

In this paper, the primary focus is on the mechanism's contact state at the end of grasping process and not the intermediate contact states or force application capabilities (as was explored in [2]).

3 Results

Using the linear underactuated system presented in section 2, we explored how the grasping performance of a single-actuator system varied as the number of degrees of freedom increased from $n = 2$ to $n = 6$. We also explored how the grasping performance of a $n = 6$ degree of freedom system varies as it is actuated by a differing number of actuators (from $m = 1$ to $m = 5$), different mechanism implementations, and different control policies. The location of each world prong was randomly sampled from a uniform distribution over the unit length of the degree of freedom (five hundred samples), and system performance was averaged across all the instances.

We assumed that all the actuators begin with two units of force, where a single unit of force is sufficient to produce joint motion equal to the length of the joint while overcoming the unit joint stiffness. We verified that at the end of the grasping process, all the joints could no longer move and all the actuators had some remaining actuation force capability, indicating that the mechanism was constrained by external contacts.

Interestingly, we noticed that the different control policies (available for a transmission mechanism with more than one actuator) produced the same number of contacts after the grasping process. Thus, the control policies did not make a difference in the final contact state of the system. The rest of the results section will focus only on how the number of actuators and the transmission mechanism influence system performance.

3.1 *The Single-Actuator System: Performance Variation with Increasing Degrees of Freedom*

Fig. 3 shows the variation in performance of a single-actuator system as the number of degrees of freedom increases from $n = 2$ to $n = 6$. We notice that with two and

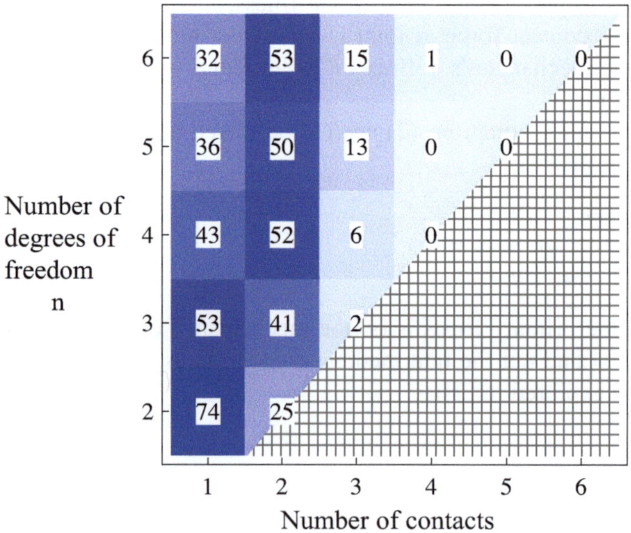

Fig. 3 Expected number of contacts for a single-actuator linear underactuated system with varying degrees of freedom. The numbers in the grid represent the probability (percentage, error in estimates less than 0.8% in all cases) of occurrence of the particular contact mode. Events in the dense checked region are not possible.

three degrees of freedom, a single contact is the most likely contact mode. However, for four and higher degrees of freedom, the double-contact mode has highest likelihood. A weighted average of these results indicate that the expected average number of contacts marginally increases as the number of degrees of freedom increases (see Table 2).

Table 2 Single-Actuator System: Performance Variation with Degrees of Freedom

Number of degrees of Freedom (n)	Expected number of contacts*
2	1.2
3	1.5
4	1.6
5	1.8
6	1.9

*Standard error is less than 0.04 in estimates.

3.2 Fixed Degree-of-Freedom System: Variation with Number of Actuators and Transmission Mechanism

Fig. 4 shows how the performance of a six degree of freedom system varies as the number of actuators that control it are increased from $m = 1$ to $m = 5$ and different

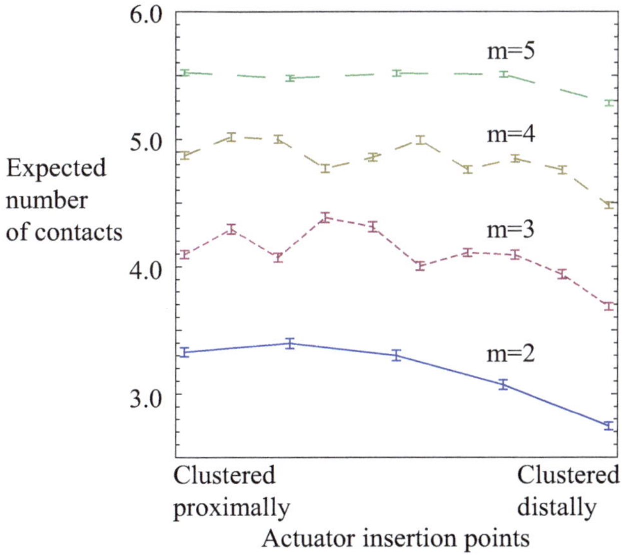

Fig. 4 Expected number of contacts for a $n = 6$ degree-of-freedom linear underactuated system as the number of actuators is increased from $m = 2$ to $m = 5$. The X-axis represents mechanism routing where the insertion points become more distal.

transmission mechanisms are used. As expected, we notice that the expected number of contacts increases with more actuators. However, the transmission system influences performance as well. Specifically, if the actuators are inserted most distally, the expected number of contacts is significantly smaller than the expected number of contacts if the actuators are inserted most proximally. However, for the $m = 3$ and $m = 4$ actuator case, there is some variability in system performance as the insertion point moves distally, and this is discussed in section 4.2.

3.3 *Fixed Degree-of-Freedom System: Best Performance Variation with Increasing Number of Actuators*

Fig. 5 shows how the best performance (across all possible transmission mechanisms) of a six degree of freedom system changes as the number of actuators are increased from $m = 1$ to $m = 5$. We notice that the most likely contact mode is $m + 1$ contacts, except for the five actuator case where the most likely number of contacts is five. Again, these results can be averaged to compute the expected number of contacts with a generic object as the number of actuators increase (see Table 3).

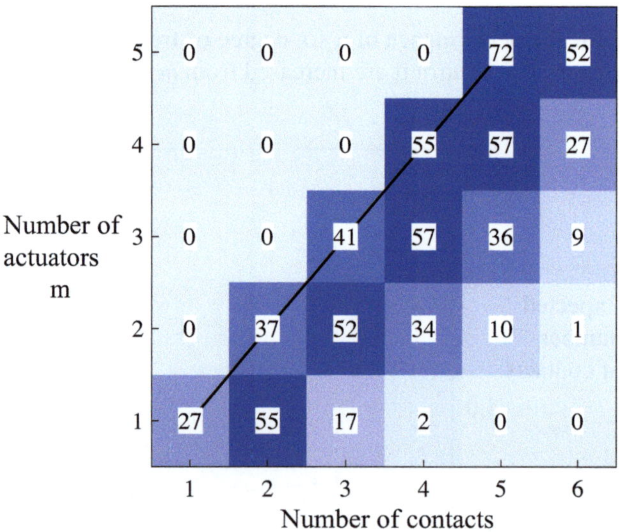

Fig. 5 Best-case expected number of contacts (across all transmission mechanisms) for a $n = 6$ degree-of-freedom linear underactuated system as the number of actuators is increased from $m = 1$ to $m = 5$. The numbers in the grid represent probability (percentage, error in estimates less than 0.42% in all cases) of occurrence of the particular contact mode. The black solid line represents the line where number of contacts equals number of actuators.

Table 3 Fixed Degree-Of-Freedom System: Best Performance Variation with Number of Actuators

Number of actuators (m)	Expected number of contacts*
1	1.9
2	3.2
3	4.1
4	4.8
5	5.4

*Standard error is less than 0.02 in estimates.

4 Discussion

4.1 Limitations of Single-Actuator Systems

From the analysis of the linear underactuated system, we notice that each additional degree of freedom provides only marginal advantage in a single-actuator system (see Table 2 and Fig. 3). A similar analysis is also required for the planar underactuated systems prevalent in robot hand research. While prior research has shown that power grasps with multiple contacts between the finger and the object are possible (by carefully designing joint compliances) even when the single actuator controls many degrees of freedom, they include assumptions about hand placement (for example, object pushed against the hand's palm) and object shape (for example, spherical shape). The expected performance of these robot hands in terms of the number of contacts the mechanism makes on average across all possible objects is still unclear. To our knowledge, this is the first work that explores the variation in expected number of contacts for a single actuator serial chain with increasing degrees of freedom.

We would also like to point out a key difference between the linear underactuated system and the planar underactuated systems that utilize revolute joints. Since the linear underactuated system has only one dimension, a contact on one joint can nullify the actuation force on another joint. In contrast, actuation forces can still be transferred to the revolute joints after the planar systems make contact with the environment because of the rotational joint kinematics. This depends on the specific contact modes (rolling versus sliding) which will determine the mechanism's ability to reconfigure [1]. Thus, these factors that influence the mechanism's adaptability must be kept in mind when analyzing the performance of planar underactuated systems.

4.2 Multi-actuator Systems

As expected, a system with more actuators has greater adaptability irrespective of how the actuation is routed (see Fig. 4). However, the control policy, or the sequence of actuator utilization, does not make a difference in terms of the number of contacts the mechanism makes with the object after the grasping process. This is a surprising and useful result which indicates that even with multiple actuators, the robotic hand can use the actuators in any order to produce an identical final contact mode. However, it was still useful to explore the various control policies in section 2.2 to show the completeness of our methods. However, the contact sequence does indeed depend on the sequence of actuator use. If the contact sequence is important (for example, making contact sequentially from the proximal link to the distal link to ensure an enveloping grasp), then the robot must plan the sequence of actuator use as well.

In terms of mechanism routing, we notice that a system where the actuator insertion points are spread across the serial chain improve mechanism adaptability. This is particularly noticed in the poor performance of the systems where all the actuators are inserted most distally. In these cases, contacts at the distal joints would completely lock the system. Interestingly, with $m = 3$ and $m = 4$ actuators, as the routing becomes more distal there is some variability in performance. This is because of the poor performance of mechanisms where the insertion points of the higher-order actuators are adjacent when compared with mechanisms where the insertion points of the higher-order actuators are spread apart. We also notice that the best performance with m actuators increases linearly until saturating at $n-1$ actuators. Thus, there is a linear advantage with each additional actuator. Future work includes an analysis of a multi-actuator system's ability to individually control the contact forces and permit object manipulation.

References

1. Balasubramanian, R., Belter, J.T., Dollar, A.M.: External disturbances and coupling mechanisms in underactuated hands. In: Proc. of ASME Internat. Design Engineering Technical Conf. and Computers and Information in Engineering Conf. (2010)
2. Balasubramanian, R., Dollar, A.M.: A comparison of workspace and force capabilities between classes of underactuated mechanisms. In: Proc. of the IEEE Internat. Conf. on Robotics and Automation, pp. 3489–3496 (2011)
3. Balasubramanian, R., Xu, L., Brook, P., Smith, J.R., Matsuoka, Y.: Human-guided grasp measures improve grasp robustness on physical robot. In: Proc. IEEE Internat. Conf. on Robotics and Automat., pp. 2294–2301 (2010)
4. Birglen, L., Laliberté, T., Gosselin, C.: Underactuated Robotic Hands. Springer (2008)
5. Carrozza, M.C., Suppo, C., Sebastiani, F., Massa, B., Vecchi, F., Lazzarini, R., Cutkosky, M.R., Dario, P.: The spring hand: Development of a self-adaptive prosthesis for restoring natural grasping. In: Autonomous Robots, pp. 125–141 (2004), doi:10.1023/B:AURO.0000016863.48502.98
6. Crisman, J.D., Kanojia, C., Zeid, I.: Graspar: A flexible, easily controllable robotic hand. IEEE Robotics and Automation Magazine, 32–38 (1996)
7. Crowder, R.M., Dubey, V.N., Chappell, P.H., Whatley, D.R.: A multi-fingered end effector for unstructured environments. In: Proc. IEEE Internat. Conf. on Robotics and Automation, pp. 3038–3043 (1999)
8. Dollar, A.M., Howe, R.D.: The highly adaptive SDM Hand: Design and performance evaluation. Internat. J. Robotics Res. 29(5), 585–597 (2010)

9. Hanafusa, H., Asada, H.: Stable prehension by a robot hand with elastic fingers. In: Robot Motion: Planning and Control, pp. 323–336. MIT Press, Cambridge (1982)
10. Hirose, S., Umetani, Y.: The development of soft gripper for the versatile robot hand. Mechanism and Machine Theory 13, 351–359 (1978)
11. Kamikawa, Y., Maeno, T.: Underactuated five-finger prosthetic hand inspired by grasping force distribution of humans. In: Proc. IEEE/RSJ Internat. Conf. on Intell. Robots and Sys., pp. 717–722 (2008)
12. Kaneko, M., Higashimori, M., Takenaka, R., Namiki, A., Ishikawa, M.: The 100G capturing robot too fast to see. ASME/IEEE Trans. on Mechatronics 8(1), 37–44 (2003)
13. Rakic, M.: Multifingered hand with self-adaptability. Robotics and Computer-Integrated Manufact 3(2), 269–276 (1989), doi:10.1016/0736-5845(89)90074-4
14. Rovetta, A.: On functionality of a new mechanical hand. ASME J. Mechanical Design 103, 277–280 (1981)
15. Shimoga, K.B.: Robot grasp synthesis algorithms: A survey. Internat. J. Robotics Res. (1996), doi:10.1177/027836499601500302

Low-Cost Multi-robot Localization

Amanda Prorok, Alexander Bahr, and Alcherio Martinoli

Abstract. Localization is an enabling technology, and a prerequisite for a wide range of robotic tasks. Despite the large amount of work already done in this domain, to date, the solution to the localization problem for fully decentralized, large-scale multi-robot systems is still an open question. In this chapter, we contribute to this particular problem outline by proposing a low-cost method: we describe a fully decentralized algorithm, particularly designed for resource-limited robotic platforms in large-scale systems. In the following sections, we elaborate the components of our method, and demonstrate the utility of our low-cost localization algorithm on groups of up to ten real mobile robots. This chapter is rounded off by bringing our approach into a larger perspective, and by discussing its potential as well as its limitations.

1 Introduction

A variety of tasks performed by multi-robot systems such as search and rescue [12, 13], environmental monitoring [5, 25], and construction of real structures [16, 32] need accurate localization to succeed. Due to the intrinsic nature of such tasks, the individual agents are often confined to a small size and weight, which sets hard limits on on-board resources. Simultaneously, a large portion of the robot's resources may be dedicated to the task at hand, especially when this task requires high-frequency perception-to-action loops, leaving little room for solving the localization problem. These compounding problems pose the challenge of designing systems and algorithms that can flexibly accommodate given restrictions, without compromising performance.

Amanda Prorok · Alexander Bahr · Alcherio Martinoli
Distributed Intelligent Systems and Algorithms Laboratory, School of Architecture, Civil and Environmental Engineering, Ecole Polytéchnique Fédérale de Lausanne (EPFL), Switzerland
e-mail: {amanda.prorok,alexander.bahr, alcherio.martinoli}@epfl.ch

D. Milutinović & J. Rosen (Eds.): Redundancy in Robot Manipulators, LNEE 57, pp. 15–33.
DOI: 10.1007/978-3-642-33971-4_2

This chapter presents a concise solution to the localization problem for a collaborative team of mobile robots. Various strategies have been followed in past works on collaborative localization—our work distinguishes itself by respecting the following design goals:

Low-cost: The time/energy spent on the localization algorithm must be inferior to that spent on the actual task at hand. Thus, we try to minimize the overall complexity of our algorithm, and simultaneously relax the communication requirements.

Full decentralization: Each robot carries responsibility for its own localization, and runs an independent localization algorithm on-board.

Any-time relative observations: Robot-to-robot observations can be made asynchronously, at any given time. This simultaneously means that there are no connectivity constraints on the robot team, and that the computational time of fusing relative observations with proprioceptive sensing is bounded.

Mobility: Since our system is decoupled and decentralized, we do not constrain mobility by making use of any methods that rely on motion agreements among the robots.

Independence of the environment: In order for our method to be equally suited for indoor and outdoor applications, in structured as well as unstructured environments, it should be self-contained and robust. Thus, we rely only on inter-robot relative sensing, and on the possibility of an initial localization (of one of the robots).

Given its efficiency in solving localization problems for unknown initial conditions and its efficiency in accommodating arbitrary probability density functions, our method of choice is the particle filter. We thus build on the general probabilistic framework of Monte-Carlo Localization (MCL) presented in [7]. In particular, our collaboration strategy exploits associated, inter-robot relative range and bearing observations. In order to accommodate the noise characteristics of typical relative range and bearing measurements, we develop a robot detection model, which is introduced into our localization algorithm. This combination forms the basis of our collaborative paradigm. Given this foundation, the key element of our approach consists of an additional routine, namely a *reciprocal* particle sampling routine, mainly designed to accelerate the convergence of a robot's position estimate (to the correct value), and to mitigate overconfidence. A collaborative localization algorithm composed of the aforementioned robot detection model jointly with the reciprocal sampling routine is very efficient with respect to its non-collaborative counterpart. However, due to the computational overhead induced by the detection model and the reciprocal sampling routine (which scale to the square of the number of particles), such an algorithm may run into real-time running constraints. This can turn out to be particularly prohibitive for platforms with hard limits on available resources. For this reason, we further extend our approach with a *particle clustering* method that reduces the complexity of the overall localization algorithm and also reduces the amount of data to be communicated. This clustering routine is especially designed to accommodate the characteristics of the range and bearing robot detection model, and does not impose an additional computational burden on the localization algorithm as a whole.

1.1 Related Work

A synopsis of currently available work on multi-robot localization promotes a division of the various approaches into two main categories: a *multi-centralized* approach, and a *decentralized* approach. The *multi-centralized* approach distinguishes itself formally from the *decentralized* approach by imposing that each robot in the team maintains a state vector containing the poses of all robots—in other words, each robot maintains a full-system state estimate, versus an estimate of only its own pose.

The *multi-centralized* [23] approach is indeed the more popular (and classical) approach, as it enables the robots to directly take account of inter-robot dependencies and to estimate correlations. However, it entails some inconveniences. In an early work, Roumeliotis et al. [30] enable the distribution of a Kalman estimation scheme by constructing communicating filters, which allows team-members to propagate their state and covariance estimates independently. Yet, as covariance matrix updates occur during each update step and require information exchange between all robots and a centralized processor, the method is particularly vulnerable to single-point failures. In particular, the requirement to update the information in all robots after a single observation of an individual robot assumes a communication infrastructure without any packet loss. The method scales in $O(N^3)$ with respect to the number of robots, and thus limits its scalability due to the high computational cost. In [19], Martinelli et al. propose an extension to [30], which relaxes the assumptions on relative observations, but without further improving the algorithm's scalability and cost. Howard et al. [10] propose an algorithm based on maximum likelihood estimation, and validate it on a team of four real robots. Their method relies on periodical information broadcasts, and it is unclear how the method scales and how sensitive it is to local minima. In a recent work, Nerurkar et al. [24] address the reduction of computational complexity and single-point failures by implementing a maximum a posteriori estimation method. Nevertheless, the $O(N^2)$ computational cost is significant. Also, the proposed method requires synchronous communication among the robots, and its feasibility still remains to be validated on real robots. Mourikis et al. [21] consider the problem of resource-constrained collaborative localization with the goal of deriving optimal sensing frequencies. Yet, as exteroceptive data is dealt with in a centralized way, the sensing frequencies inevitably decrease with an increasing number of robots, thus limiting the scalability of the approach. Cristofaro et al. [6] present a localization algorithm that arguably alleviates the problems described above. The approach is based on an extended information filter, whose implementation is distributed over the robot team members. However, its computational cost increases for each new observation made and it assumes bidirectional synchronous communication, the feasibility of which remains to be evaluated on real robots. Finally, Leung et al. [15] develop a framework based on

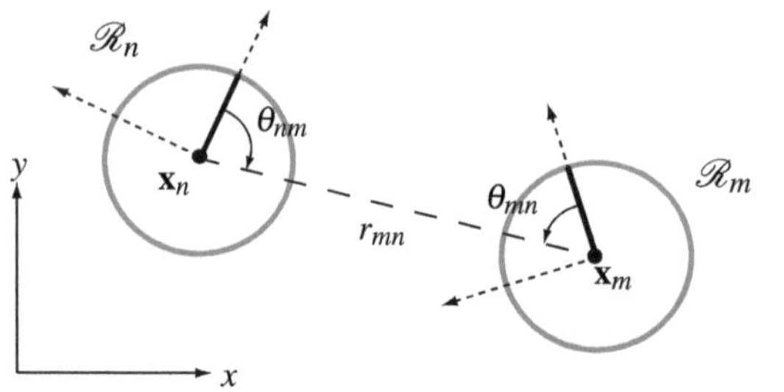

Fig. 1 System of two robots $\mathscr{R}_n$ and $\mathscr{R}_m$ at positions $\mathbf{x}_n$ and $\mathbf{x}_m$, respectively, sharing a common localization frame. The figure illustrates the robots' relative range ($r_{nm} = r_{mn}$) and bearing values (θ_{nm} and θ_{mn}).

'checkpoints' which facilitates decentralization of a given localization algorithm. Their method, however, still aims to maintain full-system state estimates on all robots, and remains to be evaluated on real robots.

The category of work representing the *decentralized* approach has an alternative take on the collaborative localization problem: each robot maintains an estimate of only its own pose, and fuses relative observations in an opportunistic fashion. Fox et al. [7] first introduced a multi-robot Monte-Carlo localization algorithm for *global* localization, that also relaxes noise assumptions as well as inter-robot dependencies. They propose a method in which robots mutually synchronize their position beliefs upon detection, and show successful global localization on two real robots. However, the method has limited scalability due to overconfidence occurring upon multiple robot detections, and no analysis is provided of the algorithm's processing requirements. Bahr et al. [1] develop a decentralized localization algorithm, based on the extended Kalman filter framework, that is especially well suited for autonomous underwater vehicles with very low data rates. This method, however, allows cyclic updates and, thus, may suffer from overconfidence. In an addition to this work [2], the authors remedy the overconfidence problem, but at the cost of a computationally expensive solution (in particular for a large number of robots and a high frequency of relative observations).

1.2 Problem Formulation

Let us consider a multi-robot system of N robots $\mathscr{R}_1$, $\mathscr{R}_2$, ..., $\mathscr{R}_N$, in a 2D space, where the number N does not necessarily need to be known by the robots (see Figure 1 for a schematic illustration of a two-robot system). For a robot $\mathscr{R}_n$, at time t, the pose $\mathbf{x}_{n,t}$ is given by the Cartesian coordinates $x_{n,t}, y_{n,t}$ and orientation $\phi_{n,t}$. Also, at time t, a robot $\mathscr{R}_m$ is in the set of neighbors $\mathscr{N}_{n,t}$ of robot $\mathscr{R}_n$ if robot $\mathscr{R}_m$ is able to take a range measurement $\tilde{r}_{mn,t}$ and bearing measurement $\tilde{\theta}_{mn,t}$ of robot $\mathscr{R}_n$. Thus, at every moment in time, the neighborhood topology is defined by the physical constraints given by the relative observation sensors deployed on the robots. Also, if $\mathscr{R}_m \in \mathscr{N}_{n,t}$, we make the assumption that the robot $\mathscr{R}_m$ can communicate with the robot $\mathscr{R}_n$. Apart from a sensing modality that enables the robots to observe inter-robot range and bearing (including a unique robot identifier), they are also equipped with a dead-reckoning self-localization module (e.g., odometry), but do not make use of any exteroceptive sensors capable of feature recognition.

As introduced earlier in this text, every robot runs its own, self-contained, collaborative particle filter, with the goal of localizing itself without any prior knowledge of the initial state or previous measurements. In practice, we assume that one of the robots is localized at the start of an exercise. It turns out that, as time evolves, our method 'propagates' the correct position belief from robot to robot with help of the relative positioning sensors, and that at some point in time all robots are localized (with respect to an upper error bound). The belief of a robot's pose is formulated as

$$\mathbf{Bel}(\mathbf{x}_{n,t}) \sim \{\langle \mathbf{x}_{n,t}^{[i]}, w_{n,t}^{[i]} \rangle | i = 1, ..., M\} = X_{n,t} \tag{1}$$

where M is the number of particles, $\mathbf{x}_{n,t}^{[i]}$ is a sample of the random variable $\mathbf{x}_{n,t}$ (the pose), and $w_{n,t}^{[i]}$ is its weight (or importance factor). The symbol $X_{n,t}$ refers to the set of particles $\langle \mathbf{x}_{n,t}^{[i]}, w_{n,t}^{[i]} \rangle$ at time t belonging to robot $\mathscr{R}_n$. This context formalizes the scope of this chapter: the method that we detail in the following sections solves the localization problem for large robot teams by exploiting collaboration.

We note that the nature of this problem scenario relates well to current real-world scenarios. In particular, in environments where it is hard or even impossible to get a GPS position update, such as underwater or inside buildings, it is always possible to exploit the mobility of one of the team mates to move into a GPS-friendly environment. In underwater robot teams [1], a robot can surface to get a GPS update. Or similarly, in search and rescue robot teams [12], a robot can navigate to the exit of a building. Several authors also comment on the advantages of heterogeneity in robot teams. Bahr et al. [1] note that for optimal localization, it is advantageous to have a few team members that are able to maintain an accurate estimate of their position through sophisticated dead-reckoning sensors, thus enabling a much larger group of robots with less sophisticated sensors to maintain an accurate position. In the same line of thought, Madhavan et al. [18] argue that when the quality of the measurements from absolute positioning sensors deteriorates for certain robots in the team, or if some of the team members do not possess absolute positioning capabilities, those robots can take advantage of other team members with complementary positioning capabilities.

1.3 Case Study

To give the reader a feel for our algorithms, we perform several experiments on a team of Khepera III robots[1] [27]. The Khepera III robot (see Figure 2) has a diameter of 12 cm, making it appropriate for multi-robot experiments in controlled environments. It has a KoreBot extension board providing a standard embedded Linux operating system on an Intel XSCALE PXA-255 processor running at 400 MHz, and uses a communication infrastructure enabled through an IEEE

---

[1] `http://www.k-team.com/`

Fig. 2 Fleet of ten Khepera III robots. The robots are all equipped with an inter-robot relative range and bearing module, which is composed of a ring of 16 infrared light emitting diodes (LEDs).

802.11b wireless card which is installed in a built-in CompactFlash slot. In order to measure the ground truth positioning to evaluate our algorithms, we installed an overhead camera system as detailed in [27], in combination with the open source tracking software SwisTrack [17]. This system allows us to monitor our robots in real-time with a mean error of about 1 cm and a maximum error below 3 cm. The robots are equipped with wheel encoders and use odometry for self-localization. Each robot also uses a relative range and bearing module [29], which provides the relative observations used by the robot detection model. Figure 2 shows ten robots equipped with a relative positioning module. In our experimental space, the boards have a proportional, additive Gaussian range noise with a standard deviation of $\sigma_r = 0.15 \cdot r_{mn}$, and a bearing noise of $\sigma_\theta = 0.15$ rad. In the following, we will discuss the localization performance in terms of the mean positioning error of all particles in the robots' beliefs with respect to the ground truth positions obtained from the overhead camera system. This metric implicitly includes the spread of the particle positions, and thus also represents the uncertainty of the position estimate.

2 Collaborative Localization

In this section, we elaborate our collaborative localization algorithm [26], which, together with the Monte-Carlo Localization (MCL) method presented in [7], forms the baseline for our work. For convenience, the complete localization algorithm is shown in Algorithm 1.

2.1 Multi-robot Monte Carlo Localization

Let us from here on consider a robot $\mathscr{R}_n$ that is detected by robot $\mathscr{R}_m$, and simultaneously receives localization information from robot $\mathscr{R}_m$. If we make the assumption that individual robot positions are independent, we can formulate the update of the belief of robot $\mathscr{R}_n$ at time t with

$$\mathbf{Bel}(\mathbf{x}_{n,t}) = p(\mathbf{x}_{n,t}|u_{n,0..t}) \cdot \int p(\mathbf{x}_{n,t}|\mathbf{x}_{m,t}, r_{mn,t}, \theta_{mn,t})\mathbf{Bel}(\mathbf{x}_{m,t})\,\mathrm{d}\mathbf{x}_{m,t} \tag{2}$$

where $u_{n,0..t}$ is the sequence of motion control actions up to time t. For such a collaboration to take place, robot $\mathscr{R}_m$ needs to communicate its range and bearing measurements $\tilde{r}_{mn,t}$, $\tilde{\theta}_{mn,t}$ and $\mathbf{Bel}(\mathbf{x}_{m,t})$ to robot $\mathscr{R}_n$. Thus a communication message is composed as $d_{mn,t} = \langle \tilde{r}_{mn,t}, \tilde{\theta}_{mn,t}, X_{m,t} \rangle$. If several robots in a neighborhood $\mathscr{N}_{n,t}$ communicate with robot $\mathscr{R}_n$, the received information is the set of all relative observations made by those robots of robot $\mathscr{R}_n$ at time t, as well as the belief representations $X_{m,t}$ of all detecting robots $\mathscr{R}_m \in \mathscr{N}_{n,t}$. We denote this data set as $D_{n,t} = \{d_{mn,t} | \mathscr{R}_m \in \mathscr{N}_{n,t}\}$. We note that the collaborative aspect of this formalism lies in the integration of robot $\mathscr{R}_m$'s belief into that of robot $\mathscr{R}_n$ (this update step is shown in Algorithm 1 in line 5). As previously discussed in [7], there are certain limitations to this approach. Due to the fact that robot $\mathscr{R}_m$ integrates its position belief into that of robot $\mathscr{R}_n$ upon detection, subsequent detections would induce multiple integrations of this belief, ultimately leading to an overconfident (and possibly erroneous) belief of the actual pose. Fox et al. remedy this shortcoming by considering two rules: *(i)* their approach does not consider negative sights (no detection) of other robots, and *(ii)* they define a minimum travel distance which a robot has to complete before detecting a same robot again. Although rule *(i)* is a practical consideration, rule *(ii)* limits the scalability and robustness of the approach. In fact, it does not respect our design goals of full mobility and any-time observations (see Sec. 1). We will see in the following sections how our approach tackles this problem by exploiting a reciprocal sampling method.

Algorithm 1. `MultiRob_Recip_MCL`$(X_{n,t-1}, u_{n,t}, z_{n,t}, D_{n,t})$

1: $\overline{X}_{n,t} = X_{n,t} = \emptyset$
2: **for** $i = 1$ to M **do**
3: $\quad \mathbf{x}_{n,t}^{[i]} \leftarrow \texttt{Motion\_Model}(u_{n,t}, \mathbf{x}_{n,t-1}^{[i]})$
4: $\quad w_{n,t}^{[i]} \leftarrow \texttt{Measurement\_Model}(\mathbf{x}_{n,t}^{[i]})$
5: $\quad w_{n,t}^{[i]} \leftarrow \texttt{Detection\_Model}(D_{n,t}, \mathbf{x}_{n,t}^{[i]}, w_{n,t}^{[i]})$
6: $\quad \overline{X}_{n,t} \leftarrow \overline{X}_{n,t} + \left\langle \mathbf{x}_{n,t}^{[i]}, w_{n,t}^{[i]} \right\rangle$
7: **end for**
8: **for** $i = 1$ to M **do**
9: $\quad r \sim \mathscr{U}(0,1)$
10: $\quad$ **if** $r \leq (1-\alpha)$ **then**
11: $\quad\quad \mathbf{x}_{n,t}^{[i]} \leftarrow \texttt{Sampling}(\overline{X}_{n,t})$
12: $\quad$ **else**
13: $\quad\quad \mathbf{x}_{n,t}^{[i]} \leftarrow \texttt{Reciprocal\_Sampling}(D_{n,t}, \overline{X}_{n,t})$
14: $\quad$ **end if**
15: $\quad X_{n,t} \leftarrow X_{n,t} + \left\langle \mathbf{x}_{n,t}^{[i]}, w_{n,t}^{[i]} \right\rangle$
16: **end for**
17: return $X_{n,t}$

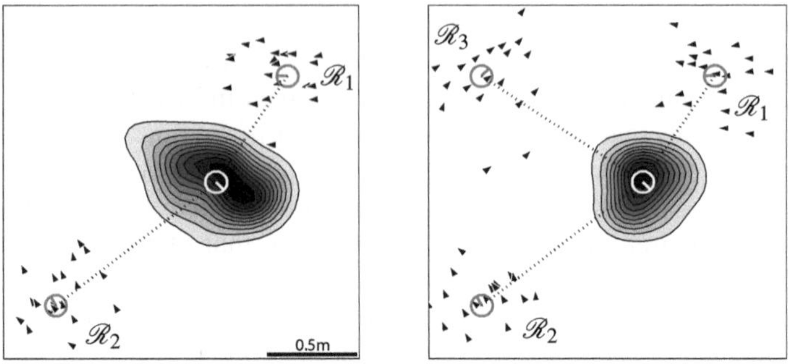

(a) (b)

Fig. 3 Example application of the detection model for multiple detecting robots (a) for two robots and (b) for three robots. Here, a set of 20 particles is shown, represented by oriented triangles superimposed over the detecting robots $\mathscr{R}_1$, $\mathscr{R}_2$, and $\mathscr{R}_3$. The detected robot is shown in white. The model's probability density is superimposed on the detected robot. The dotted line and the orientation of the robots show the actual relative range and bearing. The particle positions were generated randomly from a normal distribution ($\sigma_x = \sigma_y = 0.2$ m, and $\sigma_\phi = 0.2$ rad), and range values are perturbed by an additive Gaussian noise with $\sigma_r = 0.15$ and for the bearing values with $\sigma_\theta = 0.15$ rad.

2.2 *Range and Bearing Detection Model*

The detection model $p(\mathbf{x}_n|d_{mn})$ describes the probability that robot $\mathscr{R}_m$ detects robot $\mathscr{R}_n$ at pose $\mathbf{x}_n = [x_n, y_n, \phi_n]^\top$, given the detection data d_{mn}. This probability density function is applied to the ensemble of particles in the belief of robot $\mathscr{R}_n$, in order to adjust their weights to current relative observations. Given the nature of relative observations, we make use of a locally defined polar coordinate system. Hence, we define the transformation from Euclidean to polar coordinates $\mathbf{T}_e^p(\mathbf{x}_q, \mathbf{x}_p)$ as

$$\mathbf{T}_e^p(\mathbf{x}_q, \mathbf{x}_p) = \begin{bmatrix} r_{qp} \\ \theta_{qp} \end{bmatrix} \tag{3}$$

where

$$r_{qp} = \sqrt{(x_p - x_q)^2 + (y_p - y_q)^2} \tag{4}$$

$$\theta_{qp} = \text{atan2}((y_p - y_q), (x_p - x_q)) - \phi_q \tag{5}$$

and $\mathbf{x}_q$ defines the center of the local polar coordinate system. Thus, assuming Gaussian noise and knowledge of the range and bearing standard deviations (σ_r and σ_θ, respectively), and the independence of range and bearing measurements, the detection model is

$$p(\mathbf{x}_n|d_{mn}) = \eta \cdot \sum_{\left\langle \mathbf{x}_m^{[i]}, w_m^{[i]} \right\rangle \in X_m} \Phi\left(\mathbf{T}_e^p(\mathbf{x}_m^{[i]}, \mathbf{x}_n); \boldsymbol{\mu}, \Sigma\right) \cdot w_m^{[i]} \tag{6}$$

where $\Phi(\cdot;\boldsymbol{\mu},\Sigma)$ is a multivariate normal probability density function with mean $\boldsymbol{\mu} = [\tilde{r}_{mn,t}, \tilde{\theta}_{mn,t}]^\top$ and where η is a normalization constant. The covariance matrix is $\Sigma = \text{diag}([\sigma_r^2, \sigma_\theta^2])$ (the work in [8] provides experimental evidence for our platform showing that a range and bearing measurement behaves like two independent Gaussian variables). As seen in [26], the detection model can easily be augmented by an additional component in case robot $\mathscr{R}_n$ reciprocally detects robot $\mathscr{R}_m$. Here, for the purpose of our case-study, we use a simple Gaussian distribution in polar coordinates, but all reasonings are valid for completely arbitrary distributions. Indeed, since we use a particle filter, we can keep the same framework for any possible underlying range and bearing hardware not fulfilling the Gaussian noise assumption.

Finally, the detection model incorporating the detection data from multiple detecting robots can be formulated as the update equation shown in Algorithm 2. Figure 3 illustrates the probability density function resulting from the detection model, (a) for two detecting robots, and (b) for three detecting robots. We notice that when detection data from multiple robots is integrated into the range and bearing model, the detection precision increases.

Algorithm 2. `Detection_Model`$(D_{n,t}, \mathbf{x}_t^{[i]}, w_t^{[i]})$

1: $w \leftarrow w_t^{[i]} \cdot \prod_{d_{mn} \in D_{n,t}} p(\mathbf{x}_t^{[i]} | d_{mn})$
2: return w

2.3 *Reciprocal Sampling*

In addition to using a robot detection model for updating the belief representation $\mathbf{Bel}(\mathbf{x}_{n,t})$, our approach relies on a *reciprocal* sampling method. Let us refer to the iterative process described in Algorithm 1: instead of sampling a new particle pose $\mathbf{x}_{n,t}^{[i]}$ from $\mathbf{Bel}(\mathbf{x}_{n,t-1}^{[i]})$ in line 11, the reciprocal MCL routine in line 13 samples from the detection model $p(\mathbf{x}_n | d_{mn})$, according to Eq. 6. Thus, samples are drawn at poses which are probable given reciprocal robot observations, and which are independent of the previous belief $\mathbf{Bel}(\mathbf{x}_{n,t-1})$. By defining a reciprocal sampling proportion α, particles are sampled from the robot's own belief with a probability $1-\alpha$, and with a probability of α from the probability density function proposed by the detection model. The advantages of this procedure are twofold. Firstly, as the reciprocal sampling method exploits the information available in a whole robot team, it continuously creates particles in areas of the pose space which are likely to be significant, and thus it allows for very small particle set sizes (also shown in [26]). Secondly, by sampling new particles from the detection model, the method introduces a variance proportional to that of the relative detection sensors into the belief of the detected robot (this proportion can be tuned by varying α), and effectively mitigates overconfidence. Algorithm 3 shows the routine where line 4 represents the sampling step. There are a multitude of methods which can be applied to sample from a given distribution. In our particular case (multi-modal Gaussians), sampling from the detection model $p(\mathbf{x}_n|d_{mn})$ is cheap. For more complex probability density functions, sophisticated and efficient methods such as slice sampling [22] can be employed.

The idea of extending standard MCL with additional sampling methods was first shown in [31]. The resulting algorithm named *Mixture* MCL was shown to increase the robustness of single-robot global localization. Our method differs from that one in that it extends to collaborative multi-robot localization algorithms by sampling from the detection model of one or several mobile robots (whose positions are initially unknown) as opposed to sampling from the detection model of a potentially large set of static environmental features (whose positions have to be known or mapped a priori). Indeed, for complex environments, the method in [31] must be preceded by a fingerprinting process.

Algorithm 3. `Reciprocal_Sampling`$(D_{n,t}, \overline{X}_{n,t})$

```
1: if D_{n,t} = ∅ then
2:     x ← Sampling(X̄_{n,t})
3: else
4:     x ~ ∏_{d_mn ∈ D_{n,t}} p(x|d_mn)
5: end if
6: return x
```

We illustrate the effect of reciprocal robot detections by performing a short experiment involving two Khepera III robots, one of which is initially localized. Figure 4 shows the localization error for the second, initially unlocalized robot: In comparison to the standard sampling algorithm (Algorithm 1 with $\alpha = 0$), we see that the reciprocal sampling algorithm (Algorithm 1 with $\alpha > 0$) reduces the localization error by taking better advantage of information available on the localized teammember. Additionally, in this case where the first robot is well localized during this short time span, an increased reciprocal sampling proportion α is more efficient due to the higher probability of drawing accurate reciprocal samples.

Figure 5 shows results obtained in an experiment of 3.5 minutes duration involving ten robots (with one of the robots initially localized). The plots discuss the sensitivity of our algorithm with respect to the number of particles M, as well as its robustness with respect to communication failures. Figure 5(a) shows the localization performance (averaged over time and robots) for a variable number of particles. Larger particle sets contribute to an improved localization accuracy. Yet, an 8-fold increase in the number of particles produces a reduction of only 25% of the localization error. This result coincides with the conclusions made in [26], where it was shown that by increasing the number of particles, the performance converges to that of an ideal localization filter with an infinity of particles. Figure 5(b) shows the localization performance for variable message failure rates. Increasing failure rates induce a graceful degradation of the localization performance. This result confirms the algorithm's robustness with respect to communication failures, which ultimately reinforces the underlying asynchronous nature of our collaborative paradigm.

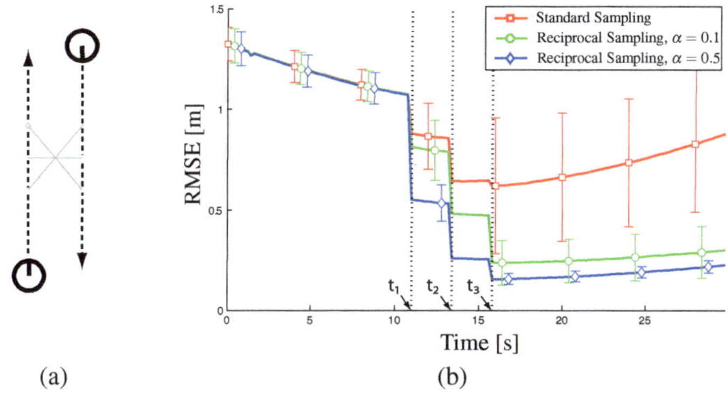

Fig. 4 (a) Schematic illustration of two robots driving past each other. Three detections are made. (b) Localization error for an initially unlocalized robot. It detects a localized robot three times along its path. The standard and reciprocal sampling algorithms (employing 50 particles) are tested 1000 times on the data set. The times at which the observations are made are marked by dotted lines (11.2s, 13.6s, 16s).

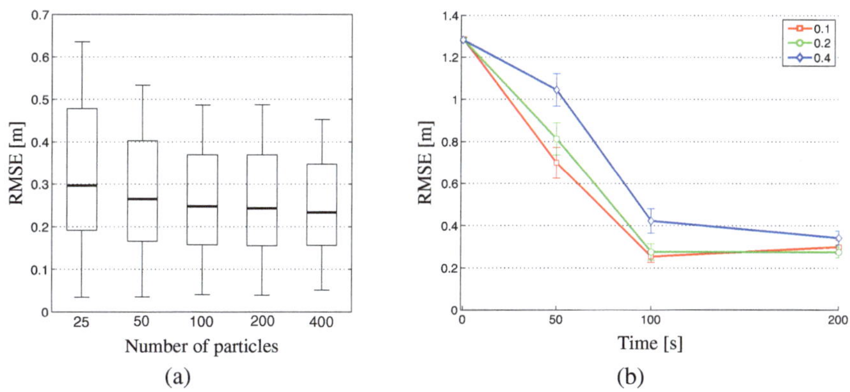

Fig. 5 Localization error for 100 evaluations of the reciprocal sampling algorithm, employing 100 particles per robot and a reciprocal sampling rate $\alpha = 0.06$. (a) Boxplots show the 25th, 50th and 75th percentile, with whiskers containing 85% of the data (for all robots and time). The algorithm is tested employing $\{25, 50, 100, 200, 400\}$ particles per robot. (b) Average error over all robots. Detection data messages are corrupted by a failure rate of $\{0.1, 0.2, 0.4\}$. The errorbars show 95% confidence intervals.

3 Particle Clustering

The algorithm complexity of the detection model $p(\mathbf{x}_n|d_{mn})$ (Eq. 6) leads to $O(M^2)$ for Algorithm 1. This cost can be prohibitive for a large number of particles M (i.e., large with respect to available computational resources). Also, a multi-robot system may have communication constraints that make sending large particle sets infeasible. Hence, even though the method applied in this paper allows for very small particle sets [26], we resort to a clustering method to further reduce the computational and communication overhead.

Let us consider a case where robot $\mathscr{R}_m$ detects robot $\mathscr{R}_n$. For better clarity in the following derivations, we will assume that $|\mathscr{N}_{n,t}| = 1$. The goal of the clustering method is to reduce the number of operations needed to compute the probability density function $p(\mathbf{x}_n|d_{mn})$. Thus, for every detection that it makes, robot $\mathscr{R}_m$ resorts to a clustering method which summarizes its set X_m composed of M particles to a set $\hat{X}_m$ composed of K cluster abstractions (or centroids), reducing the overall computational cost to $O(MK)$ (this clustering routine is detailed later, in Algorithm 4 of Section 3.1). The resulting partition of the particle set is denoted $\mathscr{C}_m$, with $|\mathscr{C}_m| = K$. An individual cluster $c_m^{[k]} \in \mathscr{C}_m$ is defined as the set of particles

$$c_m^{[k]} = \{\langle \mathbf{x}_m^{[i]}, w_m^{[i]} \rangle \mid f(\langle \mathbf{x}_m^{[i]}, w_m^{[i]} \rangle, \cdot) = k\}, \tag{7}$$

where f is a function mapping a particle to a cluster index. Also, we define $\overline{c}_m^{[k]}$ as the data abstraction of cluster $c_m^{[k]}$, representing all particles in its set by the tuple

$$\overline{c}_m^{[k]} = \langle \hat{\mathbf{x}}_m^{[k]}, \hat{w}_m^{[k]}, \hat{\boldsymbol{\mu}}_m^{[k]}, \hat{\Sigma}_m^{[k]} \rangle, \tag{8}$$

where $\hat{\boldsymbol{\mu}}_m^{[k]}$ is a two dimensional vector and $\hat{\Sigma}_m^{[k]}$ is a covariance matrix. Thus, $\hat{X}_m = \{\overline{c}_m^{[k]} \mid c_m^{[k]} \in \mathscr{C}_m\}$ is the set of K cluster abstractions. Finally, we denote the clustered detection data as $\hat{d}_{mn} = \langle \tilde{r}_{mn,t}, \tilde{\theta}_{mn,t}, \hat{X}_m \rangle$, which is sent in place of the unclustered detection data d_{mn}. Formally, given the notation introduced above, finding an optimal particle clustering is equivalent to solving the following optimization problem

$$\min_{\hat{d}_{mn}} \mathbf{D}(p(x_n|d_{mn}) \,||\, \hat{p}(x_n|\hat{d}_{mn})), \tag{9}$$

where $\hat{p}$ is an approximated detection model, and $\mathbf{D}$ a distance measure between two probability density functions. Jain et al. [11] point out that in a typical clustering task, the actual grouping (or clustering) and cluster data abstraction (or cluster representation) are separate components of the task and are commonly treated sequentially. Hence, we deal with our problem by dividing it into the two following sub-problems: *(i)* we consider the set of particles X_m and find an optimal way to create a partition $\mathscr{C}_m$, and *(ii)* we consider an arbitrary cluster $c_m^{[k]}$ in $\mathscr{C}_m$ and find an optimal way to determine its cluster abstraction $\overline{c}_m^{[k]}$. For a given set X_m, these two steps together ultimately lead to a set of cluster abstractions $\hat{X}_m$, which, instead of X_m, is included into the detection data tuple $\hat{d}_{mn}$ for every new detection made. The following paragraphs detail our low-cost clustering approach that aims to meet these specifications.

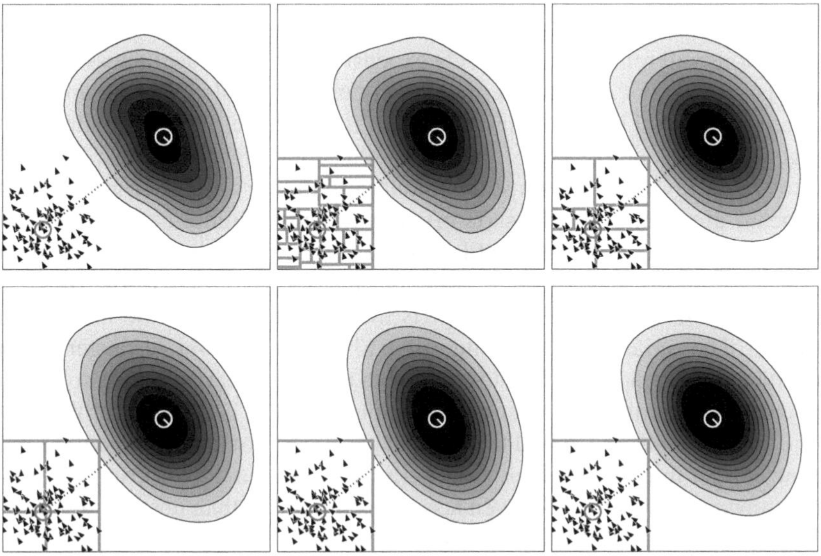

Fig. 6 The detection model (here with range and bearing noise $\sigma_r = 0.15$ and $\sigma_\theta = 0.15$) is projected on the detected robot (in white). Final cluster partitions are superimposed on the particles of the detecting robot. From left to right, top to bottom, the number of clusters K employed by the clustering algorithm is: 100, 32, 8, 4, 2, 1, for a total number of particles $M = 100$.

Algorithm 4. $\texttt{Cluster}(X_{m,t}, K)$

$\hat{X}_m \leftarrow \emptyset$
$c_m^{[1]} \leftarrow X_m$
$\mathscr{C}_m \leftarrow c_m^{[1]}$
for $k = 1$ to $K-1$ **do**
  $k_{max}, \text{dim} \leftarrow \texttt{find\_highest\_variance\_cluster}(\mathscr{C}_m)$
  $c_m^{[k_{max}]}, c_m^{[k+1]} \leftarrow \texttt{split\_cluster}(c_m^{[k_{max}]}, \text{dim})$
  $\mathscr{C}_m \leftarrow \mathscr{C}_m + c_m^{[k+1]}$
end for
for $k = 1$ to K **do**
  $\bar{c}_m^{[k]} \leftarrow \texttt{assign\_data\_abstraction}(c_m^{[k]})$
  $\hat{X}_m \leftarrow \hat{X}_m + \bar{c}_m^{[k]}$
end for
return $\hat{X}_m$

3.1 Clustering Algorithm

The optimal, combinatorial solution to the clustering problem of Equation 9 requires the evaluation of a very large number of partitions (the number of ways to partition a set of M data points into K non-empty clusters is given by Stirling number of the second kind). Even though efficient approaches have been proposed [14], combinatorial solutions still remain prohibitively expensive. Given the usefulness of clustering in a large range of disciplines, many non-combinatorial clustering approaches have been proposed [11]. Yet, since our goal is to reduce the final complexity of our algorithm, the complexity of the actual clustering algorithm must be at most equal to $O(MK)$. One of the most commonly used low-cost clustering methods is the k-means algorithm [20]. It starts off with a random initial cluster assignment and iteratively reassigns clusters until a convergence criterion is met or a maximum number of iterations L is attained. Although the algorithm has a low time complexity $O(MKL)$, its main disadvantage is that it is sensitive to the initial cluster assignment. The variant ISODATA algorithm [3] is also an iterative clustering algorithm with a time complexity of $O(MKL)$, with the additional capability to split and merge clusters according to predefined threshold values. It is therefore more flexible than the k-means and able to find the optimal partition, provided that the user is able to define correct threshold values. Non-iterative, incremental clustering algorithms have the advantage that they are even less time consuming than iterative algorithms. The leader algorithm [9] is the simplest of that kind. Data points are incrementally assigned to existing clusters based on a distance metric, with new clusters being created if all distance measures exceed a predefined criterion. Yet, given the algorithms incremental nature, the final clustering result is dependent on the order of the assignments made.

We take inspiration from the methods described above to develop a non-iterative, order-independent, non-parametric approach that produces a predefined number of K clusters. Our solution is inspired by the construction of multidimensional binary trees [4], and consists of a 2-dimensional sorting algorithm which repetitively separates the particle set along the mean of the dimension producing the highest variance, until the predefined maximum number of clusters K is attained. We note that splitting along the median instead of the mean incurs a higher complexity. A description of this algorithm is shown in Algorithm 4. The function in line 5 has a complexity $O(M)$, the function in line 6 has a complexity $O(|c_m^{[k_{max}]}|)$, and function in line 10 has a complexity $O(|c_m^{[k]}|)$. Hence, the total algorithm cost is $O(MK)$. Figure 6 shows examples of final cluster partitions for six different total numbers of clusters, performed on an identical set of 100 particles. We note that, even for maximal clustering ($K = 1$), the detection model is well approximated.

3.2 Cluster Abstraction

For an arbitrary cluster $c_m^{[k]}$, we have the non-summarized detection data $d_{mn}^{[k]} = \langle \tilde{r}_{mn,t}, \tilde{\theta}_{mn,t}, c_m^{[k]} \rangle$. The problem of finding an optimal cluster abstraction $\overline{c}_m^{[k]}$ can, thus, be formalized as

$$min\mathbf{D_{KL}}(p||\hat{p}) = \int_{-\infty}^{\infty} p(\mathbf{x}_n|\hat{d}_{mn}^{[k]}) \log \frac{p(\mathbf{x}_n|\hat{d}_{mn}^{[k]})}{\hat{p}(\mathbf{x}_n|\hat{d}_{mn}^{[k]})} \mathrm{d}\mathbf{x}_n \tag{10}$$

where $\mathbf{D_{KL}}$ is the Kullback-Leibler divergence, and $\hat{d}_{mn}^{[k]} = \langle \tilde{r}_{mn,t}, \tilde{\theta}_{mn,t}, \overline{c}_m^{[k]} \rangle$ is the summarized detection data. In [28], we showed the following. Given a point $\hat{\mathbf{x}}_{\mathbf{m}}^{[\mathbf{k}]} = [\hat{x}_m^{[k]}, \hat{y}_m^{[k]}, \hat{\phi}_m^{[k]}]^\mathsf{T}$, and the probability density function

$$\hat{p}(\mathbf{x}_n | \hat{d}_{mn}^{[k]}) = \Phi\left(\mathbf{T}_e^p(\hat{\mathbf{x}}_m^{[k]}, \mathbf{x}_n); \hat{\boldsymbol{\mu}}_m^{[k]}, \hat{\Sigma}_m^{[k]}\right), \tag{11}$$

the Kullback-Leibler divergence between p and $\hat{p}$ is minimal if

$$\hat{\boldsymbol{\mu}}_m^{[k]} = \frac{1}{|c_m^{[k]}|} \sum_{\mathbf{x}_m^{[i]} \in c_m^{[k]}} \mathbf{v}_m^{[k,i]}, \tag{12}$$

$$\hat{\Sigma}_m^{[k]} = \frac{1}{|c_m^{[k]}| - 1} \sum_{\mathbf{x}_m^{[i]} \in c_m^{[k]}} \left(\mathbf{v}_m^{[k,i]} - \hat{\boldsymbol{\mu}}_m^{[k]}\right)\left(\mathbf{v}_m^{[k,i]} - \hat{\boldsymbol{\mu}}_m^{[k]}\right)^\mathsf{T} \tag{13}$$

are the mean and covariance of $\mathbf{v}_m^{[k,i]} = \mathbf{T}_e^p(\hat{\mathbf{x}}_m^{[k]}, \check{\mathbf{x}}_m^{[i]})$, with

$$\check{x}_m^{[i]} = x_m^{[i]} + r_{mn}\cos(\theta_{mn} + \phi_m^{[i]}) \tag{14}$$

$$\check{y}_m^{[i]} = y_m^{[i]} + r_{mn}\sin(\theta_{mn} + \phi_m^{[i]}). \tag{15}$$

We note that the above equations do not take into account the uncertainty of the range and bearing observations. Thus, we propose a variant detection model $\hat{p}$ (cf. Equation 6) that explicitly takes into account noise. We have

$$\hat{p}(\mathbf{x}_n | \hat{d}_{mn}) = \eta \cdot \sum_{\overline{c}_m^{[k]} \in \hat{X}_m} \Phi\left(\mathbf{T}_e^p(\hat{\mathbf{x}}_m^{[k]}, \mathbf{x}_n); \hat{\boldsymbol{\mu}}_m^{[k]}, \hat{\Sigma}_m^{[k]} + \Sigma\right) \cdot \hat{w}_m^{[k]} \tag{16}$$

where $\hat{\boldsymbol{\mu}}_m^{[k]}$ and $\hat{\Sigma}_m^{[k]} + \Sigma$ approximate the true mean and covariance, respectively, in the presence of noise (we remind the reader that $\Sigma = \text{diag}([\sigma_r^2, \sigma_\theta^2])$). Indeed, finding a closed form solution for the true values is intractable. However, if the set of particles $c_m^{[k]}$ is densely populated, our approximation is very good. Moreover, if the particle positions coincide, and if for a given cluster $c_m^{[k]}$ the point $\hat{\mathbf{x}}_m^{[k]}$ is its center of mass, the solution is optimal. Hence, we complete the data abstraction $\overline{c}_m^{[k]} = \langle \hat{\mathbf{x}}_m^{[k]}, \hat{w}_m^{[k]}, \hat{\boldsymbol{\mu}}_m^{[k]}, \hat{\Sigma}_m^{[k]} \rangle$ (cf. Equation 8) with $\hat{\mathbf{x}}_m^{[k]}$ as the weighted center of mass, and $\hat{w}_m^{[k]}$ the cumulative weight

$$\hat{\mathbf{x}}_m^{[k]} = \frac{1}{\hat{w}_m^{[k]}} \cdot \sum_{\langle \mathbf{x}_m^{[i]}, w_m^{[i]} \rangle \in c_m^{[k]}} w_m^{[i]} \cdot [x_m^{[i]}, y_m^{[i]}, \phi_m^{[i]}]^\mathsf{T} \tag{17}$$

$$\hat{w}_m^{[k]} = \sum_{\langle \mathbf{x}_m^{[i]}, w_m^{[i]} \rangle \in c_m^{[k]}} w_m^{[i]}. \tag{18}$$

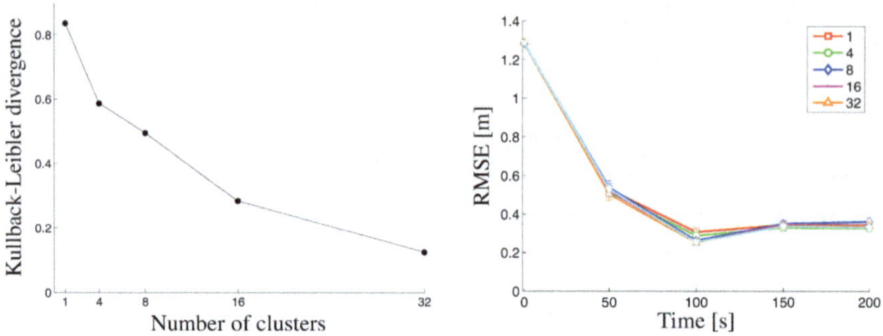

(a) (b)

Fig. 7 (a) The Kullback-Leibler divergence between the full and approximated detection models, as a function of the number of clusters employed by the clustering method. (b) Average localization error over 100 evaluations. The localization algorithm is tested, employing the clustering method using $\{1,4,8,16,32\}$ clusters. The errorbars show 95% confidence intervals.

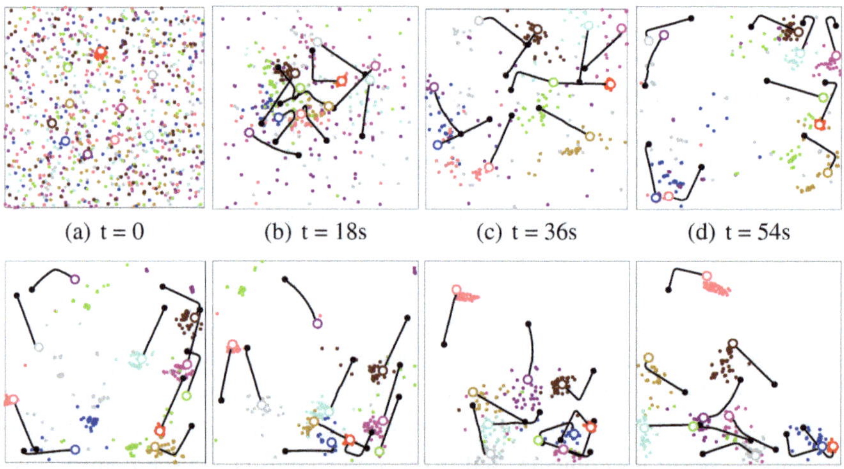

(a) t = 0 (b) t = 18s (c) t = 36s (d) t = 54s

(e) t = 72s (f) t = 90s (g) t = 108s (h) t = 126s

Fig. 8 The figure shows eight snapshots with 18s intervals of an experimental run on the team of ten Khepera III robots. Each robot employed 100 particles with a reciprocal proportion $\alpha = 0.06$, and used the clustering routine with $K = 1$. The black lines show the trajectories completed in the time intervals between snapshots, with the filled black dots representing the robot positions at the end of the previous snapshots. The red robot was initially localized.

Finally, we note that the constraints given by our approximated detection model $\hat{p}$ motivate the choice of a clustering algorithm which clusters densely located particles into common clusters (a condition which is satisfied by Algorithm 4).

Figure 7(a) shows the Kullback-Leibler divergence between the full and the approximated detection models p and $\hat{p}$, calculated from a data set gathered by ten robots. The more clusters we employ in the clustering method, the smaller the divergence to the true probability density function. This shows that our clustering method produces a valid representation of the original probability density functions. Figure 7(b) shows the localization performance when employing the clustering method for a variable number of clusters K. We note that the difference of performance between maximal clustering ($K = 1$) and modest clustering ($K = 32$) is very small. Finally, to illustrate the localization process, Figure 8 shows eight snapshots based on real data from an experiment performed over an interval of 126s during which one robot (in red) is initially localized. Each robot employed 100 particles with a reciprocal proportion $\alpha = 0.06$, and used the clustering routine with maximal clustering ($K = 1$). This experiment concludes the validation of our approach by showing how ten robots are able to converge to correct position estimates in a nevertheless simple, but effective demonstration scenario.

4 Conclusion

In this chapter, we presented a fully scalable, probabilistic, multi-robot localization algorithm based on the Monte Carlo method. Its maximal overall complexity is $O(|\mathcal{N}|MK)$, where $|\mathcal{N}|$ is the number of neighboring robots (at a given time, for a given robot in the system), M the number of particles, and K an adjustable number of clusters produced by the clustering algorithm. This clustering method has shown to produce increasingly accurate probability density function representations for large K, and when employed in practice, has shown to perform well even for very small K. Furthermore, given the asynchronous paradigm of our collaboration strategy, the algorithm's update rate is much higher than the inter-robot message communication rate. Thus, the number of detected neighbors $|\mathcal{N}|$ is in practice no higher than 1, and the complete routine complexity is reduced to $O(MK)$. Thus, the algorithm is fully scalable with respect to the number of robots in the system. In addition, the algorithm poses no communication constraints and shows a graceful performance degradation in case of message failures. Our approach was experimentally validated on a team of ten real robots.

Finally, we note that a continuation of this work should consider the following aspects in particular. We evaluated our approach on a baseline experimental setup, where the belief of a robot's position is well represented by a single particle cluster. Hence, more complex scenarios, including obstacles and multi-modal sensor models, may exhibit a significant spread of performance when clustering. In such cases, a trade-off between the number of clusters K and accuracy must be determined. Also, in severely multi-modal distributions, the construction of the cluster centroid must be revisited. In the same line of thought, more work needs to be done to explore arbitrarily distributed, non-Gaussian detection models as an extension to our generalizable framework.

Acknowledgements. The work presented in this chapter was supported by the National Competence Center in Research on Mobile Information and Communication Systems (NCCR-MICS), a center supported by the Swiss National Science Foundation under grant number 51NF40-111400.

References

[1] Bahr, A., Leonard, J.J., Fallon, M.F.: Cooperative Localization for Autonomous Underwater Vehicles. Int. Journal of Robotics Research 28(6), 714–728 (2009)
[2] Bahr, A., Walter, M.R., Leonard, J.J.: Consistent cooperative localization. In: IEEE International Conference on Robotics and Automation (ICRA), pp. 3415–3422 (2009)
[3] Ball, G.H., Hall, D.J.: ISODATA, a novel method of data analysis and classification. Technical report (1965)
[4] Bentley, J.L.: Multidimensional binary search trees used for associative searching. Commun. ACM 18(9), 509–517 (1975)
[5] Bulusu, N., Heidemann, J., Estrin, D.: GPS-less Low-Cost Outdoor Localization for Very Small Devices. IEEE Personal Communications 7(5), 28–34 (2000)
[6] Cristofaro, A., Renzaglia, A., Martinelli, A.: Distributed Information Filters for MAV Cooperative Localization. In: Martinoli, A., Mondada, F., Correll, N., Mermoud, G., Egerstedt, M., Hsieh, M.A., Parker, L.E., Støy, K. (eds.) Distributed Autonomous Robotic Systems. STAR, vol. 83, pp. 133–146. Springer, Heidelberg (2013)
[7] Fox, D., Burgard, W., Kruppa, H., Thrun, S.: A Probabilistic Approach to Collaborative Multi-Robot Localization. Autonomous Robots 8, 325–344 (2000)
[8] Gowal, S., Martinoli, A.: Bayesian Rendezvous for Distributed Robotics Systems. In: IEEE/RSJ International Conference on Intelligent Robots and Systems (IROS), pp. 2765–2771
[9] Hartigan, J.A.: Clustering Algorithms. John Wiley and Sons, Inc., New York (1975)
[10] Howard, A., Mataric, M.J., Sukhatme, G.S.: Localization for mobile robot teams: A distributed MLE approach. In: Experimental Robotics VIII, pp. 146–155 (2003)
[11] Jain, A.K., Murty, M.N., Flynn, P.J.: Data Clustering: A Review. ACM Computing Surveys 31(3), 265–323 (1999)
[12] Jennings, J.S., Whelan, G., Evans, W.F.: Cooperative search and rescue with a team of mobile robots. In: International Conference on Advanced Robotics (ICAR), pp. 193–200 (1997)
[13] Kantor, G., Singh, S., Peterson, R., Rus, D., Das, A., Kumar, V., Perreira, G., Spletzer, J.: Distributed Search and Rescue with Robot and Sensor Teams. Springer Tracts in Advanced Robotics 24, 529–538 (2006)
[14] Koontz, L.G.W., Narendra, P., Fukunaga, K.: A Branch and Bound Clustering Algorithm. IEEE Transactions on Computers C-24(9), 908–915 (1975)
[15] Leung, K.Y.K., Barfoot, T.D., Liu, H.: Decentralized Localization of Sparsely-Communicating Robot Networks: A Centralized-Equivalent Approach. IEEE Transactions on Robotics 26(1), 62–77 (2010)
[16] Lindsey, Q., Mellinger, D., Kumar, V.: Construction of cubic structures with quadrotor teams. In: Durrant-Whyte, H., Roy, N., Abbeel, P. (eds.) Robotics: Science & Systems VII, pp. 198–205. MIT Press, Cambridge (2012)

[17] Lochmatter, T., Roduit, P., Cianci, C., Correll, N., Jacot, J., Martinoli, A.: SwisTrack - A Flexible Open Source Tracking Software for Multi-Agent Systems. In: Proceedings of the 2008 IEEE/RSJ International Conference on Intelligent Robots and Systems, pp. 4004–4010 (2008)
[18] Madhavan, R., Fregene, K., Parker, L.E.: Distributed Cooperative Outdoor Multirobot Localization and Mapping. Autonomous Robots 17(1), 23–39 (2004)
[19] Martinelli, A., Pont, F., Siegwart, R.: Multi-Robot Localization Using Relative Observations. In: IEEE International Conference on Robotics and Automation (ICRA), pp. 2797–2802 (2005)
[20] McQueen, J.: Some methods for classification and analysis of multivariate observations. In: Proceedings of the Fifth Berkeley Symposium on Mathematical Statistics and Probability, pp. 281–297 (1967)
[21] Mourikis, A.I., Roumeliotis, S.I.: Optimal sensor scheduling for resource-constrained localization of mobile robot formations. IEEE Transactions on Robotics 22(5), 917–931 (2006)
[22] Neal, B.: Slice Sampling. The Annals of Statistics 31, 705–757 (2003)
[23] Nerurkar, E., Roumeliotis, S.: Asynchronous Multi-Centralized Cooperative Localization. In: IEEE/RSJ International Conference on Intelligent Robots and Systems (IROS), pp. 4352–4359 (2010)
[24] Nerurkar, E., Roumeliotis, S., Martinelli, A.: Distributed maximum a posteriori estimation for multi-robot cooperative localization. In: IEEE International Conference on Robotics and Automation (ICRA), pp. 1402–1409 (2009)
[25] Pathirana, P.N., Bulusu, N., Savkin, A.V., Jha, S.: Node localization using mobile robots in delay-tolerant sensor networks. IEEE Transactions on Mobile Computing 4(3), 285–296 (2005)
[26] Prorok, A., Martinoli, A.: A Reciprocal Sampling Algorithm for Lightweight Distributed Multi-Robot Localization. In: Proceedings of the 2011 IEEE/RSJ International Conference on Intelligent Robots and Systems (IROS), pp. 3241–3247 (2011)
[27] Prorok, A., Arfire, A., Bahr, A., Farserotu, J.R., Martinoli, A.: Indoor navigation research with the Khepera III mobile robot: An experimental baseline with a case-study on ultra-wideband positioning. In: International Conference on Indoor Positioning and Indoor Navigation, IPIN (2010), doi:10.1109/IPIN.2010.5647880
[28] Prorok, A., Bahr, A., Martinoli, A.: Low-Cost Collaborative Localization for Large-Scale Multi-Robot Systems. In: IEEE International Conference on Robotics and Automation (ICRA), pp. 4236–4241 (2012)
[29] Pugh, J., Raemy, X., Favre, C., Falconi, R., Martinoli, A.: A Fast On-Board Relative Positioning Module for Multi-Robot Systems. IEEE Transactions on Mechatronics 14(2), 151–162 (2009)
[30] Roumeliotis, S.I., Bekey, G.A.: Distributed Multirobot Localization. IEEE Transactions on Robotics and Automation 14(5), 781–795 (2002)
[31] Thrun, S., Fox, D., Burgard, W., Dellaert, F.: Robust Monte Carlo Localization for Mobile Robots. Artificial Intelligence 128(1-2), 99–141 (2000)
[32] Werfel, J., Bar-Yam, Y., Rus, D., Nagpal, R.: Distributed construction by mobile robots with enhanced building blocks. In: IEEE International Conference on Robotics and Automation (ICRA), pp. 2787–2794 (2006)

Using Torque Redundancy to Optimize Contact Forces in Legged Robots

Ludovic Righetti, Jonas Buchli, Michael Mistry,
Mrinal Kalakrishnan, and Stefan Schaal

Abstract. The development of legged robots for complex environments requires controllers that guarantee both high tracking performance and compliance with the environment. More specifically the control of contact interaction with the environment is of crucial importance to ensure stable, robust and safe motions. In the following, we present an inverse dynamics controller that exploits torque redundancy to directly and explicitly minimize any combination of linear and quadratic costs in the contact constraints and in the commands. Such a result is particularly relevant for legged robots as it allows to use torque redundancy to directly optimize contact interactions. For example, given a desired locomotion behavior, it can guarantee the minimization of contact forces to reduce slipping on difficult terrains while ensuring high tracking performance of the desired motion. The proposed controller is very simple and computationally efficient, and most importantly it can greatly improve the performance of legged locomotion on difficult terrains as can be seen in the experimental results.

Ludovic Righetti · Stefan Schaal
University of Southern California, Los Angeles, USA
Max-Planck Institute for Intelligent Systems, Tübingen, Germany
e-mail: ludovic.righetti@a3.epfl.ch, sschaal@usc.edu

Jonas Buchli
Italian Institute of Technology, Genoa, Italy
e-mail: jonas@buchli.org

Michael Mistry
University of Birmingham, Birmingham, UK
e-mail: m.n.mistry1@bham.ac.uk

Mrinal Kalakrishnan
University of Southern California, Los Angeles, USA
e-mail: kalakris@usc.edu

D. Milutinović & J. Rosen (Eds.): Redundancy in Robot Manipulators, LNEE 57, pp. 35–51.
DOI: 10.1007/978-3-642-33971-4_3

1 Introduction

We are interested in developing legged robots able to perform difficult tasks in challenging environments. In order to achieve such a goal, we need controllers that can guarantee at the same time good motion tracking performance and an adequate control of the contact interactions with the environment. Tracking performance is important for tasks requiring agility such as climbing or walking on very rough terrain, where the robot feet must be placed at very precise location. A certain degree of compliance is desirable to handle unexpected disturbances. But more importantly, the controller should also be able to directly optimize contact forces. For example, a walking robot should minimize tangential contact forces to avoid slipping while it should control contact forces to redistribute its weight among the different limbs during climbing tasks to increase the range of possible motions. In this chapter, we present an inverse dynamics controller for high tracking performance and compliance that ensures an optimal distribution of contact forces by using the redundancy available in the commands.

Model-based approaches such as inverse dynamics or operational space control offer an interesting framework for the control of legged robots. Indeed, they can greatly improve tracking performance while allowing more compliant control since they require lower error feedback gains. While these methods are standard for manipulators [13], they are not yet widely used for legged robots. Indeed legged robots are different from manipulators fixed to the ground because they are under-actuated due to their floating base[1] and they are subject to changing contact interaction with the environment as their legs move. These differences make the design of inverse dynamics controllers for legged robots more complex.

Recently, several inverse dynamics controllers were proposed for floating-base robots subject to contact constraints. These methods compute required torques without measuring contact forces by assuming idealized contact constraints and projecting the dynamics into a constraint free space [1, 7, 11]. The methods proposed in [7] is of special interest since it does not require a structured representation of the dynamics (i.e. no need to compute individual components like the inertia matrix, Coriolis, and gravity terms) and mainly relies on kinematic quantities, which makes it particularly robust to uncertainties in parameter estimation and, additionally, computationally very effective.

Interestingly when there are more than six contact constraints with the environment, the inverse dynamics problem is under-determined in the sense that there is an infinite number of torque commands for a constraint-consistent desired motion. It is the case, for example, when a biped has its two feet on the ground or when a quadruped robot with point feet has more than 2 feet on the ground. There are more degrees of freedom for the actuation than for the possible motions due to the constraints imposed by the contacts. The over-constrained case is very interesting because there is an infinite number of possible choice of commands to realize a desired motion. In general, redundancy is resolved by minimizing a cost criterion, e.g.,

[1] We refer to floating base robots that are not fixed to the ground due to their 6 non-actuated DOFs that describe the position and orientation of the robot relative to an inertial frame.

a quadratic cost in the commands as in [7, 10]. We will see in the following that we can use torque redundancy to directly manipulate the contact forces instead. It will allow us to create tracking controllers optimal with respect to any combination of linear and quadratic costs of the contact forces and the commands.

The problem of contact force distribution during locomotion has already been addressed in the literature [4,6]. Force control approaches have been proposed [3,14] to directly control the contact forces of a legged robot. These approaches have been successfully used for the control of biped robots. In general, the manipulation of contact forces is done as a primary goal, for example to create balance controllers. A potential desired motion is then treated as a secondary goal. It is in contrast with our approach that aims at creating tracking controllers able to manipulate contact forces using torque redundancy, i.e. trajectory tracking is the primary goal of the controller.

The manipulation of contact forces in the context of inverse dynamics or operational space control has also been addressed in [12]. In this case, desired contact forces are explicitly controlled using the torques acting in the nullspace of the motion. Such an approach is interesting if one has a precise objective for the contact forces. However it does not provide any optimality result on the force distribution.

Recently we have been investigating how we could use torque redundancy in inverse dynamics controllers and operational space controllers to optimize contact forces [8,9]. In this chapter, we present some of the latest developments of our work. We show how we can create inverse dynamics controllers that are optimal with respect to any combination of linear and quadratic cost in the constraint forces and in the commands. The controller is computationally simple and robust to parameter estimation errors, which make it well suited for high performance control of complex robots with a large number of degrees of freedom, such as humanoid robots. We present several applications of the controller to legged robots. Experimental results show that for the same desired motion, the use of torque redundancy can significantly improve the performance of the robot during locomotion. It is worth mentioning that several additional applications and technical details of this work can be found in [8].

2 Problem Formulation

We first consider the general problem of inverse dynamics for floating-base robots under constraints. First we present the constraint model and its assumptions. Then we present the control law recently derived by Mistry et al. [7] for legged robots and that we will use in subsequent sections.

2.1 Rigid Body Dynamics Model

Assuming that the robot in contact with its environment obeys rigid body dynamics, its equations of motion are

$$\mathbf{M}\ddot{\mathbf{q}} + \mathbf{h} = \mathbf{S}^T \tau + \mathbf{J_c}^T \lambda \tag{1}$$

under the k constraints

$$\mathbf{J}_c\ddot{\mathbf{q}} = \mathbf{b}(\mathbf{q}, \dot{\mathbf{q}}) \tag{2}$$

where $\mathbf{q} = [\mathbf{x}_j^T \ \mathbf{x}_b^T]^T$ is the vector of joint positions ($\mathbf{x}_j \in \mathbb{R}^n$) and base positions and orientations ($\mathbf{x}_b \in \mathrm{SE}(3)$), $\mathbf{M} \in \mathbb{R}^{(n+6)\times(n+6)}$ is the rigid body dynamics inertia matrix, $\mathbf{h} \in \mathbb{R}^{n+6}$ is a generalized force vector containing all the modeled forces, including the Coriolis, centrifugal and gravitational forces as well as friction forces in the joints. $\tau \in \mathbb{R}^n$ is the actuation vector and $\mathbf{S} = [\mathbf{I}_{n\times n} \quad 0_{n\times 6}] \in \mathbb{R}^{n\times(n+6)}$ is the joint selection matrix that characterize the under-actuation. $\mathbf{J}_c \in \mathbb{R}^{k\times(n+6)}$ is the Jacobian of the k constraints with the $\lambda \in \mathbb{R}^k$ Lagrange multipliers that correspond to the constraint forces.

Following the ideas from [15] we expressed the constraints in acceleration form (i.e. as given in Eq. 2). Holonomic constraints can be expressed by differentiating them twice and non-holonomic constraints by differentiating them once. We assume in the following, without loss of generality, that $\mathbf{J}_c$ is full row rank, in the sense that all constraints are linearly independent. If it is not the case then one can easily find a reduced number of independent constraints, for example by using the SVD decomposition of $\mathbf{J}_c$.

Example 1. If we assume that the position of the point feet of a legged robot are given by $\mathbf{x}_c$, then the constraints that the feet do not move relative to the ground can be written as $\mathbf{x}_c = \textit{constant}$ or equivalently by $\dot{\mathbf{x}}_c = 0$. Relating this to the motion of the joints of the robot using the Jacobian of $\mathbf{x}_c$ we have $\mathbf{J}_c\dot{\mathbf{q}} = \dot{\mathbf{x}}_c = 0$, which we differentiate once again to get $\mathbf{J}_c\ddot{\mathbf{q}} = -\dot{\mathbf{J}}_c\dot{\mathbf{q}}$.

We assume that the movement plan of the robot is expressed by desired joint accelerations that are constraint consistent, i.e., Equation (2) with $\ddot{\mathbf{q}} = \ddot{\mathbf{q}}_d$ holds. These accelerations will be satisfied if and only if they are of the form

$$\ddot{\mathbf{q}}_d = \mathbf{J}_c^G\mathbf{b} + (\mathbf{I} - \mathbf{J}_c^G\mathbf{J}_c)\ddot{\mathbf{q}}_0 \tag{3}$$

where $\mathbf{J}_c^G$ can be any generalized inverse [2] of $\mathbf{J}_c$, i.e. a matrix such that $\mathbf{J}_c\mathbf{J}_c^G\mathbf{J}_c = \mathbf{J}_c$. Here, $\ddot{\mathbf{q}}_0$ is an arbitrary acceleration vector and $(\mathbf{I} - \mathbf{J}_c^G\mathbf{J}_c)$ projects these accelerations into the null space of the constraints.

The general problem of inverse dynamics is then to compute the torques τ such that they will achieve the desired accelerations $\ddot{\mathbf{q}}_d$. The range of unconstrained movements lies in a $n+6-k$ dimensional space while the dimension of the control vector is n. Therefore we can distinguish three cases for inverse dynamics depending on the number of constraints:

- $k < 6$, the system is underactuated since there are more dimensions of movement than dimensions of actuation. There is at most one solution to the inverse dynamics problem: for a solution to exist, the desired accelerations must not only be constraint consistent, they moreover need to be consistent with the dynamics of Eq. (1). For example one can think of the case of no constraints, when a cat is falling and cannot orient its body independently from moving its joints.

- $k = 6$, the system is fully actuated. There is exactly one solution provided that the desired accelerations are constraint consistent. This case is similar to the inverse dynamics problem of a manipulator fixed to the ground.
- $k > 6$, the system is overconstrained. There is an infinite number of solutions for τ that will achieve perfect tracking of $\ddot{\mathbf{q}}_d$. It is the case, for example, when a humanoid has both feet flat on the ground, or with one foot and one hand in flat contact with the environment or when a quadruped with point feet has more than two feet on the ground.

In the following, we only consider the overconstrained case ($k > 6$) since it is the only case where torque redundancy can be used to optimize constraint forces.

2.2 Inverse Dynamics Solution Using Orthogonal Projections

By using orthogonal projections, Mistry et al. [7] proposed recently an efficient way to compute the inverse dynamics of a constrained under-actuated system without the need to measure contact forces. More precisely the authors use the QR decomposition of the constraint Jacobian $\mathbf{J_c}^T = \mathbf{Q}\begin{bmatrix}\mathbf{R}^T & \mathbf{0}\end{bmatrix}^T$, where $\mathbf{Q} \in \mathbb{R}^{(n+6)\times(n+6)}$ is an orthogonal matrix (i.e. $\mathbf{Q}\mathbf{Q}^T = \mathbf{I}$) and $\mathbf{R} \in \mathbb{R}^{6\times 6}$ is an upper triangular invertible matrix. If we decompose $\mathbf{Q} = [\mathbf{Q}_c \; \mathbf{Q}_u]$ into the constrained, $\mathbf{Q}_c \in \mathbb{R}^{(n+6)\times k}$, and unconstrained, $\mathbf{Q}_u \in \mathbb{R}^{(n+6)\times(n+6-k)}$, components, the general solution for the inverse dynamics torques given desired accelerations $\ddot{\mathbf{q}}_\mathbf{d}$ can be written as

$$\tau(\mathbf{W}, \tau_0) = \overline{\mathbf{Q}_u^T \mathbf{S}^T} \mathbf{Q}_u^T (\mathbf{M}\ddot{\mathbf{q}}_\mathbf{d} + \mathbf{h}) + (\mathbf{I} - \overline{\mathbf{Q}_u^T \mathbf{S}^T} \mathbf{Q}_u^T \mathbf{S}^T) \mathbf{W}^{-1} \tau_0 \tag{4}$$

with generalized inverse

$$\overline{\mathbf{Q}_u^T \mathbf{S}^T} = \mathbf{W}^{-1} \mathbf{S} \mathbf{Q}_u (\mathbf{Q}_u^T \mathbf{S}^T \mathbf{W}^{-1} \mathbf{S} \mathbf{Q}_u)^{-1} \tag{5}$$

where $\mathbf{W} \in \mathbb{R}^{n\times n}$ is a symmetric positive definite matrix and τ_0 is an arbitrary internal torque – τ_0 is premultiplied by a projection matrix that guarantees that τ_0 can only create internal forces, but no movement.

Moreover the resulting constraint forces can be predicted by

$$\lambda = \mathbf{R}^{-1} \mathbf{Q}_c^T (\mathbf{M}\ddot{\mathbf{q}}_d + \mathbf{h} - \mathbf{S}^T \tau) \tag{6}$$

Remark 1. In the case where $k = 6$, there is only one solution and $\overline{\mathbf{Q}_u^T \mathbf{S}^T} = (\mathbf{Q}_u^T \mathbf{S}^T)^{-1}$ and the nullspace is empty. When $k < 6$, there is at most one solution and $\overline{\mathbf{Q}_u^T \mathbf{S}^T} = (\mathbf{Q}_u^T \mathbf{S}^T)^+$, where $()^+$ denotes the Moore-Penrose generalized inverse and the nullspace is also empty.

We note that the torque control law $\tau(\mathbf{W}, \tau_0)$ is parametrized by a weight matrix and an internal torque vector. It is this parameterization that can be exploited to resolve torque redundancy. One result that is already known is that $\tau_0 = \mathbf{0}$ leads to the minimization of the cost $\tau^T \mathbf{W} \tau$ at each instant of time [7]. We show in the following how we can do the same for costs in the *constraint forces*.

3 Minimization of Constraint Forces

In the previous section, we have presented an idealized way of describing constraints using acceleration equalities. However using equalities offers a limited representational capability and cannot capture important aspects such as physical limitations on the constraint forces that can be generated. For example, in the context of a constraint that enforces a foot to stay on the ground, we cannot represent the fact that ground reaction forces should be inside the cone of friction to avoid slipping, which means that the ratio between forces tangential and normal to the contact must satisfy an inequality. We can immediately see that including such inequalities for the inverse dynamics problem will have the consequence that we will not be able to solve the inverse dynamics problem without a (possibly complex) iterative optimization algorithm. This is in contrast to the simple analytical solution presented in Eq. (4).

Another solution would be to directly minimize a cost in the constraint forces that would take into account those inequalities implicitly. If we can get the torque distribution that minimizes such a cost, then, implicitly, it will try to find a solution that tends to fulfill the inequalities. Another advantage is that one does not need to know the exact model of the contact, e.g., the friction cone. For example minimizing a cost that penalizes the tangential forces during contact will ensure that the robot minimizes slipping for all sizes of friction cones, i.e., the controller will act as conservative as possible towards slipping. While it is possible that such an approach finds a solution that violates the constraints even if a correct solution exists, we will see in the next section that it is rarely a problem in practice. A detailed discussion on limitations and advantages of this approach can be found in [8].

3.1 General Result

We now present the general result used to construct controllers that minimize any combinations of linear and quadratic costs in the contacts and in the commands.

Theorem 1. *The inverse dynamics controller (Eq.4) that minimizes the cost*

$$\frac{1}{2}\tau^T \mathbf{W}_\tau \tau + \mathbf{b}_\tau^T \tau + \frac{1}{2}\lambda^T \mathbf{W}_\lambda \lambda + \mathbf{b}_\lambda^T \lambda \tag{7}$$

is chosen by setting

$$\mathbf{W} = \mathbf{W}_\tau + \mathbf{S}\mathbf{W}_c\mathbf{S}^T \tag{8}$$

$$\tau_0 = -\mathbf{b}_\tau + \mathbf{S}\mathbf{W}_c(\mathbf{M}\ddot{\mathbf{q}} + \mathbf{h}) + \mathbf{S}\mathbf{b}_c \tag{9}$$

where

$$\mathbf{W}_c = \mathbf{Q}\begin{bmatrix} \mathbf{R}^{-T}\mathbf{W}_\lambda \mathbf{R}^{-1} & 0 \\ 0 & \mathbf{I} \end{bmatrix}\mathbf{Q}^T, \tag{10}$$

$$\mathbf{b}_c = \mathbf{Q}\begin{bmatrix} \mathbf{R}^{-T} \\ 0 \end{bmatrix}\mathbf{b}_\lambda, \tag{11}$$

$\mathbf{J}_c^T = \mathbf{Q}\begin{bmatrix}\mathbf{R}\\0\end{bmatrix}$ *is the QR decomposition of the constraint Jacobian and* $\mathbf{W}_\lambda$ *is such that* $\mathbf{W}_\tau + \mathbf{S}\mathbf{W}_c\mathbf{S}^T$ *is symmetric positive definite.*

We omit the proof as it can be found in [8] with more technical details. The result provides us with a way to directly and explicitly optimize at the same time both command costs and constraint forces using torque redundancy. The result is very general as it can be applied for any type of constraints expressed in the form of Equation (2) and for any combination of linear and quadratic costs of those constraints. The result applies to any control system whose equations of motion can be written as Equation (1) and is therefore not limited to legged robots.

Remark 2. In addition to its very nice computational properties, the formulation of the inverse dynamics control as proposed in [7] is very convenient as we can use the same orthogonal decomposition $\mathbf{Q}$ to derive the torque parametrization for optimal distribution of constraint forces.

Now cost functions can be designed depending on the desired application in order to manipulate the generated constraint forces and torques.

3.2 Examples of Optimization

In this section we discuss a few controllers that can be derived using the results of the previous section and some of their properties relevant for applications.

3.2.1 Minimization of Tangential Contact Forces

To ensure proper contact with the ground, one has to guarantee that the ground reaction forces stay within the friction cones. The friction cone is a purely geometric constraint that is defined by a friction constant and the orientation of the contact surface with the contacting foot. In the case of locomotion, to avoid slipping, one would like to have the reaction forces as orthogonal to the constraint surface as possible. In other words, the tangential forces should be minimized. Moreover in the case of a flat foot on the ground, the resulting moment around the foot should be as small as possible. The cost to optimize should therefore take into account the orientation of the ground in order to redirect contact forces in a more desirable direction.

In order to minimize the tangential forces and the moments around the foot, we propose the following cost

$$\mathbf{W}_\lambda = \begin{bmatrix} \mathbf{R}_{leg_1}^T \mathbf{W}_{leg_1} \mathbf{R}_{leg_1} & & 0 \\ & \ddots & \\ 0 & & \mathbf{R}_{leg_n}^T \mathbf{W}_{leg_n} \mathbf{R}_{leg_n} \end{bmatrix} \tag{12}$$

with

$$\mathbf{W}_{leg_i} = \mathrm{diag}(K_{tx}, K_{ty}, 1, K_{mx}, K_{my}, K_{mz}) \tag{13}$$

where $\mathbf{R}_{leg_n}$ is a rotation matrix that corresponds to the orientation of the surface with respect to the inertial frame (i.e. it aligns the reaction forces and moments of a foot with the orientation of the surface) . K_i are gains associated with the tangential forces in x,y directions and the moments in x,y,z directions. Here we assume that the z direction is normal to the ground.

Remark 3. In the case of point feet the matrix $\mathbf{W}_{leg}$ is reduced to its 3×3 upper-left sub-matrix. The gains K_t and K_m must be chosen with appropriate units such that the summations in the cost are unit consistent. The exact values of the gains need to be designed specific to the application.

3.2.2 Tracking Desired Contact Forces

Instead of only minimizing constraint forces, we can also manipulate the value of these forces explicitly in order to create more interesting contacts. For example, when one wants to manipulate the center of pressure of the feet during walking (in a case where the feet are not co-planar) or to explicitly regulate force interaction in specified directions. The previous results allow us to do that very easily. Assume we have a desired interaction force vector λ_d and we want the contact forces to follow this desired force vector as close as possible. We can then minimize a weighted square error that measures the performance in tracking

$$\frac{1}{2}(\lambda - \lambda_d)^T \mathbf{K}(\lambda - \lambda_d) \tag{14}$$

where $\mathbf{K}$ is symmetric positive definite. Then, noticing that

$$\frac{1}{2}(\lambda - \lambda_d)^T \mathbf{K}(\lambda - \lambda_d) = \frac{1}{2}\lambda^T \mathbf{K}\lambda + \frac{1}{2}\lambda_d^T \mathbf{K}\lambda_d - \lambda_d^T \mathbf{K}\lambda \tag{15}$$

we can create a controller that will optimally track those constraint forces in the sense that it will minimize the tracking error by choosing the following controller parameters

$$\mathbf{W}_\lambda = \mathbf{K} \tag{16}$$

$$\mathbf{b}_\lambda = -\lambda_d^T \mathbf{K} \tag{17}$$

It is interesting since we can use the torque redundancy to explicitly manipulate contact forces and track desired forces. The tracking performance will obviously be dependent on the desired forces and the redundancy left in the nullspace of the motion. It means that it will not be possible to track arbitrary forces, however we are guaranteed to have an optimal performance in terms of the tracking cost.

4 Experimental Results

We now present some experiments in order to illustrate the advantages of using the previously proposed redundancy resolution scheme. In the following, we apply our

Fig. 1 The robots used in the simulations: left, SARCOS humanoids, right, Boston Dynamics Little Dog

approach on two different simulated robots (Figure 1) using different optimization criteria for the controller as well as a preliminary application on a real quadruped robot.

4.1 Control of a Humanoid Robot

Our first results are shown on a simulation of the Sarcos humanoid robot (Figure 1) which is a 34-DOFs human size torque controlled humanoid robot. The main objective of the experiment is to illustrate in simple examples how the inverse dynamics controller can manipulate contact forces.

In the experiment, we keep the robot in a desired posture while minimizing tangential ground reaction forces together with the moment generated around the feet (as described in Section 3.2.1). We compare the resulting contact forces with the original inverse dynamics controller that minimizes the total command cost $\tau^T \tau$. In this static case, the contact forces are only generated through actuation and gravity.

We tested both controllers in two contexts: in a symmetric posture on a flat terrain and on a terrain where the robot is stepping on a 10 cm box. The results of the experiments are shown in Figure 2. We notice that on the flat surface, both controllers generate similar contact forces and moments. However in the asymmetric case, when the robot is on a step, we clearly see a difference in force and moment distributions. The original controller generates unnecessary tangential forces and moments to keep its posture while the other controller minimizes the contact constraints. We can clearly see the advantage of such a controller when stepping for example on an object that is not bolted to the ground and could potentially move.

4.2 Simulation of Quadruped Locomotion

Next, we show the performance of our torque controller in a more dynamic situation where we used a simulation of the LittleDog robot for quadruped locomotion. The

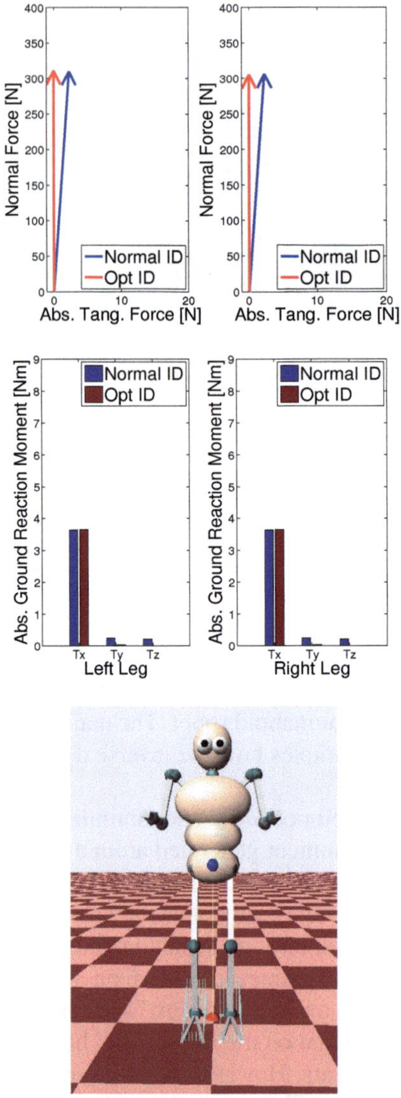

(a) Flat terrain experiments

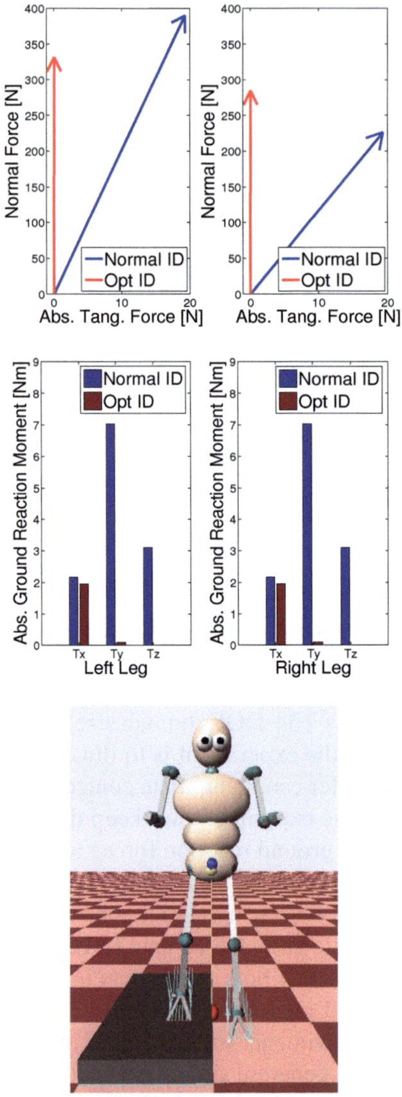

(b) Step experiments

Fig. 2 Results of the first experiment with the humanoid simulation. We depict the contact forces split into normal forces and absolute tangential forces (upper graphs) and show the 3 moments created around the feet (middle graph) and a snapshot of the posture of the robot in each experiment. The regular inverse dynamics controller is plotted in blue and the controller minimizing contact forces is plotted in red. For each experiment we set $K_{tx} = K_{ty} = 1000N^{-2}$ and $K_{mx} = K_{my} = K_{mz} = 2000Nm^{-2}$.

planning of desired joint positions, velocities and accelerations is done following the method proposed in [5]. We use a ZMP-based algorithm to plan the desired center of gravity (COG) motion of the robot. The COG plan is completed with a world space target for the position of the foot of the swing leg. These kinematic trajectories are converted into joint space reference trajectories via an analytical inverse kinematics model, which exists for this robot due to its 3 DOF legs.

In order to achieve asymptotically stable tracking of trajectories in joint space, an error feedback command in joint-space is added to the feedforward command computed by the inverse dynamics law. The resulting reference command, given desired $\mathbf{q}_d$, $\dot{\mathbf{q}}_d$ and $\ddot{\mathbf{q}}_d$, is therefore

$$\tau = \tau(\mathbf{W}, \tau_0) + \mathbf{PID}(\mathbf{q}_d, \dot{\mathbf{q}}_d) \tag{18}$$

where **PID** corresponds to a joint space PID error feedback controller.

We use a physical simulation of the Little Dog robot (Fig. 1). In order to stay as close as possible to reality, we simulate the controller as it would be executed on the real robot. The real robot has an on-board controller running at 400 Hz that generates the PID commands and can add a feedforward torque command (i.e. the inverse dynamics torque in our case). The desired positions, velocities and feedforward commands are generated on a host computer in a different controller that is running at 100 Hz. It must be noted that the inverse dynamics controller is therefore running at a relatively slow bandwidth for torque control and is much slower than the PID control loop. Therefore it can have a negative effect on the actual performance of the simulated controllers as opposed to what would be produced by an idealized or perfect model.

In order to show the performance of our method, we tested the locomotion of the robot with the controller we proposed as compared to two other controllers. First the original inverse dynamics controller that minimizes the total torque command $\tau^T \tau$, and second a PID controller without inverse dynamics but with gains twice as high as the other controllers. The higher gains are required to maintain sufficient nominal tracking performance.

We systematically tested the performance of locomotion for these three controllers on a flat, level surface with different coefficients of static friction and on a 0.25 radians sloped surface. For each of the experiments, we measured the tracking performance by computing the root mean square (RMS) tracking error in joint space. Furthermore, the distribution of ground reaction forces at each leg is recorded and the average amount of leg slipping per stance phase is computed.

In Figures 3 and 4, we illustrate the results of these experiments. We show the results only for the front left leg since the results for the other legs are qualitatively the same. We can notice that in all the experiments, the controller we proposed always achieves the best performance in terms of both low tracking error and low slipping. As expected, we see better tracking performance for both inverse dynamics controllers with low PID gains as compared to the high-gain PID controller. We must also note that, in addition, we realized a *compliant control* of the robots, which is not possible for the PID controller.

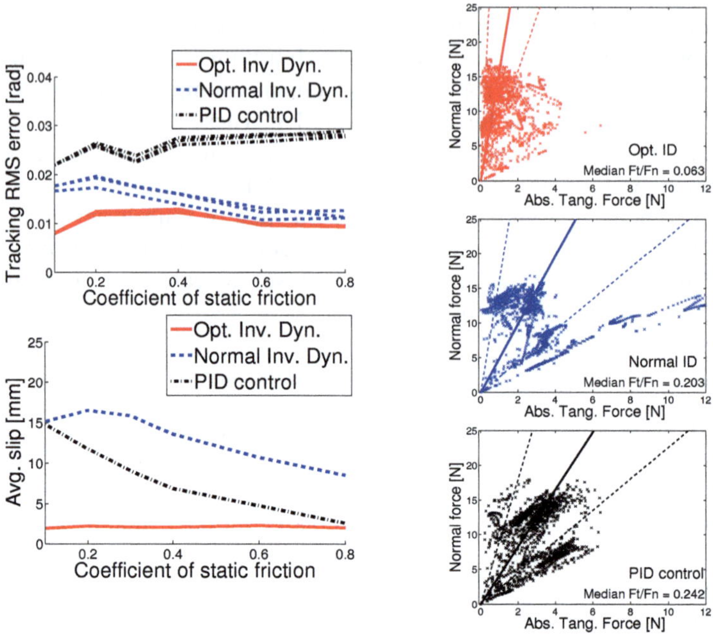

Fig. 3 Experiments on flat terrain. The figures on the left show the performance measures for the 3 different controllers (cf. text). The figures on the right show the distribution of the contact forces normal to the terrain as a function of the absolute value of the corresponding tangential contact forces. The solid lines indicates the median of the distribution and the dashed line the interquartile range of the distribution (i.e. 50 % of the distribution lies between the dashed lines). For the controller minimizing contact forces we used $K_{tx} = K_{ty} = 10N^{-2}$ in all experiments.

On flat ground, our controller minimizing tangential forces leads to very little slip (approx. 2mm), which for practical purposes can be viewed as negligible. There are two sources that lead to this tracking error despite a perfect rigid body model used in the controller and a deterministic simulation. First, the assumption of holonomic ideal constraints is not fulfilled since the simulation uses a penalty method (i.e. a spring-damper model) to model ground contact and friction. Second, there might be some numerical inaccuracies building up due to numerical integration, even though this can generally be assumed to be very small.

It should be noted that the amount of slip for the other controllers increases as the friction coefficient is lowered and goes up to more than 1.5 cm for the lowest chosen friction coefficient of 0.1. It seems that the PID controller degrades less than the normal inverse dynamics when lowering friction.

We also see that the typical distribution of ground reaction forces is more vertical for the controller optimizing the tangential forces and that the variation of this distribution is also lower. We can therefore conclude that the controller generates contact forces that are oriented more suitably for walking without slipping.

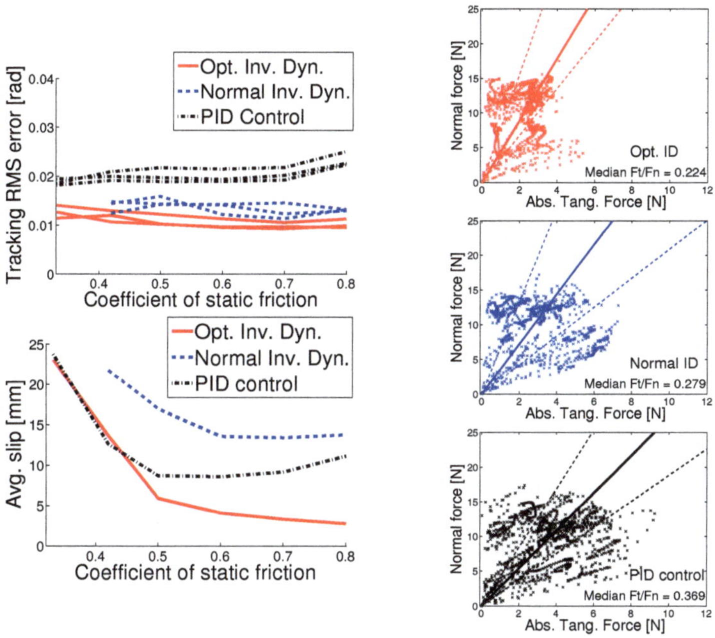

Fig. 4 Experiments on the slope. The figures on the left show the performance measures for the 3 different controllers (cf. text). The figures on the right show the distribution of the contact forces normal to the terrain as a function of the absolute value of the corresponding tangential contact forces. The solid lines indicates the median of the distribution and the dashed line the interquartile range of the distribution (i.e. 50 % of the distribution lies between the dashed lines). For the controller minimizing contact forces we used $K_{tx} = K_{ty} = 10N^{-2}$ in all experiments.

On the slope, the results of the performance measured by the slip are even more distinct. The controller having optimal distribution of forces slips less than 6 mm for a friction coefficient higher than 0.5, while the other controllers slip up to 2 to 3 times more for the same simulation conditions. We note that the robot controlled using the original inverse dynamics controller was not able to climb slopes with static friction lower than 0.42 (i.e. the robot would slip and eventually fall) while the other controllers could climb a slope with static friction as low as 0.33. The amount of slipping is still lower for the proposed controller but we note that the high gain PID controller has similar performance for the lowest friction coefficient – again we would like to point out that, however, the PID controller is not compliant and rather stiff.

Again, the distribution of forces on the ground varies less and is more vertically oriented in the case of the controller using the optimal distribution of forces.

4.3 Application to Real Robot

In this section we present experiments with the real Little Dog robot. We ran the locomotion controller with the original inverse dynamics controller minimizing the torque command and with the new controller optimizing tangential ground reaction forces. The terrain was composed of a level flat board and a slope of 0.46 radians, which is higher than the experiments done in simulation – the actual robot turned out to be more capable than our physical simulator. The controller was run at 3 different speeds with a stepping period between 3 to 4 seconds. We were not able to see significant behavioral differences, i.e. we could not see a case where one controller was able to make the robot go up the slope while the other would not. However, we consistently noticed that when using the controller minimizing the tangential forces the robot would reach the top of the slope faster as can be seen in Figure 5. We observed this behavior in all the experiments we ran. It is interesting since the planned desired trajectories were the same for both controllers.

When looking at the amount of slipping, we can see a consistent decrease in the amount of slipping with the controller using optimal distribution of contact forces, as we show in Figure 6. The amount of slip is computed accurately thanks to a motion capture system that tracks the position of the robot. We notice that the robot is slipping on average 30% less when using the new controller.

We also looked at the distribution of contact forces when the robot was walking on level ground (Figure 6). While this data has to be interpreted cautiously in view of the high level of noise in the force measurements, we clearly see a trend for a better distribution of forces when using the new proposed controller.

While we see a clear trend of improvement of locomotion on the real robot, this improvement was not sufficient enough to be able to see the behavioral differences

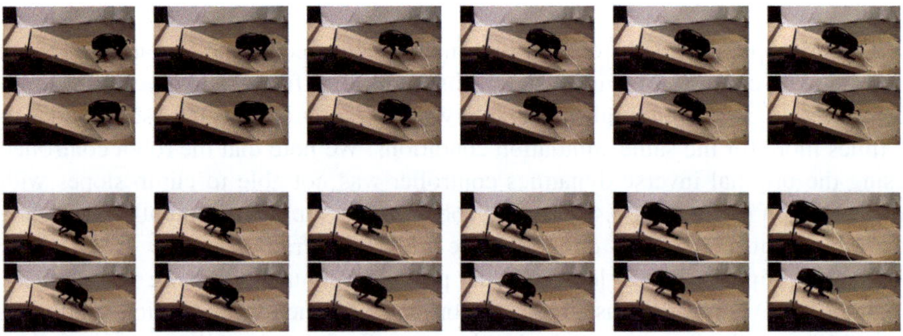

Fig. 5 Snapshots of a typical experiment. The figure is organized as two rows of snapshot, time going from left to right. There is approximately 2.5 seconds between each frame. For each frame, the upper graph is the experiment using the controller with minimization of tangential forces and the lower one corresponds to the controller minimizing the command cost. We notice that due to a reduced amount of slipping, the robot using the minimization of tangential forces reaches the top of the slope faster. The planned desired trajectories are the same for both robots.

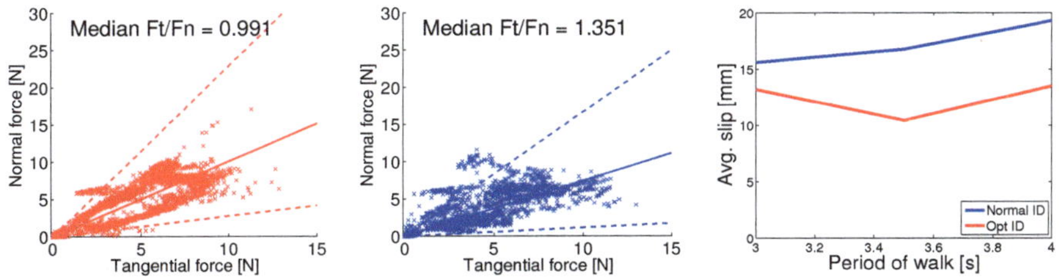

(a) Opt. ID (b) Normal ID (c) Slope climbing performance

Fig. 6 Real robot on 0.25 radians slope, in red the optimal controller in blue the normal ID controller

observed in simulation. However a few limitations of the robotic platform can explain the quality of the results:

- As we discussed in the previous section, our inverse dynamics controller is running at a 100Hz bandwidth on a host computer. Such a low bandwidth of control clearly limits the possible performance of the robot. For example, on a modern torque controlled humanoid robot such as the Sarcos humanoid, one can expect a 1kHz bandwidth of control.
- There are no torque sensors on the robot to close a torque feedback loop – torque control is inferred from current control. Therefore any error in the model converting torques into motor currents will have a negative impact on the actual torques applied to the robot compared to the desired ones.
- The quality of the dynamics model can also play a role. We evaluated the dynamics model of the robot in a similar way we did in the previous section, however it turns out that the dynamics is mainly dominated by friction in the joints, which is a local (decentralized) effect that cannot be re-distributed in a way as suggested by our controller that exploits actuation redundancy. We note that in our experiment roughly 40 to 50% of the total command is due to the PD controller, which is rather high.

These limitations are not fundamental ones and can easily be overcome in more advanced robotic platforms with joint-level force sensing and high control bandwidth. Our current results then suggest that the proposed controller should perform even better on these platforms.

5 Conclusion

In this chapter we showed how torque redundancy could be exploited to optimize contact forces in inverse dynamics controllers. The proposed redundancy resolution scheme allows to minimize any combination of linear and quadratic costs in the contact constraints and the commands. Given a desired trajectory, the controller is optimal at each instant of time. The resulting controller is surprisingly simple as it merely involves the inclusion of a weighted pseudo-inverse and an internal torque vector in the nullspace of the motion. It can therefore be implemented even on real-time computing hardware with modest computational power. Moreover it can be shown that the same controller can be used for whole-body controllers based on the operational space control framework [8].

We proposed to use this result either to track desired contact force trajectories or to minimize tangential contact forces during legged locomotion. Simulation results show that, given desired trajectories, we can exploit torque redundancy to achieve high tracking performance while guaranteeing a better distribution of contact forces and therefore better locomotion on difficult terrains. This constitutes an interesting complement to planning algorithms from the control point of view. Our results on the Little Dog robot, which is not an ideal platform for torque control, also show that the proposed controller is not a pure theoretical result but is realistic enough to be used on real systems. Such results are very encouraging as we expect to see much more improvement on a properly torque-controlled platform.

Acknowledgements. This research was supported in part by National Science Foundation grants ECS-0326095, IIS-0535282, IIS-1017134, CNS-0619937, IIS-0917318, CBET-0922784, EECS-0926052, CNS-0960061, the DARPA program on Advanced Robotic Manipulation, the Army Research Office, the Okawa Foundation, the ATR Computational Neuroscience Laboratories, and the Max-Planck-Society.

References

1. Aghili, F.: A unified approach for inverse and direct dynamics of constrained multibody systems based on linear projection operator: Applications to control and simulation. IEEE Transactions on Robotics 21(5), 834–849 (2005)
2. Ben-Israel, A., Greville, T.: Generalized inverses: theory and applications. Springer-Verlag New-York Inc. (2003)
3. Hyon, S., Hale, J.G., Cheng, G.: Full-body compliant human-humanoid interaction: Balancing in the presence of unknown external forces. IEEE Trans. on Robotics 23(5), 884–898 (2007)
4. Jiang, W.Y., Liu, A.M., Howard, D.: Optimization of legged robot locomotion by control of foot-force distribution. Transactions of the Institute of Measurement and Control 26(4), 311–323 (2004)
5. Kalakrishnan, M., Buchli, J., Pastor, P., Mistry, M., Schaal, S.: Learning, planning, and control for quadruped locomotion over challenging terrain. International Journal of Robotics Research 30, 236–258 (2011)
6. Klein, C., Kittivatcharapong, S.: Optimal force distribution for the legs of a walking machine with friction cone constraints. IEEE Transactions on Robotics and Automation 6(1), 73–85 (1990)
7. Mistry, M., Buchli, J., Schaal, S.: Inverse dynamics control of floating base systems using orthogonal decomposition. In: Proc. of the 2010 International Conference on Robotics and Automation (2010)
8. Righetti, L., Buchli, J., Mistry, M., Kalakrishnan, M., Schaal, S.: Optimal distribution of contact forces with inverse dynamics control (submitted)

9. Righetti, L., Buchli, J., Mistry, M., Schaal, S.: Control of legged robots with optimal distribution of contact forces. In: 2011 11th IEEE-RAS International Conference on Humanoid Robots (Humanoids), pp. 318–324 (2011)
10. Righetti, L., Buchli, J., Mistry, M., Schaal, S.: Inverse Dynamics Control of Floating-Base Robots with External Constraints: a Unified View. In: Proceedings of the 2011 IEEE International Conference on Robotics and Automation (ICRA 2011), pp. 1085–1090 (2011)
11. Sentis, L.: Synthesis and control of whole-body behaviors in humanoid systems. Ph.D. thesis, Stanford University (2007)
12. Sentis, L., Park, J., Khatib, O.: Compliant control of multi-contact and center of mass behaviors in humanoid robots. IEEE Transactions on Robotics 26(3), 483–501 (2010)
13. Siciliano, B., Sciavicco, L., Villani, L., Oriolo, G.: Robotics: Modelling, Planning and Control. Advanced Textbooks in Control and Signal Processing. Springer, London (2009), doi:10.1007/978-1-84628-642-1
14. Stephens, B., Atkeson, C.: Dynamic balance force control for compliant humanoid robots. In: International Conference on Intelligent Robots and Systems, IROS (2010)
15. Udwadia, F., Kalaba, R.: On the foundations of analytical dynamics. Int. J. of Non-Linear Mechanics 37, 1079–1090 (2002)

Exploiting Heterogeneity in Robotic Networks

Nicola Bezzo, R. Andres Cortez, and Rafael Fierro

Abstract. In this chapter we consider the problem of coordinating robotic systems with different dynamics, sensing and vision capabilities, to achieve a unison mission goal. An approach that makes use of a heterogeneous team of agents has several advantages when cost, integration of capabilities, or possible large search areas need to be investigated. A heterogeneous team allows for the robots to become "specialized" in their abilities and therefore accomplish sub-goals more efficiently which in turn makes the overall mission more efficient. We first propose a prioritized search algorithm combined with communication constraints to provide a decentralized prioritized sensing control algorithm for a heterogenous sensor network that maintains network connectivity. By specifying particular edge weights in the proximity graph, we provide a technique for biasing particular connections within the heterogenous sensor network. Then in the second part of the chapter we show a hierarchical approach to optimally allocate the tasks of a mission to specific agents. We develop a decentralized algorithm based on artificial physics and potential functions to guide the heterogeneous robotic network in the environment while maintaining connectivity constraints. Simulations and an experimental result are provided to show the applicability of the proposed frameworks.

1 Introduction

In recent years we have witnessed an increase in the use of mobile robots for different applications spanning from military to civilian operations. Search and rescue

Nicola Bezzo · Rafael Fierro
Department of Electrical & Computer Engineering, University of New Mexico, Albuquerque, NM 87131, USA
e-mail: {nicbezzo,rfierro}@ece.unm.edu

R. Andres Cortez
Los Alamos National Laboratory, Los Alamos, NM 87545, USA
e-mail: racortez31@gmail.com

D. Milutinović & J. Rosen (Eds.): Redundancy in Robot Manipulators, LNEE 57, pp. 53–75.
DOI: 10.1007/978-3-642-33971-4_4

missions, disaster relief operations, and surveillance are just few examples of scenarios where the use of autonomous and intelligent robotic systems is preferred over the use of human first responders. In such operations wireless communication needs to be reliable over the robotic network to maneuver the unmanned vehicles and transmit information. We are interested in heterogeneous robotic systems with agents having different dynamics, sensing behaviors, and functionalities. For instance, aerial vehicles have the capability to cover an area faster, while flying over obstacles, but cannot have a detailed view of mines, caves, or buildings where line-of-sight is lost. On the other hand, ground vehicles, like wheeled or multi-legged robotic platforms, can explore a limited area but with much more accuracy. In this work we exploit the cooperative coordination of ground and aerial robots to perform operations impossible by using only a single type of agents. Specifically, here we consider two scenarios. In the first case study, we create a theoretical approach to build the connectivity links between communication relay and sensor agents. Then we combine a prioritized search algorithm to send sensor agents to areas with a high probability of having good information. In the second part of this chapter, we envision a scenario, depicted in Fig. 1, in which an aerial vehicle carries a crawling robot for fast deployment in a target place. After the crawling robot is deployed, it explores an area inaccessible by the quadrotor (*i.e.,* a cave, a mine, small target, etc.). The aerial transportation system together with other mobile relays act as communication transceivers to tether the signal between a fixed base station and the exploring crawler on the ground.

When dealing with such systems, the communication within the network is fundamentally important due to the uncertainties of the wireless channels. In this chapter we pursue strategies to enhance the connectivity of a network made of heterogeneous robotic agents and a fixed base station. We are motivated by those situations where the communication infrastructure is not available or suddenly discontinued (*e.g.,* earthquakes, hurricanes, tornados, *etc.*) and in which the environment is partially known. In such scenarios, it is necessary to have a fast deployment of communication nodes to cover and search unknown and possibly hazardous areas.

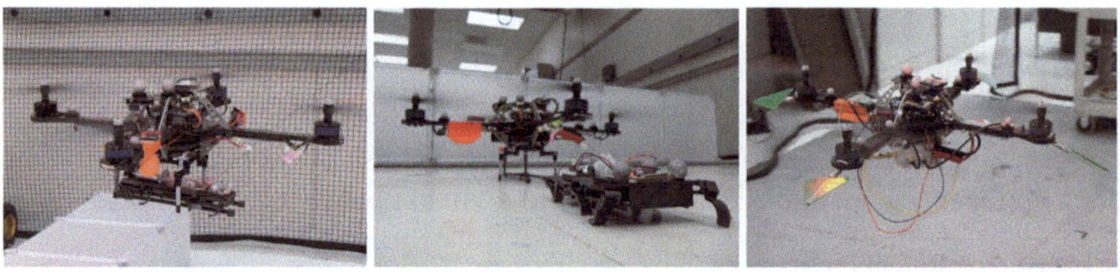

(a) (b) (c)

Fig. 1 (a) A Hummingbird quadrotor [1] transporting a OctoRoACH crawling robot [2]; (b) Deployment of the crawler using a quadrotor; and (c) An example of aerial mobile relay with four directional antennas.

1.1 Related Work

The use of heterogeneity in robotic applications is a recent topic of research that is attracting several researchers because of the challenges created when dealing with multi-agent systems having different dynamics, sensing, and manipulation capabilities. Authors in [3] consider formally a heterogeneous system and analyze its properties based on graph coloring techniques to assign colors to different types of agents. Similar to the work presented in this chapter, the authors in [4] use agents with different dynamics and capabilities to execute multiple missions in a decentralized fashion considering task sequencing and a consensus-based technique.

Wireless communication in robotics is one of the main aspects to guarantee autonomous and decentralized swarming behavior. The robotics and control community are very active in investigating the integration of communication in robotics applications, because the uncertainties found in wireless channels can compromise the performance of the entire multi-agent system. For instance, the authors in [5, 6] formally analyze the communication channel properties and use them to optimally navigate autonomous agents to improve some communication performances such as the Signal-to-Noise Ratio (SNR), Bit Error Rate (BER) and Capacity. In [7], the authors propose a modified Traveling Salesperson Problem to navigate an underwater vehicle in a sensor field, using a realistic model for acoustic communication. In [8], a Rician fading model for the communication channel is utilized in a pursuit-evasion game with two mobile agents moving in a cluttered environment. In [9], the authors optimize routing probabilities to ensure desired communication rates while using a distributed hybrid approach. In [10], we tether a chain of mobile routers to keep line of sight communication between a base station and a user that moves in a concave environment. Authors in [11] show extensive experimental results in which the communication throughput is optimized by making small variations in the positions of the mobile agents in the environment. A stop-time policy is proposed in [6] to control a robot to adjust its speed based on wireless link quality. Authors in [12] use aerial vehicles as relays to build the communication infrastructure for a team of ground vehicles by designing a gradient based controller that incorporates the *Signal to Interference Ratio* (SIR). In [13], the authors present a decentralized framework to maintain a graph connected by using a gradient-based control. Also, to demonstrate that the graph is connected the authors use the notion of algebraic connectivity. Similarly to the work presented in this chapter, the authors in [14] present a multi-agent system with interaction between aerial and ground vehicles based on task assignment for complex missions. In [15], the researchers present a set of algorithms to repair connectivity within a network of mobile routers and then show outdoor experimental results to validate their algorithms, while, in [16] an agent attempts to estimate the network topology by using only local information. Work in [17] shows that a group of UAVs, by exploiting antenna diversity, follows the gradient of the signal available in a certain environment.

In [18], the authors utilize a spring-mass system analogy together with graph theory in order to demonstrate stability of a swarm of robots that move in a decentralized fashion to cover an area while maintaining connectivity. Similar work using graph theory is proposed in [19], in which the author use the notion of Neighbor-Every-Theta (NET) graph to build connections among the nodes of a homogeneous network of mobile agents. More recently, the authors in [20] provide a survey about graph connectivity in mobile robot swarms, discussing different approaches and algorithms to maintain and optimize connectivity among mobile robot networks, from a graph-theoretical point of view.

Finally, from a sensing perspective, authors in [21] present an optimization framework to maneuver aerial vehicles equipped with cameras to perceive a certain area based on field of view properties. In this chapter we use a similar optimization approach to decide which agent or combination of agents need to be used to accomplish a certain task, based on the environment knowledge, sensors and type of vehicles available at the beginning of a mission.

The remainder of this chapter is organized as follows. In Section 2 we define the heterogeneous system and formulate the connectivity problem considering relay and sensor agents. Then we present the algorithms necessary to move the heterogeneous system using a prioritized search method. In Section 3 we analyze a similar scenario but we use a task allocation algorithm and decentralize control laws based on artificial physics and potentials to optimally move the heterogeneous system in a partially known environment. In both case studies, we present simulation results and for the second scenario also an experiment to validate the proposed framework. Finally, we draw conclusions in Section 4.

2 Heterogeneous Connected Robotic System

In this section we formally give a definition of heterogeneous robotic network followed by the problem formulation and connectivity constraints used to minimize motion constraints.

Definition 1. (Heterogeneous System): A network of $\mathcal{N}$ robots is called *heterogeneous* if the members of the network are interconnected, act together toward a common objective and produce results impossible to obtain by the action of homogeneous agents, and if the following conditions hold:

- one or more agents in the network have different motion dynamics with respect to other agents in the system;
- one or more agents in the network have different sensing constraints with respect to other agents in the systems, but all agents have wireless communication capabilities.

2.1 Problem Formulation

We begin by considering a heterogeneous team of $\mathcal{N}$ agents consisting of $\mathcal{N}_g$ sensing agents, which we will consider for this application as Unmanned Ground Vehicles (UGVs), and $\mathcal{N}_r$ relay agents, which we will assume are Unmanned Aerial Vehicles (UAVs), in two and three dimensions. Assume the $\mathcal{N}_g$ sensing agents are equipped with sensors capable of sensing an environmental phenomena within a finite radius $\mathcal{S}_s$ and communicating within a finite radius $\mathcal{C}_c(\mathbf{x}) \leq \mathcal{C}_{c_{max}}$ that depends on the position of the robots. Also let us assume that the $\mathcal{N}_r$ relay agents are capable of communicating over a finite radius $\mathcal{C}_{rc}$ such that $\mathcal{C}_{rc} > \mathcal{C}_{c_{max}}$ *i.e.*, the relay robots are better equipped for communication than the sensing agents and the relay robots' communication range is not dependent on location. Let $\mathcal{W}$ denote the area of interest and assume that $\mathcal{W}$ is a simple convex polygon with boundary $\partial\mathcal{W}$, including its interior. Define C_{obs} as the union of all obstacles in the region $\mathcal{W}$, and let $\mathcal{W}_{\text{ofree}} = \mathcal{W} \setminus C_{\text{obs}}$ be the area within $\mathcal{W}$ that is free of obstacles. We define the probability of detection map $M(\mathbf{x})$, which reflects the probability of detecting an environmental phenomena over the area to be searched [22]. The goal here is to find the set of feasible policies for the heterogeneous network such that the sensing agents are able to cover $M(\mathbf{x})$ while maintaining connectivity among the network.

2.2 Communication Constraints

We formulate the *connectivity constraint set* for each particular communication link possibility of our heterogeneous network based on the geometry of the communication radii. For the following definitions we will use $\mathcal{B}(\mathbf{x}, r)$ to denote a closed ball of radius r centered at $\mathbf{x}$ in $\mathbb{R}^3$.

Definition 2. (*Relay/Sensor or Relay* connectivity constraint set) Consider two agents, one relay agent i located at position $\mathbf{x}_i$ and one sensing or relay agent j located at position $\mathbf{x}_j$ such that the euclidean distance $||\mathbf{x}_i - \mathbf{x}_j|| \leq \mathcal{C}_{rc}$. Then the connectivity constraint set of agent i with respect to agent j is

$$\Upsilon_{\text{d}_{rs}}(\mathbf{x}_i, \mathbf{x}_j) = \Upsilon_{\text{d}_{rr}}(\mathbf{x}_i, \mathbf{x}_j) = \mathcal{B}\left(\frac{\mathbf{x}_i + \mathbf{x}_j}{2}, \frac{\mathcal{C}_{rc}}{2}\right). \tag{1}$$

Definition 3. (*Sensor/Sensor* connectivity constraint set) Consider two sensing agents, one agent i located at position $\mathbf{x}_i$ and one agent j located at position $\mathbf{x}_j$ such that $||\mathbf{x}_i - \mathbf{x}_j|| \leq \mathcal{C}_c$. Then the connectivity constraint set of agent i with respect to agent j is

$$\Upsilon_{\text{d}_{ss}}(\mathbf{x}_i, \mathbf{x}_j) = \mathcal{B}\left(\frac{\mathbf{x}_i + \mathbf{x}_j}{2}, \frac{\mathcal{C}_c}{2}\right). \tag{2}$$

Definition 4. (Connectivity constraint set for relay agent w.r.t. heterogeneous network) Consider a group of agents containing both sensing and relay agents located at $\mathcal{X} = \{\mathbf{x}_1, \mathbf{x}_2, \ldots, \mathbf{x}_{\mathcal{N}}\}$. Then the connectivity constraint set of relay agent i with respect to all other agents in the group is

$$\Upsilon_{\text{d}_{hr}}(\mathbf{x}_i, \mathcal{X}) = \{p \in \Upsilon_{\text{d}_{rr}}(\mathbf{x}_i, \mathbf{x}_j) | q \in \mathcal{X} \setminus \{\mathbf{x}_i\} \text{ s.t. } ||q - \mathbf{x}_i|| \leq \mathcal{C}_{rc}\}. \tag{3}$$

Definition 5. (Connectivity constraint set for sensor agent w.r.t. heterogeneous network) Consider a group of agents containing both sensing and relay agents located at $\mathcal{X} = \{\mathbf{x}_1, \mathbf{x}_2, \ldots, \mathbf{x}_{\mathcal{N}}\}$. Then the connectivity constraint set of a sensor agent i with respect to all other agents in the group is

$$\Upsilon_{\mathrm{d}_{hs}}(\mathbf{x}_i, \mathcal{X}) = \Lambda_{ss} \cap \Lambda_{sr}. \tag{4}$$

where

$$\Lambda_{ss} = \cap_{j=1}^{n} \Upsilon_{ss}(\mathbf{x}_i, \mathbf{x}_j), \text{ where } \mathbf{x}_j \in \text{ sensors}, \tag{5}$$

$$\Lambda_{sr} = \cap_{k=1}^{m} \Upsilon_{sr}(\mathbf{x}_i, \mathbf{x}_k), \text{ where } \mathbf{x}_k \in \text{ relays}. \tag{6}$$

The connectivity constraint sets of Definitions 2 - 5 describe the set of allowable positions that each robot may take such that the communication network will remain connected. Thus the connectivity constraint sets define the feasible motion for each individual robot to remain connected with the network.

2.3 *Heterogeneous Proximity Graph*

Due to the heterogeneity of our sensor network, we must define an appropriate proximity graph. As a reminder, a proximity graph describes connections between a set of vertexes based on their relative distances.

Definition 6. (Proximity Graph, [23]) Let $Q \subset \mathbb{R}^n$. A proximity graph $\mathcal{G}$ associates to a set of distinct points $\mathcal{X} = \{\mathbf{x}_1, \ldots, \mathbf{x}_{\mathcal{N}}\} \subset Q$, an undirected graph with vertex set $\mathcal{X}$ and whose edge set is given by $E_{\mathcal{G}}(\mathcal{X}) \subseteq \{(\mathbf{x}_i, \mathbf{x}_j) \in \mathcal{X} \times \mathcal{X} | \mathbf{x}_i \neq \mathbf{x}_j\}$.

We see that due to the heterogeneity of our network, the edge set of our proximity graph should depend on the agent type. The following definition describes how the edge set should be created for our heterogeneous proximity graph.

Definition 7. (Heterogeneous r($\mathbf{x}$)-disk graph)Two agents $\mathbf{x}_i$ and $\mathbf{x}_j$ are neighbors if they are located within a distance $r(\mathbf{x}) = \mathcal{C}_c$ if both $\mathbf{x}_i$ and $\mathbf{x}_j$ are sensing agents or $r(\mathbf{x}) = \mathcal{C}_{rc}$ if one of the agents is a relay agent, *i.e.*,

$$(\mathbf{x}_i, \mathbf{x}_j) \in E_{\mathcal{G}_{\mathrm{disk}(r(\mathbf{x}))}}(\mathcal{X}) \tag{7}$$

$$\text{if} \begin{cases} ||\mathbf{x}_i - \mathbf{x}_j|| \leq \mathcal{C}_c \text{ and } \mathbf{x}_i, \mathbf{x}_j \text{ both sensing agents} \\ ||\mathbf{x}_i - \mathbf{x}_j|| \leq \mathcal{C}_{rc} \text{ and } \mathbf{x}_i \text{ or } \mathbf{x}_j \text{ is a relay agent.} \end{cases} \tag{8}$$

In the $\mathcal{G}_{\mathrm{disk}(r(\mathbf{x}))}(\mathcal{X})$ graph, edges depend on the agent distances as well as agent connection combinations.

The heterogeneous r($\mathbf{x}$)-disk proximity graph, $\mathcal{G}_{\mathrm{disk}(r(\mathbf{x}))}(\mathcal{X})$, allows us to represent the network topology of our heterogeneous system. It is seen that depending on the configuration of the network there may exist heavy redundancy in the connections. This redundancy comes at the cost of more constraints on each agent, therefore reducing the size of the set of possible inputs that guarantee connectivity. This reduction stems from the fact of the intersection of multiple sets.

Let us now define the weighted complete graph which we will denote as $\mathcal{G}$ throughout the rest of this section.

Definition 8. (Weighted Complete Graph, $\mathcal{G}$) Let $Q \subset \mathbb{R}^n$. The weighted complete graph $\mathcal{G}$ associates to a set of distinct points $\mathcal{X} = \{\mathbf{x}_1, \ldots, \mathbf{x}_{\mathcal{N}}\} \subset Q$ an undirected graph with vertex set $\mathcal{X}$ and whose edges $e = (\mathbf{x}_i, \mathbf{x}_j) \in E_{\mathcal{G}}(\mathcal{X})$ have the following weights $w(e)$,

$$w(e) = \begin{cases} ||\mathbf{x}_i - \mathbf{x}_j|| + C_{rc} & \text{if } \mathbf{x}_i,\ \mathbf{x}_j \text{ both sensing agents} \\ ||\mathbf{x}_i - \mathbf{x}_j|| & \text{if } \mathbf{x}_i \text{ or } \mathbf{x}_j \text{ is a relay agent.} \end{cases} \quad (9)$$

2.4 *Minimizing Motion Constraints*

Given a formal way of representing the motion constraints for each agent with respect to the heterogeneous group, we are now left with trying to minimize the constraints (links) in such a way that we expand the input set the agents can choose from, that still guarantees connectivity at the next time step. One solution is to take $\mathcal{G}_{\text{disk}(\mathbf{x}))}(\mathcal{X})$ and run a minimum spanning tree algorithm to determine a subgraph of the r($\mathbf{x}$)-disk graph that has the minimum number of connections needed to remain connected [24, 25].

2.4.1 Shaping the Network Configuration

To help bias *relay/sensor* connections over *sensor/sensor* connections with respect to the Minimum Spanning Tree (MST) we now formulate a weighting factor for *sensor/sensor* connections. From definitions (2) and (3) we see that the motion constraint set for *relay/sensor* connections is larger than *sensor/sensor* connections due to a bigger communication radius. With the help of Fig. 2(a) we look at the scenario of one relay and two sensing agents. In terms of the MST, all connections that have a possibility of being biased can be broken down in this way. For ease of notation we will refer to the MST $\mathcal{G}_{\text{disk}(r(p))}$ as just the MST.

Let $||\mathbf{x}_i - \mathbf{x}_j|| = l$, $||\mathbf{x}_i - \mathbf{x}_k|| = l_1$ and $||\mathbf{x}_j - \mathbf{x}_k|| = l_2$. Let us assume that $l < l_1 \leq C_{rc}$ and $l_2 \leq C_c$. From the construction of the MST, the solid edges between $\mathbf{x}_i$, $\mathbf{x}_j$, and $\mathbf{x}_k$ in Figs. 2(a-b) will be chosen since $l + l_2 < l_1 + l_2$ and $l + l_2 < l + l_1$.

Let $\varepsilon > 0$ denote the minimum distance between two sensing agents, *i.e.*, physical footprint. Under other circumstances ε can also be considered the threshold distance where two sensing agents should communicate directly. To bias the *relay/sensor* connection (dotted line) a weighting factor ξ_1, must be constructed such that when $l_2 = \varepsilon$, $\xi_1 l_2 \geq C_{rc}$. Defining $\xi_1 = \left(\frac{C_{rc}}{\varepsilon} + \delta_1\right)$ with $\delta_1 \geq 0$ we get the following

$$\xi_1 l_2 = \left(\frac{C_{rc}}{\varepsilon} + \delta_1\right) l_2,$$
$$\xi_1 l_2 = C_{rc} + \delta_1 \varepsilon,$$
$$\xi_1 l_2 \geq C_{rc}.$$

Therefore, with the connection weighting factor ξ_1 we now have that $l + l_1 \leq l + \xi_1 l_2$ and $l + l_1 \leq l_1 + \xi_1 l_2$.

Weighting the *sensor/sensor* connection (edge) by a factor of ξ_1 allows us to bias the MST to choose the *relay/sensor* connections.

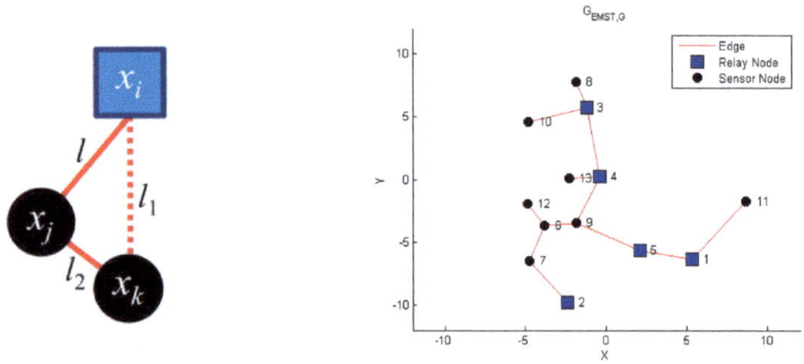

(a) (b)

Fig. 2 (a) One relay agent (square) and two sensing agents (circle) used to formulate weighting factor for *sensor/sensor* connections and (b) Example of the MST for thirteen agents with no connection weights.

In a similar fashion we can bias *relay/relay* connections. This is advantageous for certain mission objectives or when large amounts of data need to be transferred directly to a relay node. It may not be efficient or even possible to send large amounts of data through a sensing node to reach another relay node.

Using Fig. 3(a), as previously stated let us assume the minimum distance between any two agents is $\varepsilon > 0$. Let us also assume that from Fig. 3(a) that $l, l_1, l_2 < C_{rc}$ and for convenience assume $l_1 < l_2 < l$. From the point of view of the MST the solid edges between $\mathbf{x}_i$, $\mathbf{x}_j$, and $\mathbf{x}_k$ in Fig. 3(a) will be chosen since $l_1 + l_2 < l_1 + l$ and $l_1 + l_2 < l + l_2$.

To bias direct *relay/relay* connections (Fig. 3(a) dotted line), we use a weighting factor $\xi_2 = \frac{\varepsilon}{l}$. Choosing ξ_2 in this way insures that a direct *relay/relay* connection will be chosen over the multi-hop connection by the MST algorithm in Fig. 3, *i.e.*, relay$\rightarrow$ sensor$\rightarrow$ relay. This is seen from the fact that,

$$\xi_2 l = \frac{\varepsilon}{l} l,$$

$$\xi_2 l = \varepsilon.$$

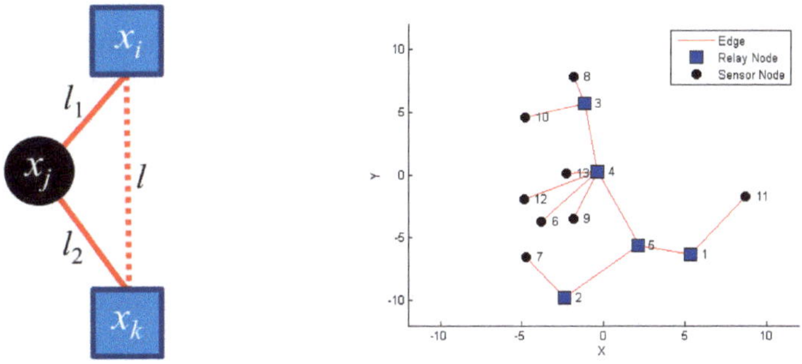

(a) (b)

Fig. 3 (a) Figure of two relay agents (squares) and one sensing agent (circle) used to formulate weighting factor for *relay/relay* connections and (b) Example MST_{CW} for thirteen agents with both *sensor/sensor* and *relay/relay* connection weights. Notice the *relay/sensor* connections are chosen over *sensor/sensor* connections and *relay/relay* connections are chosen over *relay/sensor* connections.

Therefore, now the distance between $\mathbf{x}_i$ and $\mathbf{x}_k$ is ε from the point of view of the MST algorithm. Since the minimum distance of any two agents is ε, the MST will choose the direct *relay/relay* link. Fig. 3(b) shows the network configuration using both ξ_1 and ξ_2 as connection weights (MST_{CW}).

2.5 *Prioritized Sensing with Connectivity Constraints*

In this section we combine our prioritized sensing behavior with the heterogeneous connectivity constraints to create an algorithm that can send sensing agents to areas with the highest possibility of having good information while also guaranteeing that the heterogenous network will remain connected. We also evaluate the performance of the heterogeneous algorithm against its homogeneous equivalent to determine the usefulness of heterogeneity within our approach.

2.5.1 Feasible Motion Sets: Sensing Agents

To combine the prioritized sensing objective with the network connectivity constraints we need to merge our probability of detection map, $(M(\mathbf{x}))$, and the connectivity constraint sets we previously computed. To do this we refer back to the prioritized sensing algorithm presented in [22], particularly the set D_i, which is the set of points within robot *i's* Voronoi partition not occupied by obstacles. For ease of notation let us define Υ_i as the connectivity constraint set for robot i as described in Section 2.2. Let us now define the feasible motion set, $\mathcal{M}_i$, for robot i as the following

$$\mathcal{M}_i = D_i \cap \Upsilon_i. \tag{10}$$

The feasible motion set $\mathcal{M}_i$ for robot i is the set of points within its Voronoi partition which is not occupied by obstacles and is limited to the points where network communication links can be maintained. Now we have a set, $\mathcal{M}_i$, which we can optimize over, that will guarantee network connectivity throughout the search process. The updated prioritized search algorithm which takes into account network connectivity is outlined in Algorithm 1.

Algorithm 1. Prioritized Sensing with Connectivity Constraints

while $t < t_{\text{final}}$ **do**
  for $x_i = 1, \dots, \mathcal{N}_g$ **do**
    Calculate Υ_i from $\mathcal{G}_{MST_{CW}}$ (Equations (1)-(4))
    Determine $V_i \in Q$
    Calculate $\mathcal{M}_i = D_i \cap \Upsilon_i$
    Optimize over $\mathcal{M}_i$ to determine an approximate maximum $\tilde{g}^*$ in $\mathcal{M}_i$ of $M(\mathbf{x})$
    if $\tilde{g}^*$ is reachable **then**
      $g_{\mathbf{x}_i} \Leftarrow \tilde{g}^*$
    else
      $\tilde{g}_i^* = \max(Y_1, \dots, Y_{N-1})$ excluding $\tilde{g}^*$ that was previously calculated
    end if
    Calculate $f_i(\mathbf{x}_i, g_{\mathbf{x}_i}, d)$ in $\mathcal{M}_i$ with g_{xi} set as the goal point.
    $\mathbf{u}_i = -k\nabla f_i(\cdot)$ where $k = |\mathbf{x}_i - g_{\mathbf{x}_i}|$
    $\forall \mathbf{x} \in \mathcal{S}_s, M(\mathbf{x}) = \beta M(\mathbf{x})$
    Exchange map information with neighbors in $\mathcal{G}_{MST_{CW}}$ graph
  end for
end while

In Algorithm 1 it is seen that when network connectivity is taken into consideration, the set over which the probability of detection (POD) map is being optimized may "shrink" due to the intersection of the two sets D_i and Υ_i. This makes sense since now the algorithm has to come to a compromise between the sensing objective and the network connectivity objective.

2.5.2 Feasible Motion Sets: Relay Agents

Our approach to addressing connectivity maintenance of a sensor network is to convert the network to a heterogeneous one by adding relay agents capable of better communication capabilities. For this particular application we are assuming that sensing agents are UGVs and relay agents are UAVs flying at a constant altitude. By using these assumptions, relay agent collisions with sensing agents in the network are not considered. Also, the relay agents communication range is considered to be its projection on the two dimensional space.

For the motion planning of relay agents in the network, Algorithm 2 is used to compute the centroid of each relay agents motion constraint set (MCS_i). Each relay agent i will move towards its respective centroid of its feasible motion set. By construction, the centroid c_i^* of each agents motion constraint set (MCS_i) lives in the interior of its motion constraint set. Therefore by setting c_i^* as the goal point for each relay agent, the heterogeneous team should remain connected when each goal point is reached by the respective agents. Algorithm 2 produces behavior of the relay agents that acts to balance the network in terms of the distances between its connected links. This is a desirable behavior in sensing/search problems because it allows the network to "stretch" as sensing agents move towards the outer regions of the search space. In essence, as a sensing agent moves towards uninvestigated areas it "pulls" a relay agent with it in order to maintain the connectivity of the network.

Algorithm 2. Centroidal Behavior ($g_{\mathbf{x}_i} = c_i^*$)

```
while t < t_final do
    for i = 1,...,N_r do
        Calculate c_i^* from MCS_i (Equations (1)-(4))
        g_{x_i} <= c_i^*
        while Δt < T_s do
            u_i(t) = -K x_i(t)
        end while
    end for
end while
```

2.6 *Simulations*

For the following simulation (Fig. 4(a), relay agents apply Algorithm 2 while sensing agents apply Algorithm 1 .We implement the algorithms on a heterogeneous network made up of 7 sensing agents and 4 relay agents. The search space is taken to be 60m $\times$ 60m square area, with the communication ranges for the sensing agents, $C_c = 3$m and $C_{rc} = 16$m for the relay agents. The sensing agents in the network are initialized in a random configuration with the relay agents situated at (-8m, -8m), (-8m, 8m), (8m, -8m), (8m, 8m) such that the initial configuration of the heterogeneous network is connected. The sensing radius of the sensing agents are taken to be 3m and the parameter that reflects the reduction of the probability of detection map, β is taken to be 0.8. Each simulation lasts for 50 iterations of the algorithm, approximately 120 seconds. Table 1 summarizes the results for five simulations.

Table 1 Heterogeneous Sensor Network (7 sensing, 4 relay agents): Prioritized Sensing

Simulation Number	Average POD per m^2	Max POD Value
1	0.0043	0.178
2	0.0104	0.337
3	0.0034	0.168
4	0.0031	0.124
5	0.0147	0.333

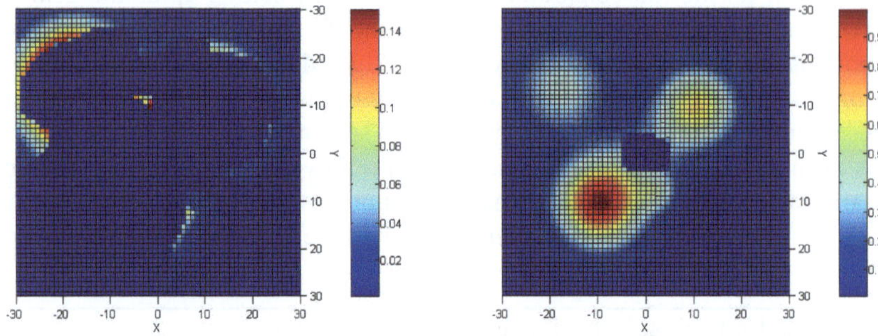

(a) (b)

Fig. 4 Probability of Detection Map after 50 iterations of the algorithm for (a) a Heterogeneous System and (b) a Homogeneous System. Notice that in (a) most of the area has been searched with exception to the upper left hand corner while in (b) just a small portion of the workspace is covered.

Fig. 4(a) shows the reduced POD map, $M(\mathbf{x})$, after 50 iterations of the algorithm with seven sensing agents and four relay agents. Notice that most of the area has already been searched and has been reduced to around 0.14. Finally in Fig. 4(b) we assume there are no relay agents in the network but only seven sensing agents. We can see that imposing the communication constraints directly on the network of sensing agents prevents the prioritized sensing algorithm significantly. The main reason is that the communication radius of the sensing agents is small compared to the area they have to explore. Thus to use a heterogeneous sensor network is advantageous when the area to explore is much larger than that of the communication range of the sensing agents.

3 Optimal Deployment of Heterogeneous Systems

In this section we present a generalized approach to deploy a number of heterogeneous robotic agents and achieve a common mission goal in a fast and efficient manner. We begin by building a task allocation function that considers the number and type of agents available to perform a specific goal, taking into account the sensing and communication constraints. Then we create a decentralized controller to move the selected agents toward the goal while maintaining connectivity through the use of mobile router relays. We assume that the heterogeneous system is composed by $\mathcal{N}_g$ sensor agents, $\mathcal{N}_a$ aerial transportation systems that can carry sensor agents, and $\mathcal{N}_r$ aerial communication relays. The total number of robots in the system is $\mathcal{N} = \mathcal{N}_g + \mathcal{N}_a + \mathcal{N}_r$. Throughout the discussion we use the notation R_a to symbolize an aerial relay and R_g for a ground sensor.

The problem we are solving in this section can be stated formally as follows:

Problem 1. Deployment of Heterogeneous Robotic Networks: Given a heterogeneous robotic network of $\mathcal{N}$ agents, L a leader agent chosen through optimization, and $\mathcal{T}$ a certain task to be accomplished within a specified area $\mathcal{W} \in \mathbb{R}^2$, find a set of feasible policies $\mathbf{u}_i \in \mathcal{U}$ for the leader and the minimum number of robots (mobile router relays) such that L is able to achieve the mission goal $\mathcal{T}$ while maintaining connectivity all of the time with a fixed base station b.

The approach considered to solve this problem can be divided into two main components:

- *Mission planning strategy* to decide which agents to be used during the mission; and
- *Communication analysis* to maintain connectivity during the operation.

3.1 Mission Planning Optimization

We desire to achieve certain mission goals j that consist of exploring and sensing a specific area $\mathcal{T}_j$ centered in c_v, to obtain as much information as possible with minimum cost. We assume that the mission and the robotic agents are initially positioned in a planar workspace $\mathcal{W} \subset \mathbb{R}^2$. Let $\mathbf{x}_i \in \mathcal{X}$ represent the state of each agent with $\mathbf{x}_i = (x_i, y_i, z_i)^T$ the position of the robot i with respect to the base station b, considered as the reference frame $b = (x_b, y_b, 0)^T$. We want to control the minimum number of agents to accomplish a mission goal quickly while minimizing the energy of the system.

If necessary, a relay agent can carry a sensor agent to help accomplishing a mission. By combining both an aerial and a ground vehicle together, like in Figs. 1(a-b), we can create a better agent to take advantage of the qualities of both entities and pass a big obstacle using the aerial vehicle and deploy the ground agent to explore in more detail a target.

In order to create a proper task assignment among the heterogeneous robotic system, we need to create an allocation function that considers the following factors:

- *the quality of execution of a certain task*. For instance a proper quality function can be the sensing accuracy, the precision in moving toward a certain goal or the quality of the communications;
- *the cost of execution of a certain task*. In this case we consider the power necessary to drive/fly the vehicle to a certain location, the maximum speed and time to accomplish a goal.

Therefore, combining these two cost functions, the main objective is to find a proper controller $\mathbf{u}^*$ that minimizes an utility function $J(\mathcal{X})$ with respect to the state space $\mathcal{X} = \{\mathbf{x}_1, \mathbf{x}_2, \ldots, \mathbf{x}_{\mathcal{N}}\}$. We consider the following utility function

$$J(\mathcal{X}) = \sum_{i,j} (\tau_{ij} c_{ij} + \lambda_{ij} h_{ij}) \tag{11}$$

subject to:

$$\begin{aligned} &\mathbf{x}_i^0 = (x_i^0, y_i^0, 0) \quad \forall i = 1 \ldots \mathcal{N} \\ &\mathbf{x}_o = (x_o, y_o, z_o) \quad \forall o = 1 \ldots \mathcal{O} \\ &\textstyle\sum_j \tau_{ij} \leq 1, \quad \sum_j \lambda_{ij} \leq 1 \ \forall i \\ &\textstyle\sum_i \tau_{ij} = 1, \quad \sum_i \lambda_{ij} = 1 \ \forall j \end{aligned} \tag{12}$$

where $\mathbf{x}_i^0$ is the initial position of the robots and $\mathbf{x}_o \in \mathscr{O}$ is the position of the obstacles that are known a priori. c_{ij} and h_{ij} are the cost functions and τ_{ij} and λ_{ij} assignment variables that can take either 0 or 1 binary values. In particular, $\tau_{ij} = 1$ if a certain task j is assigned to i considering the cost function c_{ij} and it is 0 otherwise. Similarly, $\lambda_{ij} = 1$ if the jth task is assigned to the ith agent considering the cost function h_{ij}.

The first cost function, c_{ij} is defined as follows

$$c_{ij} = \lim_{\mathbf{x}_i \to c_V} f(\mathbf{x}_i, \mathcal{T}_j, \alpha_i) \tag{13}$$

with

$$f(\mathbf{x}_i, \mathcal{T}_j, \alpha_i) = \begin{cases} k_i(\mathbf{x}_i - c_V)^2 & \text{if } ||\mathbf{x}_i - c_V|| > \sqrt{\alpha_i} \\ \alpha_i & \text{otherwise} \end{cases}, \tag{14}$$

where α_i value depends on the vision capabilities of the agent in consideration, and the type of mission to be accomplished. k_i depends on α_i and describes the perceptibility quality of an agent i. In general we assume that aerial relays R_a have a limited perceptibility capability but are able to converge to a desired position faster, while ground sensors R_g are slower but can perceive higher resolution details of the environment. This behavior is represented by properly tuning k_i and α_i associated with each agent. For instance $k_{R_a} > k_{R_g}$ and $\alpha_{R_a} > \alpha_{R_g}$.

The second cost function is defined as

$$h_{ij} = \varepsilon_i \left(\frac{\chi_{\text{ofree}}^{ij}}{\dot{\mathbf{x}}_i} \right) \tag{15}$$

with $\varepsilon_i \geq 0$ a term that considers the power consumption of agent i per unit time. χ_{ofree}^{ij} is the obstacle free path between the position $\mathbf{x}_i$ and the final position c_V of $\mathcal{T}_j$. Finally $0 \leq \dot{\mathbf{x}}_i \leq M_i$ is the velocity of agent i constrained above by a maximum value M_i that depends on each agent dynamics. For instance an aerial robot R_a such as a quadrotor is faster than a car like ground robot R_g, thus $M_{R_a} \geq M_{R_g}$.

Thus, by using this approach, we can decide the combination of agents to be used to achieve a certain mission. We will denote this combination of agents as leaders of the group as $\mathbf{x}_{Lj} = \{\tau_{1j}\mathbf{x}_1 + \lambda_{1j}\mathbf{x}_1, \ldots, \tau_{\mathcal{N}j}\mathbf{x}_{\mathcal{N}} + \lambda_{\mathcal{N}j}\mathbf{x}_{\mathcal{N}}\}$ if $\tau_{ij} \neq \lambda_{ij}$ or $\mathbf{x}_{Lj} = \{\tau_{1j}\mathbf{x}_1, \ldots, \tau_{\mathcal{N}j}\mathbf{x}_{\mathcal{N}}\}$ if $\tau_{ij} = \lambda_{ij}$. For ease of discussion, through the remaining of the chapter we will use the notation $\mathbf{x}_L$ to refer to the position of the leader L.

3.2 *Path Planning of the Heterogeneous System*

The dynamics of the ith robotic agent can be approximated using the following model

$$\begin{aligned} \dot{\mathbf{x}}_i &= \mathbf{v}_i, \quad i = [1, \ldots, \mathcal{N}] \\ \dot{\mathbf{v}}_i &= \mathbf{u}_i, \end{aligned} \tag{16}$$

where $\mathbf{x}_i = [x_i \quad y_i \quad z_i]^T \in \mathbb{R}^3$ is the position vector of the ith agent relative to the base station b, $\mathbf{v}_i \in \mathbb{R}^3$ and $\mathbf{u}_i \in \mathbb{R}^3$ denotes the velocity and acceleration (control input), respectively, for each agent $i \in \mathcal{N}$.

3.2.1 Goal Potential

Once the task allocation is completed, one or a combination of more agents L is deployed to reach the goal and accomplish the task, in a decentralized fashion. To navigate the leader L toward its goal $\mathcal{T} \in \mathcal{W}$, we can define an attractive potential function centered in the target position.

For simplicity sake, we use the following quadratic weight function to represent the goal attractive function:

$$\nu(\mathbf{x}_L) = \frac{A_\nu}{2} \|\mathbf{x}_L - c_\nu\|^2, \tag{17}$$

where $A_\nu \in \mathbb{R}^+$ determines the size of the quadratic range and $c_\nu \in \mathbb{R}^2$ is the center of the target attractive function. The relative attractive force is obtained by the negative gradient of (17):

$$F_\nu(\mathbf{x}_L) = -\nabla_{\mathbf{x}_L} \nu(\mathbf{x}_L) = -A_\nu(\mathbf{x}_L - c_\nu), \tag{18}$$

where $A_\nu > 0$ because we want an attractive field towards the center of the goal function.

3.2.2 Local Sensing Behavior

The workspace, $\mathcal{W}$, is populated with $\mathcal{O}$ known obstacles and N_o fixed polygonal obstacles $\{O_1, \ldots, O_{N_o}\}$, whose geometries and positions are assumed unknown. We imagine a scenario in which the agents have limited knowledge of the environment and the obstacles are detected locally when the robot moves to certain locations of the workspace. Once an obstacle is detected we define a local workspace repulsive potential field [26], in which the potential value approaches infinity as the robot approaches the obstacle, and goes to zero if the robot is at a distance greater than ρ_0 from the obstacle. Formally

$$W_{O,i} = \begin{cases} \frac{1}{2}\eta_i \left(\frac{1}{\rho(\mathbf{x}_i)} - \frac{1}{\rho_0}\right)^2 & \text{if } \rho(\mathbf{x}_i) \leq \rho_0 \\ 0 & \text{if } \rho(\mathbf{x}_i) > \rho_0 \end{cases}, \tag{19}$$

where $\rho(\mathbf{x}_i)$ is the shortest distance between the agent and any detected obstacle in the workspace and η_i is a constant.

Then, the repulsive force is equal to the negative gradient of $W_{O,i}$, or in formula for $\rho(\mathbf{x}_i) \leq \rho_0$ it is given by

$$F_{O,i} = \eta_i \left(\frac{1}{\rho(\mathbf{x}_i)} - \frac{1}{\rho_0}\right) \frac{1}{\rho(\mathbf{x}_i)^2} \nabla\rho(\mathbf{x}_i), \tag{20}$$

where $\nabla\rho(\mathbf{x}_i)$ is the gradient of the minimum distance between the robot and the closest detected obstacle.

3.2.3 Connectivity Maintenance

In our work the communications are simulated by spring interactions among the agents of the network.

The dynamics of the ith robotic router among the total $\mathcal{N}_r$ available, follow the model:

$$\mathbf{u}_i = \left[\sum_{j\in S_i} \kappa_{ij}\,(l_{ij}-l_0)\,\hat{\mathbf{d}}_{ij}\right] - \gamma_i\dot{\mathbf{x}}_i, \tag{21}$$

$$\text{with } i = 1,\ldots,\mathcal{N}_r \text{ and } i \neq j,$$

where S_i is the set of neighbor robots connected to the ith router and since we are using a spring-mass model these links between agents are virtual springs; $l_{ij} = \|\mathbf{x}_i - \mathbf{x}_j\|$ is the length of the spring between robot i and j and l_0 is the rest spring length; $\hat{\mathbf{d}}_{ij}$ is the unit vector indicating the direction of the force of the virtual spring between the robots and finally κ_{ij} and γ_i are the spring constant and the damping coefficient respectively, between robots i and j . Note that l_0 depends on the two agents i and j. Specifically, within this work, two situations can occur:

- if $C_i > C_j$, then $l_0 = \rho_i$ with ρ_i the maximum range of communication of agent i; and
- if $C_i = C_j$, then $l_0 = \rho_i = \rho_j$ with ρ_i (ρ_j) the maximum range of communication of agent i (j).

3.2.4 Routing Potential

The target to reach can be at a distance greater than the communication range within a fixed base station, therefore, some mobile routers may be needed to maintain connectivity and route information to the base station. While L explores the environment attracted by the goal potential, it detects some obstacles along the path and starts to build a map. Also at every step, a cumulative decreasing attractive potential is created and transmitted to the group of mobile routers. This attractive function has the following expression:

$$\varphi(\mathbf{x}_i)(t+\Delta t) = \min\left\{\varphi(\mathbf{x}_i)(t), \phi(\mathbf{x}_i)(t+\Delta t)\right\} \tag{22}$$

with:

$$\phi(\mathbf{x}_i)(t+\Delta t) = B_\phi - (t+\Delta t)\zeta\frac{A_\phi}{2}e^{-\frac{\|\mathbf{x}_i-\mathbf{x}_L\|^2}{\ell_\phi}} \tag{23}$$

and

$$\varphi(\mathbf{x}_i)(t) = \phi(\mathbf{x}_i)(t), \tag{24}$$

where $A_\phi \in \mathbb{R}^+$, $\ell_\phi \in \mathbb{R}^+$, $0 < \zeta \leq 1$ and $B_\phi \geq A_\phi$.

Depending on the mission planning, the mobile routers are chosen based on the minimum feasible distance between the leader and the base station, the type of leader deployed, and the cost and quality of the communication service of the agents available.

3.3 Controllers

By assembling together all the pieces described in the previous sections the general control law for the leader follows

$$\mathbf{u}_L = \begin{cases} F_{O,L} - \nabla_{\mathbf{x}_L} v(\mathbf{x}_L) & \text{if } ||\mathbf{x}_L - c_v|| > \varepsilon_v \\ 0 & \text{otherwise} \end{cases}, \tag{25}$$

with ε_v small to ensure that the leader gets as close as possible to the target position. Based on the optimization algorithm (11) if $\tau_{ij} \neq \lambda_{ij}$ $\forall i \in \mathcal{N}$ and a given mission goal $\mathcal{T}_j$, two agents are necessary to optimally achieve $\mathcal{T}_j$, having a transportation relay R_{a_L} moving a sensor agents R_{g_L} to the goal position. The combination relay/sensor, that we denote by $L = \{a_L, g_L\}$, will move toward the goal attracted by (18) and will stop to deploy R_{g_L} when the following constraint is satisfied.

Given $\mathcal{B}_j = \mathcal{B}(c_v, \rho_j)$ the minimum ball of radius ρ_j centered in the target position c_v that contains the object to visit, the control laws for the two agents R_{a_L} and R_{g_L} are:

$$\mathbf{u}_{a_L} = \begin{cases} \mathbf{u}_L & \text{if } \mathbf{x}_L \in \mathcal{W} \setminus \mathcal{B}_j \\ \mathbf{u}_i & \text{if } \mathbf{x}_L \in \mathcal{B}_j \end{cases}, \tag{26}$$

where $\mathcal{W} \setminus \mathcal{B}_j$ is the portion of workspace excluded $\mathcal{B}_j$, $\mathbf{u}_i$ is the control law for the mobile routers and is defined in (28), and,

$$\mathbf{u}_{g_L} = \begin{cases} 0 & \text{if } \mathbf{x}_L \in \mathcal{W} \setminus \mathcal{B}_j \\ \mathbf{u}_L & \text{if } \mathbf{x}_L \in \mathcal{B}_j \end{cases}, \tag{27}$$

The control law for the mobile routers is given by:

$$\mathbf{u}_i = \begin{cases} P_{ij} - \gamma_i \dot{\mathbf{x}}_i - \nabla_{\mathbf{x}_i} \varphi(\mathbf{x}_i) & \text{if } ||\mathbf{x}_L - c_v|| > \varepsilon_v \\ P_{ij} - \gamma_i \dot{\mathbf{x}}_i & \text{otherwise} \end{cases}, \tag{28}$$

with $P_{ij} = \left[\sum_{j \in S_i} \kappa_{ij} (l_{ij} - l_0) \hat{\mathbf{d}}_{ij}\right]$

Specifically, for the mobile router control law, if $||\mathbf{x}_L - c_v|| > \varepsilon_v$ and $||\mathbf{x}_L - b|| \geq \frac{l_0}{2}$, we create the following constraints:

- $\forall m \in (S_L \cap S_b)$, with S_L (S_b) the set of neighbor robots in communication range with L (b), $\exists$ at least one m such that

$$||\mathbf{x}_m - \mathbf{x}_L|| \leq \frac{l_0}{2}. \tag{29}$$

When such situation occurs, $\{m \text{ s.t. } ||\mathbf{x}_m - \mathbf{x}_L|| = \min_{i\in(S_L\cap S_b)}(||\mathbf{x}_i - \mathbf{x}_L||)\}$ is used as the first mobile router in the chain between the base station b and the leader L;

- if m chosen in the previous step moves a distance $||\mathbf{x}_m - \mathbf{x}_b|| \geq \frac{l_0}{2}$ then another node n is chosen with the same criterion given in the above step $\{n \text{ s.t. } ||\mathbf{x}_n - \mathbf{x}_m|| = \min_{i\in(S_m\cap S_b)}(||\mathbf{x}_i - \mathbf{x}_m||)\}$. This iteration is repeated until the leader reaches its final position and doesn't move.

Constraint (29) is imposed to achieve robustness in connectivity. Before stating formally the robustness theorem, we give the following definition:

Definition 9. (Stem Graph) A *stem* is a graph in which two special nodes (a head and a tail) are singled out. This particular tree has n total vertices of degree ≤ 2 and $n-1$ edges; it is connected and contains no closed loops (cycles).

Based on Definition 9, the following theorem can be formulated:

Theorem 1. *Given a stem graph $\mathcal{G}(V,E)$ in which each node $v_i \in V$ is a mobile agent with spring-mass-damping dynamics* (21) *and thus each edge $e_i \in E$ is a virtual spring connectivity link, if one node disappears from the graph (i.e., one agent stops working) then we can guarantee that the graph is able to stay connected and all nodes reach a new stable state.*

Proof. In [26] we showed that a spring-mass system eventually reaches equilibrium in which all agents have a null state (*i.e.*, $\dot{\mathbf{x}} = 0$). When this situation occurs, the potential of each agent of the system is

$$U_i = \sum_{j\in S_i} \frac{1}{2}\kappa_{ij}(||\mathbf{x}_i - \mathbf{x}_j|| - l_0)^2 \geq 0 \tag{30}$$

Given $\mathcal{N}_{bL}$ the number of mobile routers in the chain between b and L, the total potential of the system is given by:

$$U_{\text{tot}} = \sum_{i=1}^{\mathcal{N}_{bL}} U_i \geq 0 \tag{31}$$

If one node p, connected to m and n, disappears, because of the constraint in (29), we will have that

$$U_{\text{tot}}^{+} - U_{\text{tot}}^{-} \leq 0 \tag{32}$$

where U_{tot}^{+} is the energy after p is removed from the chain and U_{tot}^{-} is the energy before the network loses connection with p. This result is intuitive because we have assumed that m and n are in communication range. Since l_{pm} (and l_{pn}) $< l_{mn} \leq l_0$ we have that the energy after the switch is less than the energy before the switch, bringing the network to a new more stable point while maintaining connectivity.

3.4 Simulation Results

In this section we present a simulation result to demonstrate the use of the control laws and optimization algorithm developed in this section. We create a partially known environment in which we assume to know the position and shape of some obstacles in the environment while other obstacles are detected only locally as the agents move to achieve the mission goal.

We consider a heterogeneous group of mobile robots with the following classes of agents available in the environment: 5 aerial transportation systems R_{a_i} $(i=1,\ldots,5)$ with sensing footprint $\mathcal{S}_a$ and communication range area $\mathcal{C}_a$. Each of these agents can carry a ground sensor robot; 5 ground sensor robots R_{g_i} $(i=1,\ldots,5)$ with sensing footprint $\mathcal{S}_g < \mathcal{S}_a$ and communication range area $\mathcal{C}_g \leq \mathcal{C}_a$; 10 aerial communication relays R_{r_i} $(i=1,\ldots,10)$ with positioning capabilities (*i.e.*, ability to estimate their position in the workspace relative to b) and communication range area $\mathcal{C}_r = \mathcal{C}_a$.

Fig. 5(a) shows the simulation workspace $\mathcal{W}$ with the target to reach and the types of agents available, as described above. Not all the agents are deployed during the mission, but just the minimum amount necessary to achieve a certain goal, following the theoretical approach described in the previous sections.

Assumption 2. For simplicity we assume that both aerial transportation systems and aerial relays can take off and land very fast, therefore we neglect the time delay necessary for an aerial vehicle to reach a certain altitude. Also in this work we make the aerial vehicles fly at a certain predefined altitude. In this way we see the workspace as a two-layer space (*ground* and *aerial* layers) in which at the ground level all obstacles are available and at the aerial level only fewer obstacles are perceptible, as depicted in Fig. 5(b).

The goal of the simulation in Fig. 6 is to reach and explore the object located in the far right corner of Fig. 6. Twenty robotic agents constituting the heterogeneous system are available, clustered around the base station (bottom left of the figures).

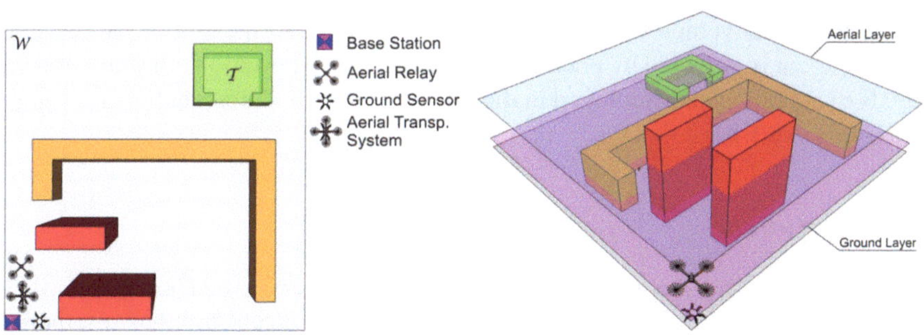

(a) (b)

Fig. 5 (a) The environment used during the simulation presented in this section and (b) A 3-D visualization of the operational layers of the heterogeneous robotic system.

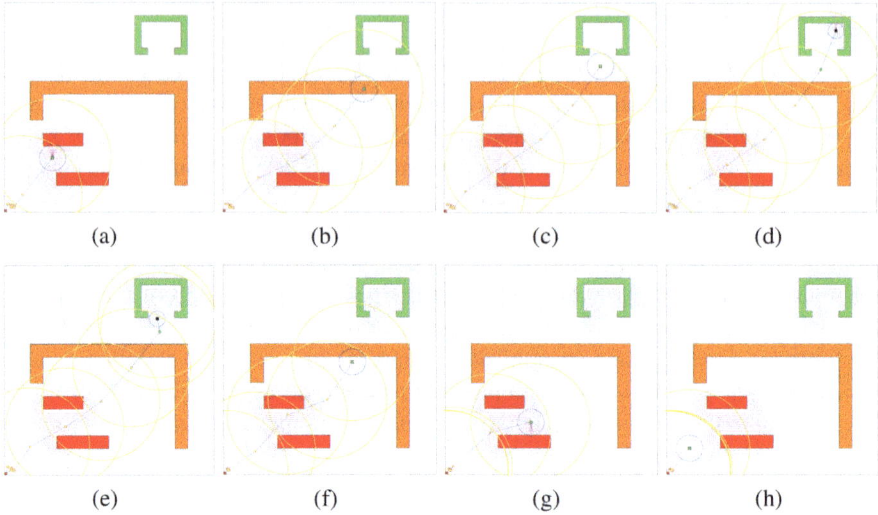

Fig. 6 Simulation result. (a)-(c) Deployment of a sensor agent R_g (bold square) using an aerial transportation relay R_a (diamond). (d) and (e) show different moments about the exploration of the target $\mathcal{T}$ before R_g is recovered by R_a.Through the mission a series of mobile routers move behind the leader to maintain connectivity with a fixed base station (bottom left of the figures). In (f-h) finally all the deployed agents return to the base station location.

After running the optimization algorithm, an aerial transportation relay R_a carrying a ground sensor R_g is chosen to achieve the mission goal. Precisely R_a moves attracted by potentials, toward $\mathcal{T}$. Upon reaching destination, the sensor agent is deployed to explore the target (Figs. 6(d) and (e)). While this operation is performed, a chain of mobile router relays R_r is built behind the leader transportation system to maintain connectivity with the base station (Figs. 6(a-e)). Note that the first two rectangular obstacles, located in the bottom left of Fig. 6 are unknown, while the large obstacle in the middle of the figure is known a priori and it is at a low altitude allowing the aerial vehicles to fly over it (see Fig. 5(b)). At the end of the mission, the sensor agent is attracted by potentials toward the transportation relay, as depicted in Fig. 6(e) and all agents deployed in the mission are pulled toward the base station position (Figs. 6(f-h)).

3.5 Experiment

An experiment has been performed using two quadrotors AscTec Hummingbird [1] and a crawling robot OctoRoACH [2, 27]. One of the quadrotors has a structure installed underneath its body allowing the crawler to sit while in flight (Fig. 1(a)) and to exit the structure when on the ground (Fig. 1(b)). The quadrotors are autonomously controlled by VIs developed in LabVIEW in which the position controllers (x, y, and z) and the yaw angle (ψ) controllers are implemented in each case

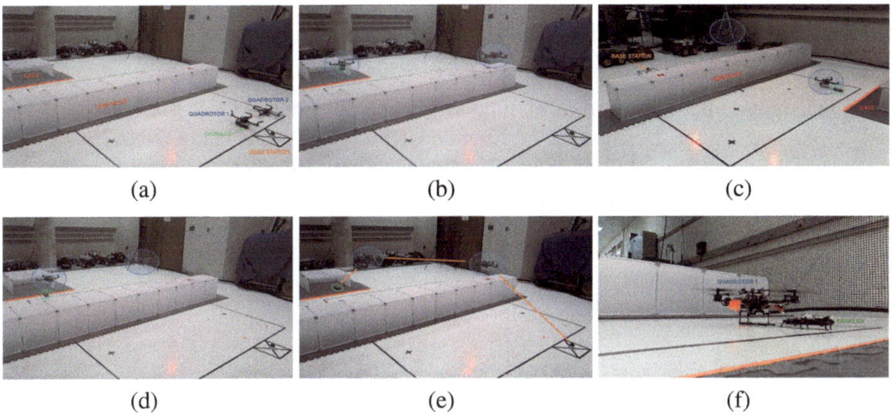

Fig. 7 Experimental result. (a)-(c) Deployment of a crawling robot via a quadrotor aerial system, while a second quadrotor is used as a communication relay. Snapshots in (d) and (e) show the quadrotors maintaining connectivity between the base station and the crawler that is exploring the cave. In (f) a quadrotor rescues the ground robot. Note that snapshots (c) and (f) belong to the same experiment but are taken from other angles for ease of viewing.

by applying a PID [28]. Figs. 7(a-c) show the deployment of a crawling robot by means of a quadrotor relay in order to allow the exploration of a cave. A second quadrotor is used to maintain connectivity between the nodes of the system during the operation (Figs. 7(d-e)). Finally, at the end of the mission the quadrotor recovers the crawler (Fig. 7(f)) and all the nodes convey to the base station location.

It is important to mention that communication connectivity is implemented by maintaining a certain distance separation between the nodes of the system. During the mission no agent gets disconnected and the nodes are able to cooperatively perform the assigned task.

4 Conclusions

In this chapter we have shown how to exploit heterogeneity to cooperatively coordinate a robotic network and perform complex missions in which a homogeneous system (*i.e.*, robotic networks with same capabilities) has a limited performance. We first address a prioritized sensing behavior with communication constraints for a network made up of sensing agents and mobile communication relays. Precisely, we show that by biasing the connections between sensing and relay agents we can obtain a better coverage and connectivity. In the second part of the chapter, we take advantage of several theoretical results to deploy the heterogeneous system in the workspace while minimizing energy and achieving a certain goal in a faster and a more accurate manner. We use optimal task assignment to select which agent or combination of agents to deploy and a chain of mobile router relays to maintain connectivity among the network. Both simulations and the experimental results demonstrate the applicability of the proposed strategies.

Acknowledgements. This work was supported by NSF grants ECCS #1027775 and IIS #0812338, and by the Department of Energy URPR Grant #DE-FG52-04NA25590. We would like to thank Prof. Luke Lester (PI Microelectronics Center) for providing the OctoRoACH crawling robots as part of the Micro Autonomous Systems & Technology (MAST) Program.

References

1. Ascending Technologies GmbH, `http://www.asctec.de/`
2. Pullin, A., Kohut, N., Zarrouk, D., Fearing, R.: Dynamic turning of 13 cm robot comparing tail and differential drive. In: IEEE International Conference of Robotics and Automation, ICRA (to appear, 2012)
3. Abbas, W., Egerstedt, M.: Distribution of agents in heterogeneous multi agent systems. In: IEEE Conference on Decision and Control (CDC), pp. 976–981 (December 2011)
4. Di Paola, D., Gasparri, A., Naso, D., Ulivi, G., Lewis, F.: Decentralized task sequencing and multiple mission control for heterogeneous robotic networks. In: IEEE International Conference on Robotics and Automation (ICRA), pp. 4467–4473. IEEE (2011)
5. Ghaffarkhah, A., Mostofi, Y.: Communication-aware surveillance in mobile sensor networks. In: American Control Conference (ACC), pp. 4032–4038. IEEE (2011)
6. Lindhé, M., Johansson, K.: Adaptive exploitation of multipath fading for mobile sensors. In: IEEE International Conference on Robotics and Automation (ICRA), pp. 1934–1939 (2010)
7. Hollinger, G., Mitra, U., Sukhatme, G.: Autonomous data collection from underwater sensor networks using acoustic communication. In: 2011 IEEE/RSJ International Conference on Intelligent Robots and Systems (IROS), pp. 3564–3570. IEEE (2011)
8. Bhattacharya, S., Basar, T., Hovakimyan, N.: Singular surfaces in multi-agent connectivity maintenance games. In: Proc. of IEEE Conference on Decision and Control and European Control Conference (CDC-ECC), pp. 261–266 (December 2011)
9. Zavlanos, M., Ribeiro, A., Pappas, G.: Distributed control of mobility & routing in networks of robots. In: Proc. IEEE Workshop on Signal Process. Advances in Wireless Commun., pp. 7545–7550 (2010)
10. Bezzo, N., Fierro, R., Swingler, A., Ferrari, S.: A disjunctive programming approach for motion planning of mobile router networks. International Journal of Robotics and Automation 26(1) (2011)
11. Vieira, M.A.M., Taylor, M.E., Tandon, P., Jain, M., Govindan, R., Sukhatme, G.S., Tambe, M.: Mitigating multi-path fading in a mobile mesh network. In: Ad Hoc Networks (2011)
12. Gil, S., Schwager, M., Julian, B., Rus, D.: Optimizing communication in air-ground robot networks using decentralized control. In: IEEE International Conference on Robotics and Automation (ICRA), pp. 1964–1971 (2010)
13. Sabattini, L., Chopra, N., Secchi, C.: On decentralized connectivity maintenance for mobile robotic systems. In: Proc. of IEEE Conference on Decision and Control and European Control Conference (CDC-ECC), pp. 988–993 (December 2011)
14. Whitten, A.K., Choi, H., Johnson, L., How, J.P.: Decentralized task allocation with coupled constraints in complex missions. In: Proc. of American Control Conference (ACC), San Francisco, CA (June 2011)

15. Derbakova, A., Correll, N., Rus, D.: Decentralized self-repair to maintain connectivity and coverage in networked multi-robot systems. In: Proc. of IEEE International Conference of Robotics and Automation (ICRA), Shangai, China, May 09-13 (2011)
16. Chowdhary, G., Egerstedt, M., Johnson, E.: Network discovery: An estimation based approach. In: Proc. of American Control Conference (ACC), San Francisco, CA (June 2011)
17. Griffin, B., Fierro, R., Palunko, I.: An autonomous communications relay in GPS-denied environments via antenna diversity. Journal of Defense Modeling & Simulation JDMS, Special Issue on Intelligent Behaviors for Tactical Unmanned Systems 9(1), 33–44 (2012)
18. Shucker, B., Murphey, T., Bennett, J.: Convergence-preserving switching for topology-dependent decentralized systems. IEEE Transactions on Robotics 24(6), 1405–1415 (2008)
19. Poduri, S., Pattem, S., Krishnamachari, B., Sukhatme, G.: Using local geometry for tunable topology control in sensor networks. IEEE Transactions on Mobile Computing 8(2), 218–230 (2009)
20. Zavlanos, M., Egerstedt, M., Pappas, G.: Graph-theoretic connectivity control of mobile robot networks. Proceedings of the IEEE (99), 1525–1540 (2011)
21. Schwager, M., Julian, B.J., Angermann, M., Rus, D.: Eyes in the sky: Decentralized control for the deployment of robotic camera networks. Proceedings of the IEEE (99), 1–21 (2011)
22. Cortez, R.A., Fierro, R., Wood, J.: Prioritized sensor detection via dynamic voronoi-based navigation. In: IEEE/RSJ International Conference on Intelligent Robots and Systems (IROS), pp. 5815–5820 (October 2009)
23. Bullo, F., Cortés, J., Martínez, S.: Distributed Control of Robotic Networks. Applied Mathematics Series. Princeton University Press (2009)
24. Cortez, R., Fierro, R., Wood, J., Lumia, R.: Heterogeneous sensor network for prioritized sensing. In: IEEE/RSJ International Conference on Intelligent Robots and Systems (IROS), pp. 2333–2339. IEEE (2011)
25. Bollobás, B.: Modern Graph Theory. Springer-Verlag New York Inc. (1998)
26. Bezzo, N., Yuan, Y., Fierro, R., Mostofi, Y.: A decentralized connectivity strategy for mobile router swarms. In: Proc. of the 18th World Congress of the International Federation of Automatic Control (IFAC), Milan, Italy (August 30, 2011)
27. Biomimetic Millisystems Lab, UC Berkeley, `http://robotics.eecs.berkeley.edu/~ronf/Biomimetics.html`
28. Bezzo, N., Griffin, B., Cruz, P., Donahue, J., Fierro, R., Wood, J.: A cooperative heterogeneous mobile wireless mechatronic system. To appear in IEEE/ASME Trasactions on Mechatronics (2012)

Variational Analysis of Snakelike Robots

Gregory S. Chirikjian

Abstract. Snakelike robots have been employed in applications ranging from search-and-rescue to minimally invasive surgical procedures, and may yet find new applications in maintaining civil infrastructure and the repair of satellites. This chapter reviews how variational methods have been used previously to analyze three classes of snakelike robots: (1) hyper-redundant manipulators guided by backbone curves; (2) flexible steerable needles; and (3) concentric tube continuum robots. Both coordinate-dependent Euler-Lagrange formulations and coordinate-free Euler-Poincaré approaches are discussed. Variational methods are used in the context of hyper-redundant manipulators to constrain degrees of freedom and to provide a means of redundancy resolution. In contrast, variational methods are used for non-holonomic steerable needles, which have fewer actuatable degrees of freedom than those in the task space, to generate optimal open-loop plans for the needle to follow, and to model needle-tissue interactions. For concentric tube robots, variational methods provide a means to determine equilibrium conformations dictated by the principles of elastic mechanics. This chapter therefore illustrates by example how variational methods are a natural tool for the analysis and planning of various kinds of snakelike robots.

1 Introduction

The focus of this chapter is a review of variational methods as applied to snakelike robots. This section provides a brief review of the literature on snakelike robots and their application areas, and provides background mathematics used in the remainder of the chapter.

Gregory S. Chirikjian
Department of Mechanical Engineering, Johns Hopkins University
e-mail: gregc@jhu.edu

D. Milutinović & J. Rosen (Eds.): Redundancy in Robot Manipulators, LNEE 57, pp. 77–91.
DOI: 10.1007/978-3-642-33971-4_5

1.1 Literature Review

The study of snakelike robots, initiated by S. Hirose and collaborators in the early 1970s and summarized in [14], has multiple different facets. Snakelike locomotion systems [20] have been designed for search and rescue [32]. Elastic-filament manipulators have been used for positioning and orienting objects, as well as for nasal and throat surgery [27]. Steerable needle concepts and associated nonholonomic planning have been proposed for minimally invasive biopsy and treatment procedures [23, 2, 30, 29]. Continuum [5, 21] and concentric elastic tube robots [26, 13, 31] have been proposed for a variety of tasks including medical procedures. And hyper-redundant manipulators have been proposed for satellite servicing and inspection in nuclear power plants.

The emphasis of this chapter is on snake robots with a fixed base. Therefore, locomotion systems are not modeled here. Similar questions arise both in the redundancy resolution of fully actuated, hyper-redundant, manipulators, and in the mechanical analysis and planning of underactuated snakelike systems (flexible needles and active cannulae). In all cases the snakelike robotic device is described by a *backbone curve* of the form

$$\mathbf{x}(t) = \int_0^t [1+\varepsilon(s)]\mathbf{u}(s)ds \tag{1}$$

for $t \in [0,1]$. Here t is a curve parameter and $\mathbf{u}(s)$ is a unit vector. When $\varepsilon(s) \equiv 0$, the curve parameter becomes arclength. This backbone curve might represent the centerline of a needle or cannula, or it might be an artificial construct to which a highly articulated (hyper-redundant) manipulator is supposed to adhere, so as to facilitate planning. For a hyper-redundant manipulator that is modular in nature, it can be broken up into segments, and each segment can be identified with a piece of backbone curve, $[t_i, t_{i+1}]$. The "fitting" of the actual device to the curve requires that information about the orientation of reference frames attached to the curve at t_i and t_{i+1} also be provided. In principle, any smoothly evolving set of reference frames of the form $\{R(t)\,|\,t \in [0,1]\}$ will suffice, where $R(t) = [\mathbf{u}(t), \mathbf{n}_1(t), \mathbf{n}_2(t)]$, and $\mathbf{u}(t)$, $\mathbf{n}_1(t)$ and $\mathbf{n}_2(t)$ are mutually orthogonal unit vectors, and $\mathbf{u}(t) \times \mathbf{n}_1(t) = \mathbf{n}_2(t)$.

Then, the combination of backbone curve and evolving set of reference frames can be written as a set of pairs

$$g(t) = (R(t), \mathbf{x}(t)).$$

The position and orientation of a reference frame attached at $t = t_{i+1}$ relative to $t = t_i$ is then

$$[g(t_i)]^{-1} \circ g(t_{i+1}) = \left([R(t_i)]^T R(t_{i+1}),\, [R(t_i)]^T(\mathbf{x}(t_{i+1}) - \mathbf{x}(t_i))\right)$$

where $\circ$ denotes the composition of rigid-body motions (or equivalently, the multiplication of corresponding homogeneous transformation matrices). As the physical robot moves, it is desirable that its joints stay near the middle of their range. For this reason, "optimal" backbone curves that locally vary as little as possible while satisfying global end constraints were investigated [10, 11]. The following subsection reviews the variational methods that are applicable to this problem, and to all of these systems.

1.2 Background Mathematics

Let $\mathbf{q} \in D \subset \mathbb{R}^n$ be a vector of local coordinates describing a region of the configuration space of a system. The classical variational problem is that of extremizing a functional of the form

$$I = \int_{t_1}^{t_2} f(\mathbf{q}, \dot{\mathbf{q}}, t)\, dt \tag{2}$$

subject to constraints of the form

$$\int_{t_1}^{t_2} h_i(\mathbf{q}, t)\, dt = H_i. \tag{3}$$

The solution to this problem results from introducing Lagrange multipliers, $\lambda = [\lambda_1, ..., \lambda_m]^T \in \mathbb{R}^m$ and defining

$$L(\mathbf{q}, \dot{\mathbf{q}}, t) = f(\mathbf{q}, \dot{\mathbf{q}}, t) + \sum_{i=1}^{m} \lambda_i h_i(\mathbf{q}, t) \tag{4}$$

so as to satisfy the *Euler-Lagrange equation* [4, 15, 16, 18]

$$\frac{d}{dt}\left(\frac{\partial L}{\partial \dot{\mathbf{q}}}\right) - \frac{\partial L}{\partial \mathbf{q}} = \mathbf{0} \tag{5}$$

as well as end constraints on $\mathbf{q}(t_1)$ and $\mathbf{q}(t_2)$, and the constraints in (3). Numerical shooting methods can be used to refine the values of $\dot{\mathbf{q}}(t_1)$ and λ until a solution is reached.

While the coordinate-dependent Euler-Lagrange approach has tremendous value in engineering applications, in some scenarios the associated singularities cause difficulties. For this reason, less-known coordinate-free approaches also have their place. When the configuration space of a system (such as a rigid-body, or reference frames attached to a snakelike robot) have the structure of a Lie group, the problem can be formulated in terms of extremizing functionals of the form

$$I = \int_{t_1}^{t_2} f(g; g^{-1}\dot{g}; t)\, dt \tag{6}$$

where $g(t)$ is an element of a matrix Lie group G (such as the group of 3×3 rotation matrices, $SO(3)$, or 4×4 homogeneous transformations, $SE(3)$) and $g^{-1}\dot{g}$ is simply the product of the matrices representing g^{-1} and $\dot{g}$, the latter of which is not an element of G.

Given a functional of the form equation (6), and constraint equations of the form

$$\int_{t_1}^{t_2} h_k(g)\,dt = H_k \tag{7}$$

The necessary conditions for extremizing (6) can then be written in terms of the functions f and h_k as

$$\frac{d}{dt}\left(\frac{\partial f}{\partial \xi_i}\right) + \sum_{j,k=1}^{n} \frac{\partial f}{\partial \xi_k} C_{ij}^{k}\,\xi_j = \tilde{E}_i^r(f + \sum_{l=1}^{m} \lambda_l h_l) \tag{8}$$

where $\xi = [\xi_1, ..., \xi_n]^T$ contains the independent nonzero entries in the matrix $g^{-1}\dot{g}$ (e.g., angular velocities when $G = SO(3)$, or infinitesimal twists when $G = SE(3)$), and $\{C_{ij}^k \,|\, i, j, k = 1, ..., n\}$ is a set of constants called the structure constants, which are fixed for any given G and choice of extracting ξ's from $g^{-1}\dot{g}$. That is, if $\xi_i = (g^{-1}\dot{g}, E_i)$ where $\{E_i\}$ is a basis for the Lie algebra of G, and $(\cdot,\cdot)$ is an inner product on this Lie algebra. (Natural basis elements for $SO(3)$ and $SE(3)$ are described in detail in [6, 7, 9].)

In this context

$$(\tilde{E}_i^r f)(g) \doteq \left.\frac{d}{dt} f(g \circ \exp(tE_i))\right|_{t=0}$$

is akin to a directional derivative. Equation (8) is a modified version of the *Euler-Poincaré* equation [1, 24, 3, 17], which is a coordinate-free version of the Euler-Lagrange equation in (5).

Both the Euler-Lagrange and Euler-Poincaré equations have been used in the modeling of snakelike robots. A point that is often glossed over when these equations are applied is that (5) and (8) provide only *necessary* conditions for optimality. And in general there is no guarantee that the resulting solution will be optimal in the sense intended by the user. However, in some applications involving snakelike robots, it can be shown that globally optimal solutions result from this formulation.

These globally optimal situations are discussed in the remainder of this chapter, which is structured as follows: Section 2 discusses optimal backbone curve shapes and reparameterizations. This forms the core of the chapter, as many of these methods are reused and applied in different ways in remaining sections. Section 3 discusses the variational modeling of concentric tube (active cannula) devices. Section 4 illustrates how these same methods can be applied in needle steering.

2 Optimal Hyper-Redundent Manipulator Backbone Curves

This section describes two kinds of variational problems related to backbone curves of hyper-redundant manipulators. First, Section 2.1 discusses the globally optimal reparameterization of given curves. Then, Section 2.2 formulates the coordinate-free variational approach to generating backbone curves. And Section 2.3 derives the corresponding variational equations. For more detailed treatments of the results presented in this section see [8, 19, 7].

2.1 Optimal Reparameterization

Suppose that an arc-length parameterized curve $\mathbf{x}(s) \in \mathbb{R}^n$ is given, and that the shape of this curve is desirable, but the temporal evolution of the position of a particle along the curve is sought such that the integral of a a cost functional $\tilde{f}(\mathbf{x}, \dot{\mathbf{x}}) = \frac{1}{2}\dot{\mathbf{x}}^T G(\mathbf{x})\dot{\mathbf{x}}$ should be minimized along the curve. From the chain rule, $\dot{\mathbf{x}} = \frac{d\mathbf{x}}{ds}\dot{s}$, and so $f(s,\dot{s}) \doteq \tilde{f}(\mathbf{x}(s), \frac{d\mathbf{x}}{ds}\dot{s}) = \frac{1}{2}\left(\frac{d\mathbf{x}}{ds}\right)^T G(\mathbf{x}(s))\left(\frac{d\mathbf{x}}{ds}\right)\dot{s}^2 = \frac{1}{2}g(s)\dot{s}^2$.

Though the Euler-Lagrange equations only provide *necessary* conditions for optimality, this is a situation in which the structure of the function $f(\cdot)$ will guarantee that the solution generated by the Euler-Lagrange equations is a globally optimal solution.

$$f(s,\dot{s},t) = \frac{1}{2}g(s)\dot{s}^2$$

and $s^*(t)$ denotes the solution obtained by the Euler-Lagrange equations of variational calculus, then the resulting cost is

$$J(s^*) = \left(\int_0^1 g^{\frac{1}{2}}(s^*(t))\dot{s}^* dt\right)^2 = \left(\int_0^1 g^{\frac{1}{2}}(s^*)ds^*\right)^2.$$

This is because $s^*(0) = 0$ and $s^*(1) = 1$ and $ds^*/dt > 0$. Note that this means that the value of the integral in the above expression for $J(s^*)$ is *independent* of the path $s^*(t)$. This does not mean that $J(s^*)$ itself is independent of s^*. Rather, it means that after the form of the candidate optimal path obtained from the Euler-Lagrange equation is substituted back into the cost functional the resulting value can be written as

$$J(s^*) = \left(\int_0^1 g^{\frac{1}{2}}(s)ds\right)^2$$

since the name of the variable of integration is irrelevant. Then from here if we substitute any $s(t)$ with $ds/dt > 0$, and use the Cauchy-Schwarz inequality

$$\left(\int_0^1 a(t)b(t)\,dt\right)^2 \leq \left(\int_0^1 |a(t)|^2\,dt\right)\left(\int_0^1 |b(t)|^2\,dt\right)$$

with $a(t) = g^{\frac{1}{2}}(s(t))ds/dt$ and $b(t) = 1$, it follows that

$$J(s^*) \leq J(s) \tag{9}$$

for any possible $s(t)$.

More generally, suppose that, for whatever reason, a globally minimal solution to a variational optimization problem with $f_1(\mathbf{q}, \dot{\mathbf{q}}, t) = \frac{1}{2}\dot{\mathbf{q}}^T M(\mathbf{q})\dot{\mathbf{q}}$ and $\mathbf{q}(0)$ and $\mathbf{q}(1)$ specified has been solved via the Euler-Lagrange equations and minimization over all resulting paths that connect the specified end points. This solution then can be used to "bootstrap" a globally optimal solution to a larger variational problem in which

$$f_2(\mathbf{q},\theta,\dot{\mathbf{q}},\dot{\theta},t) = \frac{1}{2}\dot{\mathbf{q}}^T M(\mathbf{q})\dot{\mathbf{q}} + \frac{1}{2}\|\dot{\theta} - B(\mathbf{q})\dot{\mathbf{q}}\|_W^2 \tag{10}$$

where $\|B\|_W^2 = \mathrm{tr}(B^T W B)$ is the weighted Frobenius norm where $W = W^T > 0$.

The Euler-Lagrange equations for the original variational problem are of the form

$$\frac{d}{dt}\left(\frac{\partial f_1}{\partial \dot{\mathbf{q}}}\right) - \frac{\partial f_1}{\partial \mathbf{q}} = \mathbf{0} \Longrightarrow$$
$$M(\mathbf{q})\ddot{\mathbf{q}} + \frac{1}{2}\frac{\partial}{\partial \mathbf{q}}(\dot{\mathbf{q}}^T M(\mathbf{q})\dot{\mathbf{q}}) = \mathbf{0}. \tag{11}$$

Let the solution to this system of equations subject to boundary conditions be denoted as $\mathbf{q}^*(t)$.

The Euler-Lagrange equations for the new system will be

$$\frac{d}{dt}\left(\frac{\partial f_2}{\partial \dot{\mathbf{q}}}\right) - \frac{\partial f_2}{\partial \mathbf{q}} = \mathbf{0} \Longrightarrow$$
$$M(\mathbf{q})\ddot{\mathbf{q}} - \frac{d}{dt}\left[B^T(\mathbf{q})W(\dot{\theta} - B(\mathbf{q})\dot{\mathbf{q}})\right] +$$
$$\frac{1}{2}\frac{\partial}{\partial \mathbf{q}}(\dot{\mathbf{q}}^T M(\mathbf{q})\dot{\mathbf{q}}) + \frac{\partial}{\partial \mathbf{q}}\left[\dot{\mathbf{q}}^T B^T(\mathbf{q})\right] W(\dot{\theta} - B(\mathbf{q})\dot{\mathbf{q}}) = \mathbf{0} \tag{12}$$

and

$$\frac{d}{dt}\left(\frac{\partial f_2}{\partial \dot{\theta}}\right) - \frac{\partial f_2}{\partial \theta} = \mathbf{0} \Longrightarrow$$
$$\dot{\theta} - B(\mathbf{q})\dot{\mathbf{q}} = \mathbf{a} \tag{13}$$

where $\mathbf{a}$ is a constant vector of integration. Substituting the right hand side of (13) back into (12) and using the chain rule,

$$\frac{d}{dt}(B^T(\mathbf{q})W\mathbf{a}) = \dot{\mathbf{q}}^T\frac{\partial}{\partial \mathbf{q}}\left[B^T(\mathbf{q})W\mathbf{a}\right] = \frac{\partial}{\partial \mathbf{q}}\left[\dot{\mathbf{q}}^T B^T(\mathbf{q})\right] W\mathbf{a},$$

means that (12) reduces to (11), and so the optimal solution for the "$\mathbf{q}$-part" of the problem again will be $\mathbf{q}^*(t)$. The right hand side of (13) means that the candidate optimal solution for this new problem is

$$\theta^*(t) = \mathbf{a}t + \mathbf{b} + \int_0^t B(\mathbf{q}^*(s))\dot{\mathbf{q}}^*(s)\,ds \tag{14}$$

where $\mathbf{a}$ and $\mathbf{b}$ are constant vectors that can be matched to boundary values, and $\mathbf{q}^*(t)$ is the solution to the original variational problem with cost $f_1(\mathbf{q},\dot{\mathbf{q}},t)$. The global optimality of the solution $(\mathbf{q}^*(t),\theta^*(t))$ is guaranteed by the assumption that optimal $\mathbf{q}^*(t)$ is obtained a priori, and the global optimality of $\theta^*(t)$ in (14) can be observed by substituting any $\theta(t) = \theta^*(t) + \varepsilon(t)$ where $\varepsilon(0) = \varepsilon(1) = \mathbf{0}$ into the cost function and observing that this never improves the cost.

This class of problems can be viewed in a slightly different way by rewriting (10) as

$$f(\mathbf{q},\theta,\dot{\mathbf{q}},\dot{\theta},t) = \frac{1}{2}\begin{bmatrix}\dot{\mathbf{q}}\\ \dot{\theta}\end{bmatrix}^T \begin{pmatrix} B^T(\mathbf{q})WB(\mathbf{q}) & B^T(\mathbf{q})W \\ WB(\mathbf{q}) & W \end{pmatrix} \begin{bmatrix}\dot{\mathbf{q}}\\ \dot{\theta}\end{bmatrix}. \tag{15}$$

This means that if a quadratic cost can be decomposed in this way then the solution to the larger problem will inherit the global optimality from the original problem.

As a concrete application of this class of problems, consider the problem of simultaneous *curve reparameterization and optimal roll distribution*. Start with an initially arclength-parameterized curve $\mathbf{x}(s)$ for $s \in [0,1]$, and frames defined using the Frenet-Serret apparatus. If the Frenet frames are $(Q_{FS}(s),\mathbf{x}(s))$, then a new set of smoothly evolving reference frames can be defined as $(Q(t),\mathbf{x}(t)) = (Q_{FS}(s(t))R_1(\theta(s(t))),\mathbf{x}(s(t)))$, where $R_1(\theta)$ is an added twist, or roll, of the Frenet frames about the tangent. Given a cost of the form

$$C \doteq \frac{1}{2}\int_0^1 \left\{\frac{1}{2}r^2 \mathrm{tr}(\dot{Q}\dot{Q}^T) + \dot{\mathbf{x}}\cdot\dot{\mathbf{x}}\right\} dt \tag{16}$$

$$= \frac{1}{2}\int_0^1 \left\{(r^2\kappa^2(s)+1)\dot{s}^2 + r^2(\tau(s)\dot{s}+\dot{\theta})^2\right\} dt \tag{17}$$

(where r is a specified constant with units of length, and $\kappa(s)$ and $\tau(s)$ are respectively the curvature and torsion of the curve), the goal is to find a simultaneous reparameterization $s = s(t)$, and $\theta(s(t))$ so as to minimize C.

The integrand here is of the form in (10) with s taking the place of $\mathbf{q}$ and θ taking the place of θ. Since the curve reparameterization problem (with the second term in the integral set to zero) is a one-dimensional variational problem with $f(s,\dot{s},t) = \frac{1}{2}g(s)(\dot{s})^2$, global optimality is preserved. And from the discussion above, this guarantees the global optimality of the composite problem.

As a result, the sorts of simultaneous curve reparameterization and optimal roll distribution to satisfy end constraints obtained from variational calculus in [10, 11] in the context of "hyper-redundant" (snakelike) robotic arms are in fact optimal.

2.2 Cases When the Euler-Poincaré Equation Gives Globally Minimal Solutions

Given a cost function of the form

$$f(g,\xi,t) = \frac{1}{2}\xi^T W \xi = \frac{1}{2}\sum_{i,j=1}^{n} w_{ij}\xi_i\xi_j$$

with $W = W^T > 0$, the Euler-Poincaré equations are of the form

$$\frac{d}{dt}\left(\sum_{j=1}^{n} w_{ij}\xi_j\right) + \sum_{j,k=1}^{n}\left(\sum_{l=1}^{n} w_{kl}\xi_l\right) C_{ij}^k \xi_j = 0. \tag{18}$$

Let

$$S_{lj}^{i} \doteq \sum_{k=1}^{n} w_{kl} C_{ij}^{k}. \tag{19}$$

If $S_{lj}^{i} = -S_{jl}^{i}$ then $\sum_{j,l=1}^{n} S_{lj}^{i} \xi_l \xi_j = 0$ and (18) reduces to

$$W\dot{\xi} = \mathbf{0} \implies \xi(t) = \xi(0) \implies g(t) = g(0) \circ e^{t\hat{\xi}(0)}.$$

This means that when these conditions hold the shortest path computed from variational calculus connecting $g(0) = g_0$ and $g(1) = g_1$ is

$$g(t) = g_0 \circ \exp(t \cdot \log(g_0^{-1} \circ g_1)). \tag{20}$$

Furthermore, this path will be globally optimal because of the structure of the cost function. However, if $S_{lj}^{i} \neq -S_{jl}^{i}$ then (18) does not reduce and the path generated by variational calculus is generally not this geometric one.

2.2.1 Extensible Backbone Curve Conformations

A nonuniform extensible elastic filament with unstretched length L has elastic energy of the form

$$E_1 = \int_0^L F(\xi(s), s)\, ds \quad \text{where} \quad F(\xi(s), s) = \frac{1}{2}[\xi(s) - \xi_0]^T K(s)[\xi(s) - \xi_0]. \tag{21}$$

Here $\hat{\xi}_0 \in se(3)$ defines the local shape of the minimal energy conformation at each value of curve parameter s, and $K(s)$ is the 6×6 stiffness matrix that describes resistance to change in each direction of infinitesimal motion. Off-diagonal terms in this matrix describe couplings that have been observed experimentally. Both $\hat{\xi}(s)$ and $\hat{\xi}_0$ are body-fixed quantities in the sense that $\hat{\xi}(s) = g^{-1} dg/ds$ and $\hat{\xi}_0 = g_0^{-1} dg_0/ds$ for the one-dimensional set of reference frames $g(s)$ and $g_0(s)$. Sometimes it is convenient to define

$$\mathbf{k} \doteq K\xi_0.$$

Given $\xi_0(s)$, it is possible to integrate the matrix differential equation

$$\frac{dg_0}{ds} = g_0\, \hat{\xi}_0(s)$$

subject to the initial condition $g(0) = e$ (the identity element of $SE(3)$ is the identity matrix $e = \mathbb{I}_4$) for $s \in [0, L]$ to obtain the minimal energy conformation rooted at the identity. In the case when $\xi_0(s)$ is a constant vector, this will be a helix (with circular arcs and line segments as special cases).

Note that the independent variable is now a curve parameter, s, rather than time, t. Here the curve parameter s is taken to be the curve parameter of the filament in its undeformed (referential) conformation $g_0(s)$.[1]

2.2.2 Inextensible Backbone Curves

For arc-length-parameterized backbone curves

$$\mathbf{a}(L) = \int_0^L \mathbf{u}(s)\,ds \quad \text{and} \quad \mathbf{u}(s) = R(s)\mathbf{e}_3. \tag{22}$$

This can be viewed as the inextensible filament "growing" along the direction indicated by the tangent for each value of arc length, s, up to a total length of L.

In this case the stiffness matrix, B, is 3×3, and the resulting problem becomes one of minimizing

$$I = \frac{1}{2}\int_0^L [\omega(s) - \omega_0]^T B[\omega(s) - \omega_0]\,ds \tag{23}$$

subject to the constraints (22). Here $\omega = (R^T dR/ds)^\vee$ and $\omega_0 = (R_0^T dR_0/ds)^\vee$ are angular velocities as seen in the body-fixed frame, where arclength s replaces time as the independent variable.

Unlike the extensible problem, which was an unconstrained variational minimization problem on $SE(3)$, this is a constrained variational problem on $SO(3)$, and will therefore involve the use of Lagrange multipliers. The Euler-Poincaré equations for both cases are worked out in the following section.

2.3 *Resulting Variation Equations Describing Backbone Curves*

Here the necessary conditions for coordinate-free variational minimization of the energy functionals described in the previous section are established. This is a straightforward application of the Euler-Poincaré equations. In the inextensible and shearless case the group $G = SO(3)$, and in the extensible case the group is $G = SE(3)$. In both cases there are six free degrees of freedom to specify the end position and orientation of the elastic filament. In the extensible case these degrees of freedom correspond to the six scalar components of the initial conditions $\xi(0)$, whereas in the inextensible case they correspond to three scalar initial conditions $\omega(0)$ and three scalar Lagrange multipliers (components of λ) corresponding to the three constraints that define $\mathbf{a}(L)$ in (22).

[1] In the extensible case the curve parameter s can be viewed as the arc length in the referential (undeformed) conformation of the filament, which does not necessarily mean that s will be arclength in the pushed forward (deformed) version of the filament. However, in the inextensible model, s retains its role as arclength after deformation since deformations are restricted to bending and twisting in that model.

2.3.1 Inextensible Case

Considering the case of (23) with the kinematic constraint of inextensibility (22), one writes Eq. (8) with $f = U$ for $i = 1,2,3$ together as the vector equation

$$B\dot{\omega} + \omega \times (B\omega - \mathbf{b}) = \begin{pmatrix} -\lambda^T A\mathbf{e}_2 \\ \lambda^T A\mathbf{e}_1 \\ 0 \end{pmatrix} \tag{24}$$

where a dot represents differentiation with respect to arclength s, $\lambda \in \mathbb{R}^3$ is the vector of Lagrange multipliers necessary to enforce the vector constraint in Eq. (22), and the right-hand-side of Eq. (24) results from the fact that

$$E_i^r(\lambda^T A\mathbf{e}_3) = \frac{d}{dt}\lambda^T A(\mathbb{I} + tE_i)\mathbf{e}_3\Big|_{t=0} = \lambda^T AE_i\mathbf{e}_3 = \lambda^T A(\mathbf{e}_i \times \mathbf{e}_3).$$

Eq. (24) is solved iteratively subject to the initial conditions $\omega(0) = \mu$ which are varied together with the Lagrange multipliers until $\mathbf{a}(L)$ and $A(L)$ attain the desired values. $A(s)$ is computed from $\omega(s)$ in Eq. (24) by integrating the matrix differential equation

$$\dot{A} = A\left(\sum_{i=1}^{3} \omega_i(s)E_i\right),$$

and $\mathbf{a}(L)$ is then obtained from Eq. (22). Numerical methods for updating μ and λ so as to push the position and orientation of the distal end to specified values are described in [19].

2.3.2 Extensible Rods

From (8) and (21) one can obtain the following Euler-Lagrange equation for the extensible case:

$$K\dot{\xi} + (K\xi - \mathbf{k}) \wedge \xi = \mathbf{0} \tag{25}$$

where $\wedge$ is the product of infinitesimal rigid-body motions defined by

$$\begin{pmatrix} \omega_1 \\ \mathbf{v}_1 \end{pmatrix} \wedge \begin{pmatrix} \omega_2 \\ \mathbf{v}_2 \end{pmatrix} = \begin{pmatrix} \omega_2 \times \omega_1 + \mathbf{v}_2 \times \mathbf{v}_1 \\ \omega_2 \times \mathbf{v}_1 \end{pmatrix}.$$

This wedge operator is related to the *ad* operator as

$$\xi_1 \wedge \xi_2 = -[ad(\xi_2)]^T \xi_1 \tag{26}$$

where $\xi_i = [\omega_i^T, \mathbf{v}_i^T]^T$, $i = 1,2$ and the matrix of *ad* operator is defined as

$$[ad(\xi)] = \begin{pmatrix} \widehat{\omega} & \mathbb{O} \\ \widehat{\mathbf{v}} & \widehat{\omega} \end{pmatrix}.$$

Eq. (25) is solved subject to the initial conditions $\xi(0) = \eta \in \mathbb{R}^6$. This, together with the kinematic condition

$$\dot{g} = g\left(\sum_{i=1}^{6} \xi_i E_i\right), \tag{27}$$

is integrated for $0 \leq s \leq L$ to define $g(\xi, L)$. From this point everything follows in exactly the same way as for the inextensible case. For any fixed value of $L \in \mathbb{R}^+$, (25) and (27) can together be thought of as defining a mapping from $\mathbb{R}^6$ (the initial conditions η) into $SE(3)$. This mapping can be generated by numerically solving these ordinary differential equations. It is not a one-to-one mapping, and finding all values of initial conditions that map to a specific end position and orientation of the filament is quite challenging [19].

3 Variational Methods in Concentric Tube Robots

Concentric-tube robots (or active canulae) consist of n concentric elastic tubes that have been pre-bent. Several research groups have investigated these sorts of snake-like robots, including [13, 31] and work referenced therein. One analysis method that is applicable to these robots is a variational approach, as presented in [25]. Indeed, the use of the Euler-Poincaré equations in this application follows in much the same way as for the derivation of optimal backbone curves for hyper-redundant manipulators. The tubes are essentially inextensible, and so the energy due to deformation can be written as

$$E = \frac{1}{2}\sum_{i=1}^{n} \int_0^1 [\omega_i(s) - \omega_i^*(s)]^T B_i(s) [\omega_i(s) - \omega_i^*(s)]\, ds \tag{28}$$

where $\omega_i^*(s)$ describes the equilibrium shape of the ith tube which has 3×3 stiffness matrix $B_i(s)$. All tubes have the same length and initial and final positions and orientations. As a result determining the equilibrium shape is a matter of determining $\omega_1(s)$ and the relative angles (and angular rates) that the other tubes exhibit relative to this. As explained in [25],

$$\omega_i(s) = \exp(-\theta_i(s) E_3)\, \omega_1(s) + \dot{\theta}_i(s)\, \mathbf{e}_3$$

where $\mathbf{e}_3 = [0, 0, 1]^T$ and E_3 is the skew-symmetric matrix such that $E_3\mathbf{x} = \mathbf{e}_3 \times \mathbf{x}$ and $\theta_i(s)$ is the amount of twist that each tube $i = 2, 3..., n$ undergoes relative to tube 1, consistent with the boundary conditions imposed. Since the one-dimensional rotations described by each of the θ_i are in $SO(2)$, the Euler-Poincaré formulation can be (and has been) employed successfully with

$$G = SO(3) \times SO(2) \times \cdots \times SO(2)$$

to determine the equilibrium conformations.

4 Variational Methods in Needle Steering

Two kinds of variational problems arise in steering of flexible needles in soft tissues. First, when the tissue is firm and does not deform substantially as the needle is inserted, the optimal control problem of how to push and twist the needle "as little as possible" while still reaching the goal presents itself. Second, if the tissue is soft, the question of how the potential energy in the needle and that in the tissue equilibrate becomes a significant problem. Both of these issues are variational in nature, and efforts that have already been published are reviewed, and future directions are sketched here.

4.1 *Kinematic Needle Steering*

When a flexible needle with a sharp beveled tip is inserted into tissue, it has a tendency to follow a roughly circular path. This resulted in a 3D version of a 'unicycle' model described in [23], and was extended to the 'bicycle' model in [29]. In the unicycle model the reference frame attached to the needle tip executes a trajectory of the form

$$\begin{pmatrix} \dot{R} & \dot{\mathbf{x}} \\ \mathbf{0}^T & 0 \end{pmatrix} = \begin{pmatrix} R & \mathbf{x} \\ \mathbf{0}^T & 1 \end{pmatrix} \begin{pmatrix} 0 & -\omega(t) & 0 & 0 \\ \omega(t) & 0 & -\kappa & 0 \\ 0 & \kappa & 0 & v(t) \\ 0 & 0 & 0 & 0 \end{pmatrix}$$

where κ is the constant curvature of the path that a needle naturally wants to follow, and $(\omega(t), v(t0)$ are the control inputs. See, for example, [22] for notation and [12] an up-to-date survey of the field of flexible needle steering. The variational problem then becomes that of minimizing the control cost

$$C_1 = \frac{1}{2}\int_0^1 \left\{c_1(\omega(t))^2 + c_2(v(t))^2\right\} dt$$

subject to the boundary conditions $g(0) = (\mathbb{I}, \mathbf{0})$ and $g(1) = g_d$ (the desired target frame). A practical problem that is encountered is the lack of repeatability of observed needle-steering trajectories. Building on the prior work in [33], this has been modeled as a stochastic version of the above kinematic constraint in [23], and probabilistic methods on Lie groups have been employed to solve the problem [22, 28].

4.2 *Soft-Tissue Effects*

When the tissue into which the needle is inserted is soft, a tradeoff of elastic energy takes place between the needle and the tissue. As a thought experiment, consider the following scenario. An elastic needle is deformed so as to satisfy certain end constraints. The shape of this needle will then be given by the equations in Section 2.3. Now imagine that the needle is held in place by an external field, and the ends of the needle are capped with solid balls. Imagine further that a liquid polymer is then poured around the needle and its end caps and is allowed to harden without any residual stress. If the external field is the released, an equilibrium will be reached by the needle/endcaps and the surrounding material. The details of this equilibrium will depend on the boundary conditions, material properties of the needle and surrounding environment, and the size and properties of the end caps. If the needle and end caps are constrained to remain within a deformed version of the dumbbell-shaped cavity that was formed when the imaginary polymer cured (without separation or gaps, but with slipping allowed) then in the infinitesimal strain limit, the equilibrium will be described by finding the deformation field $\mathbf{f}(\mathbf{x})$ such that

$$\frac{1}{2}\int_0^1 k\|\frac{d^2}{dt^2}\mathbf{f}(\mathbf{x}(t)) - \frac{d}{dt}\mathbf{u}(t)\|^2 dt = \frac{1}{2}\int_{\mathbb{R}^3} \|\frac{1}{2}((\nabla_{\mathbf{x}}\mathbf{f})^T + \nabla_{\mathbf{x}}\mathbf{f}) - \mathbb{I})\|_E^2 d\mathbf{x}$$

where E is a weighting matrix that contains the same information as the elasticity tensor of the medium, and k is the scalar bending stiffness of a needle with isotropic cross section. In other words, the energy due to bending of the needle from its equilibrium state should balance with the energy due to deformation of the medium. The solution to this problem is again a variational problem. And this is one which is challenging on its own, and in combination with understanding the mechanics of interaction between the needle-tip and the medium.

5 Conclusions

Variational methods as a tool for modeling and planning of highly articulated and flexible manipulators were reviewed. The broad range of applicability of these methods make them a potentially important tool for researchers studying redundant manipulators. In the context of hyper-redundant manipulators, which have many articulated degrees of freedom, variational methods are useful in defining slowly varying backbone curves, as well as an optimal distribution of reference frames that evolve along it. In the context of concentric-tube (active-cannula) robots, variational methods provide the solution to the equilibrium conformations resulting from the interaction of the tubes, and in needle-steering problems, the variational approach provides a tool for planning trajectories.

Acknowledgements. This work was supported by NIH Grant R01EB006435 "Steering Flexible Needles in Soft Tissue" (A. Okamura, PI).

References

1. Abraham, R., Marsden, J.E.: Foundations of Mechanics. Benjamin/Cummings, San Mateo, CA (1978)
2. Alterovitz, R., Lim, A., Goldberg, K., Chirikjian, G.S., Okamura, A.M.: Steering flexible needles under Markov motion uncertainty. In: Proc. IEEE/RSJ Int. Conf. on Intelligent Robots and Systems, pp. 120–125 (August 2005)

3. Bloch, A., Krishnaprasad, P.S., Marsden, J.E., Ratiu, T.S.: The Euler-Poincaré equations and double bracket dissipation. In: Communications in Mathematical Physics, vol. 175, p. 1 (1996)
4. Brechtken-Manderscheid, U.: Introduction to the Calculus of Variations. Chapman and Hall, New York (1991)
5. Chirikjian, G.S.: Hyper-Redundant Manipulator Dynamics: A Continuum Approximation. Advanced Robotics 9(3), 217–243 (1995)
6. Chirikjian, G.S., Kyatkin, A.B.: Engineering Applications of Noncommutative Harmonic Analysis. CRC Press, Boca Ratin (2001)
7. Chirikjian, G.S.: Stochastic Models, Information Theory, and Lie Groups: Volumes 1 +2. Birkhäuser, Basel (2009/2012)
8. Chirikjian, G.S.: Group Theory and Biomolecular Conformation, I.: Mathematical and computational models. J. Phys. Condens. Matter 22, 323103 (2010)
9. Chirikjian, G.S., Kyatkin, A.B.: An Operational Calculus for the Euclidean Motion Group with Applications in Robotics and Polymer Science. J. Fourier Analysis and Applications 6(6), 583–606 (2000)
10. Chirikjian, G.S.: Theory and Applications of Hyper-Redundant Robotic Manipulators, Division of Engineering and Applied Science, California Institute of Technology (June 1992), `http://thesis.library.caltech.edu/4458/1/Chirikjian_gs_1992.pdf`
11. Chirikjian, G.S., Burdick, J.W.: Kinematically optimal hyper-redundant manipulator configurations. IEEE Trans. Robot. Autom. 11, 794 (1995)
12. Cowan, N.J., Goldberg, K., Chirikjian, G.S., Fichtinger, G., Alterovitz, R., Reed, K.B., Kallem, V., Park, W., Misra, S., Okamura, A.M.: Robotic Needle Steering: Design, Modeling, Planning, and Image Guidance. In: Rosen, J., Hannaford, B., Satava, R. (eds.) Surgical Robotics – Systems, Applications, and Visions, pp. 557–582. Springer (2011)
13. Dupont, P.E., Lock, J., Itkowitz, B., Butler, E.: Design and Control of Concentric-Tube Robots. IEEE Transactions on Robotics 26(2), 209–225 (2010)
14. Hirose, S.: Biologically Inspired Robots: Serpentile Locomotors and Manipulators. Oxford University Press (1993)
15. Ewing, G.M.: Calculus of Variations with Applications. W.W. Norton and Co., New York (1969)
16. Gruver, W.A., Sachs, E.: Algorithmic Methods in Optimal Control. Pitman Publishing, Ltd., Boston (1980)
17. Holm, D.D., Marsden, J.E., Ratiu, T.S.: The Euler-Poincaré equations and semidirect products with applications to continuum theories. Adv. Math. 137, 1 (1998)
18. Kamien, M.I., Schwartz, N.L.: Dynamic Optimization: The Calculus of Variations and Optimal Control in Economics and Management. North-Holland, New York (1991)
19. Kim, J.-S., Chirikjian, G.S.: Conformational analysis of stiff chiral polymers with end constraints. Molecular Simulation 32(14) (December 15, 2006)
20. Ma, S.: Analysis of creeping locomotion of a snake-like robot. Advanced Robotics 15(2), 205–224 (2001)
21. Neppalli, S., Csencsits, M.A., Jones, B.A., Walker, I.D.: Closed-Form Inverse Kinematics for Continuum Manipulator. Advanced Robotics 23(15), 2077–2091 (2009)
22. Park, W., Wang, Y., Chirikjian, G.S.: The path-of-probability algorithm for steering and feedback control of flexible needles. International Journal of Robotics Research 29(7), 813–830 (2010)
23. Park, W., Kim, J.S., Zhou, Y., Cowan, N.J., Okamura, A.M., Chirikjian, G.S.: Diffusion-based motion planning for a nonholonomic flexible needle model. In: Proceedings of the IEEE Int. Conf. on Robotics and Automation, Barcelona, Spain, 6 pages (April 2005)

24. Poincaré, H.: Sur une forme nouvelle des equations de la mechanique. Cr. Hebd. Acad. Sci. 132, 369 (1901)
25. Rucker, C., Webster III, R.J., Chirikjian, G.S., Cowan, N.J.: Equilibrium Conformations of Concentric-Tube Continuum Robots. International Journal of Robotics Research 29(10), 1263–1280 (2010)
26. Sears, P., Dupont, P.: A Steerable Needle Technology Using Curved Concentric Tubes. In: 2006 IEEE/RSJ International Conference on Intelligent Robots and Systems, October 9-15, pp. 2850–2856 (2006)
27. Simaan, N., Taylor, R., Flint, P.: A dexterous system for laryngeal surgery. In: Proceedings of 2004 IEEE International Conference on Robotics and Automation, ICRA 2004, April 26 -May 1, vol. 1, pp. 351–357 (2004)
28. Wang, Y., Chirikjian, G.S.: Nonparametric Second-Order Theory of Error Propagation on the Euclidean Group. International Journal of Robotics Research 27(11-12), 1258–1273 (2008)
29. Webster III, R.J., Kim, J.-S., Cowan, N.J., Chirikjian, G.S., Okamura, A.M.: Nonholonomic Modeling of Needle Steering. International Journal of Robotics Research 25(5-6), 509–525 (2006)
30. Webster III, R.J., Cowan, N.J., Chirikjian, G.S., Okamura, A.M.: Nonholonomic modeling of needle steering. In: 9th International Symposium on Experimental Robotics. Springer Tracts in Advanced Robotics, vol. 21, pp. 35–44 (March 2006)
31. Webster III, R.J., Romano, J.M., Cowan, N.J.: Mechanics of Precurved-Tube Continuum Robots. IEEE Trans. Robotics 25(1), 67–78 (2009)
32. Wolf, A., Brown, H.B., Casciola, R., Costa, A., Schwerin, M., Shamas, E., Choset, H.: A mobile hyper redundant mechanism for search and rescue tasks. In: Proceedings of 2003 IEEE/RSJ International Conference on Intelligent Robots and Systems (IROS 2003), October 27-31, vol. 3, pp. 2889–2895 (2003)
33. Zhou, Y., Chirikjian, G.S.: Probabilistic Models of Dead-Reckoning Error in Nonholonomic Mobile Robots. In: ICRA 2003, Taipei, Taiwan (September 2003)

Robustness in the Presence of Task Differentiation in Robot Ensembles

M. Ani Hsieh and T. William Mather

Abstract. In the last fifteen years, much interest has been focused on the deployment of large teams of autonomous robots for applications such as environmental monitoring, surveillance and reconnaissance, and automated parts inspection for manufacturing. The objective is to leverage the team's inherent redundancy to simultaneously cover wide regions and achieve massive parallelization in task execution while remaining robust to individual failures. Despite recent successes, significant challenges remain, in part, due to the difficulties associated with managing and coordinating the various redundancies that exist in a large team of homogeneous agents. In this chapter, we present an ensemble approach towards the design of distributed control and communication strategies for the dynamic allocation of a team of robots to a set of tasks. This approach uses a class of stochastic hybrid systems to model the robot team dynamics as a continuous-time Markov jump process. The main advantage is a lower-dimensional representation of the team dynamics that is amenable to system-level analysis of the team's performance in the presence of task differentiation. We show how such analysis can be further used to design and optimize individual robot control policies through simulations and experimental validation.

1 Introduction

In the last fifteen years, much interest has been focused on the deployment of large teams of autonomous robots for applications such as environmental monitoring, surveillance and reconnaissance, and automated parts inspection for manufacturing. The objective is to leverage the team's inherent redundancy to simultaneously cover wide regions and achieve massive parallelization in task execution while remaining robust to individual failures. Despite recent successes, significant challenges remain, in part, due to the difficulties associated with managing and coordinating

---

M. Ani Hsieh · T. William Mather
Drexel University, 3141 Chestnut St, Philadelphia, PA 19104, USA
e-mail: {mhsieh1, twm32}@drexel.edu

D. Milutinović & J. Rosen (Eds.): Redundancy in Robot Manipulators, LNEE 57, pp. 93–108.
DOI: 10.1007/978-3-642-33971-4_6

the various redundancies that exist in a large team of homogeneous agents. In this chapter, we present an ensemble approach towards the design of distributed control and communication strategies for the dynamic allocation of a team of robots to a set of tasks. This approach uses a class of stochastic hybrid systems to model the robot team dynamics as a continuous-time Markov jump process. The result is a lower-dimensional representation of the team dynamics that is amenable to system-level analysis of the team's performance in the presence of task differentiation.

In our work, we consider the allocation of a team of robots to a collection of tasks distributed within a workspace. Applications include automated warehouses where inventory from the loading docks must be transported and stowed in their designated locations or deploying robots to cover different regions with their sensors for monitoring or surveillance purposes. In these examples, the team must have the ability to autonomously move from one task location to another, distribute themselves accordingly among the various locales to ensure task completion or to provide appropriate sensor coverage, all the while remaining robust to changes in the environment or individual failures.

The assignment of robots to a collection of spatially distributed tasks is a variant of the multi-task (MT) robots, single-robots (SR), time-extended assignment (TA) problem [8]. In the multi-robot domain, market-based approaches have gain much success [7, 25, 10, 5, 19, 15, 16] and can be further improved when learning is incorporated [4]. However, these methods often scale poorly in terms of team size and number of tasks. Furthermore, these approaches often depend on timely communication of the various local cost and utility functions and thus may not be suitable for situations when inter-agent wireless communication is unreliable or extremely limited [6, 9].

In this chapter, we present an ensemble approach towards the modeling, analysis, and design of distributed coordination strategies for the dynamic allocation of a team of homogeneous robots to a collection of spatially distributed tasks. Our approach is similar to existing work where macroscopic continuous models are used to describe the dynamics of a robot swarm [20, 18, 14]. The technique builds upon the representation of the individual robot controllers as probabilistic finite state machines which enables the approximation of team dynamics as a continuous-time Markov process [20, 18, 13]. In [11, 13], macroscopic models were used to synthesize stochastic agent-level control policies to enable the dynamic allocation of a team of robots to multiple locales in predefined proportions without explicit inter-agent wireless communication. Different from [18], the desired allocation was achieved through the selection of the individual robot transition rates and were extended to account for navigation delays in [1].

In our work, we assume a team of identical robots where the individual agent-level control policy is given by a sequential composition of individual task controllers. Different from existing work, we use the lower dimensional macroscopic representation of the ensemble dynamics for analysis and controller synthesis. We present two different approaches towards the design of stochastic transition rules that enable the team to autonomously achieve a desired distribution across a collection of spatially distributed task. In the first approach, mean-field models are used to analyze and mitigate the effects of unexpected inter-robot interactions on the team's ability to maintain a desired allocation [21]. In the second approach, agent-level control policies that can affect both the mean and the variance of the distribution is obtained by modeling the team dynamics as a polynomial Stochastic Hybrid System (pSHS) [17, 23].

The novelty of the contribution is a team-size invariant approach towards the design of distributed agent-level control policies that can respond to robot failures in a natural way and ensure a graceful degradation of the system. By providing a systems-level view of the team dynamics, the inter-agent communication needs of a desired coordination strategy can be more explicitly accounted for at the controller synthesis stage. The rest of the chapter is organized as follows: Section 2 presents the development of the macroscopic models for an ensemble of robots executing a collection of tasks with deterministic task execution times. Section 3 describes the analysis and design of our ensemble model derived distributed control strategies. We conclude with a brief discussion of future work in Section 4.

2 Problem Formulation

Consider the deployment of N robots to M tasks each located within a different region in the workspace. In this section, we show how continuous macroscopic models can be obtained from a collection of individual robot controllers. The goal is to use these models to design a decentralized control policy to enable the team to autonomously distribute across the M tasks and maintain the desired allocation at the various regions.

2.1 Single Robot Controller

Given a collection of $\{1,\ldots,M\}$ tasks/sites, we use a directed graph, $\mathscr{G} = (\mathscr{V},\mathscr{E})$, to model the pairwise precedence constraints between the tasks. Each task is represented by a vertex in $\mathscr{V} = \{1,\ldots,M\}$. A directed edge exists between two vertices $(i,j) \in \mathscr{V} \times \mathscr{V}$ if task i must precede task j and we denote this relation as $i \prec j$. Then, the set of edges, $\mathscr{E}$, is given by $\mathscr{E} = \{\forall(i,j) \in \mathscr{V} \times \mathscr{V} | i \prec j\}$. We assume $\mathscr{G}$ is a strongly connected graph, *i.e.*, a directed path exists for any $i,j \in \mathscr{V}$.

Given the M tasks, we denote the set of task controllers for each task as $\{U_1,\ldots,U_M\}$ and assume that the single robot controller is obtained through the sequential composition of $\{U_1,\ldots,U_M\}$ such that the precedence constraints specified by $\mathscr{G}$ are satisfied. We represent the robot controller as a finite state automaton where each automaton state i is associated with a task controller U_i. Fig. 1(a) shows an example robot controller where the arrows denote state transitions that satisfy the constraints specified in $\mathscr{G}$.

In this work, we consider the assignment/allocation of the team to M tasks/sites. The team's objective is to maintain the desired allocation of the robots across the various regions. At each site, robots execute U_i for a pre-specified amount of time τ_i. This represents the time required by a robot to complete the task at the given site.

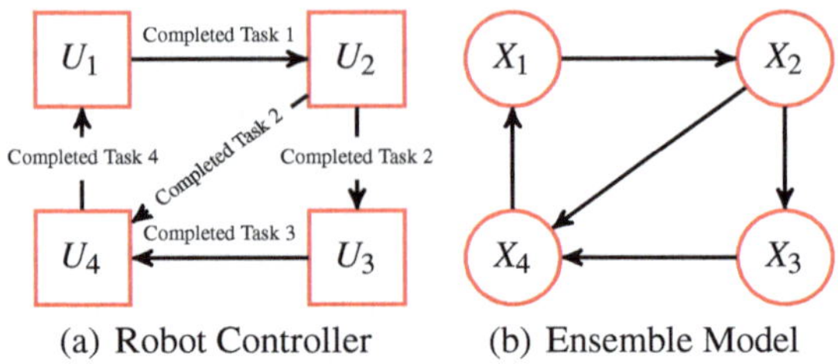

Fig. 1 (a) The robot controller. The robot changes controller states based on the guard conditions. (b) Graphical representation of the equivalent chemical reaction network for a robot ensemble executing the tasks.

Once the task has been completed, the robot must navigate to the next adjacent site based on the constraints encoded in $\mathscr{G}$. As such, we assume each robot has complete knowledge of $\mathscr{G}$, the ability to localize within the workspace, and is capable of navigating from one task/site to another while avoiding collisions with other robots in the workspace.

2.2 *The Ensemble Model*

For a team of N robots, each executing the same sequentially composed controller, *e.g.*, the one in Fig. 1(a), the ensemble dynamics can be modeled as a *polynomial stochastic hybrid system* (pSHS). This enables us to derive lower dimensional macroscopic models that describe the time evolution of the distribution of the team across the various tasks/sites.

Let $X_i(t)$ and $\bar{X}_i$ denote the number of robots executing task i (or being site at i) and the desired number of robots for task i respectively. Then the fraction of the robots at site i is given by $x_i(t) = X_i(t)/N$ with $\bar{x}_i$ denoting the desired fraction of robots. The specification in terms of fractions rather than absolute numbers provides a team size invariant formulation and is practical for scaling purposes. Since the tasks are spatially distributed, the robots will move from one task to another and must avoid collisions with other robots. The variability in robot arrival times at each site is modeled using *transition rates*. For every edge $e_{ij} \in \mathscr{E}$, we assign constant $k_{ij} > 0$ such that k_{ij} defines the transition probability per unit time for one agent from site i to go to site j. Furthermore, we assume the ensemble dynamics is Markov which will allow us to model the dynamics of the robot distribution as a set of linear differential equations. It is important to note that in general $k_{ij} \neq k_{ji}$.

2.2.1 Mean-Field Dynamics

It was shown in [13, 1] that the time evolution of the population fraction executing task i can be modeled as a continuous-time Markov process in the absence of task execution times, *i.e.*,

$$\frac{d}{dt}x_i(t) = \sum_{(j,i)\in\mathscr{E}} k_{ji}x_j(t) - \sum_{(i,j)\in\mathscr{E}} k_{ij}x_i(t). \tag{1}$$

The task execution times can be incorporated by reformulating the above equation as a delayed differential equation of the form

$$\frac{d}{dt}x_i(t) = \sum_{(j,i)\in\mathscr{E}} k_{ji}x_j(t-\tau_j) - \sum_{(i,j)\in\mathscr{E}} k_{ij}x_i(t). \tag{2}$$

We note that these models are mean-field descriptions of the team dynamics where the system state is given by $\mathbf{x}(t) = [x_1(t),\ldots,x_M(t)]^T$. In [13, 2], the desired distribution across the M sites was achieved by using (1) to optimize the k_{ij} terms to meet specific ensemble performance metrics.

2.2.2 Moment Dynamics

While the mean-field formulation provides a model of the time evolution of the fractions of robots at each task location, it is possible to provide a different macroscopic description of the ensemble dynamics by considering the rates of change of the various moments of the robot population distribution. Similar to the use of fractions in the previous section, the specification in terms of the moments of the robot population will also provide a team size invariant formulation. This is achieved by describing the ensemble dynamics using a set of transition rules of the form:

$$X_i \xrightarrow{k_{ij}} X_j \quad \forall\, e_{ij} \in \mathscr{E}. \tag{3}$$

The above expression represents a stochastic transition rule with k_{ij} as the per unit reaction rate and $X_i(t)$ and $X_j(t)$ as discrete random variables. In the robotics setting, equation (3) implies that robots at site i will transition to site j with a rate of $k_{ij}X_i$.

In this formulation, the system states are the random variables $X_i(t)$ with the state vector given by $\mathbf{X}(t) = [X_1(t),\ldots,X_M(t)]^T$. Given the set of stochastic transition rules in (3), the moment equations for the discrete random variable X_i is given by the extended generator of the system [12]. For a real-valued function $\psi(X_i)$, the extended generator is an expression for the time derivative of the expected value of ψ, *i.e.*, $\frac{d}{dt}\mathrm{E}[\psi(X_i)] = \mathrm{E}[L\psi(X_i)]$, and takes the form

$$L\psi(X_i) = \sum_j \left[(\psi(X_i-1)-\psi(X_i))w_{ji} + (\psi(X_i+1)-\psi(X_i))w_{ij}\right]. \tag{4}$$

The right hand side of (4) gives the continuous time derivatives of the system for a discrete change in the state X_i. The expression $[\psi(X_i-1)-\psi(X_i)]$ represents the change in ψ given a unit change in the discrete variable X_i, while w_{ij} represents the frequency at which the change occurs. For the system given by (3), $w_{ij} = k_{ij}X_i$. To obtain the rate of the change of the expected value of X_i, $\frac{d}{dt}\mathrm{E}[X_i]$, we let $\psi(X_i) = X_i$ in (4). Similarly, to obtain $\frac{d}{dt}\mathrm{E}[X_i^2]$, we let $\psi(X_i) = X_i^2$.

For the case when $M = 2$, the first and second moment dynamics for X_1 are given by

$$\begin{aligned}\frac{d}{dt}\mathrm{E}[X_1] &= \mathrm{E}\left[((X_1+1)-X_1)k_{21}X_2+((X_1-1)-X_1)k_{12}X_1\right]\\ &= k_{21}\mathrm{E}[X_2]-k_{12}\mathrm{E}[X_1]\\ \frac{d}{dt}\mathrm{E}[X_1^2] &= \mathrm{E}\left[((X_1+1)^2-X_1^2)k_{21}X_2+((X_1-1)^2-X_1^2)k_{12}X_1\right]\\ &= -2k_{12}\mathrm{E}[X_1^2]+2k_{21}\mathrm{E}[X_1X_2]+k_{21}\mathrm{E}[X_2]+k_{12}\mathrm{E}[X_1].\end{aligned}$$

When the w_{ij}'s are linear with respect to the system state $\mathbf{X}$, the moment equations are closed. This means that the time derivative for the first moment of X_i, $\frac{d}{dt}\mathrm{E}[X_i]$, is only dependent on the first moments of X_i for $i = 1,\ldots,M$, *i.e.*, $\mathrm{E}[X_1],\ldots,\mathrm{E}[X_M]$, the second moments are dependent on the first and second moments, and so on and so forth. This is important because when the moment equations are closed, the moment dynamics can be expressed as a linear matrix equation.

In general, the ensemble moment dynamics for the system with M tasks/sites is given

$$\begin{aligned}\frac{d}{dt}\mathrm{E}[X] &= \mathbf{K}\mathrm{E}[X]\\ \frac{d}{dt}\mathrm{E}[XX^T] &= \mathbf{K}\mathrm{E}[XX^T]+\mathrm{E}[XX^T]\mathbf{K}^T+\Gamma(\alpha,\mathrm{E}[X])\end{aligned} \tag{5}$$

where $[\mathbf{K}]_{ij} = k_{ji}$ and $[\mathbf{K}]_{ii} = -\sum_{(i,j)\in\mathscr{E}} k_{ij}$. It is important to note that $\mathbf{K}$ is a Markov process matrix and thus is negative semidefinite. This coupled with the conservation constraint $\sum_i X_i = N$ leads to exponential stability of the system given by (5) [11, 17, 23]. Each entry in the matrix of second moments is determined from the moment closure methods shown above where the entries of $\Gamma(\alpha,\mathrm{E}[X])$ are all linear with respect to the k_{ij}'s and the means $\mathrm{E}[X]$. For a system with two states, X_1 and X_2, $\Gamma(\alpha,\mathrm{E}[X])$ is defined as

$$\Gamma(\alpha,\mathrm{E}[X]) = \begin{bmatrix} k_{12}\mathrm{E}[X_1]+k_{21}\mathrm{E}[X_2] & -k_{12}\mathrm{E}[X_1]-k_{21}\mathrm{E}[X_2]\\ -k_{12}\mathrm{E}[X_1]-k_{21}\mathrm{E}[X_2] & k_{12}\mathrm{E}[X_1]+k_{21}\mathrm{E}[X_2]\end{bmatrix}.$$

Similar to the mean-field description, the k_{ij}'s can be chosen to enable a team of robots to autonomously maintain some desired mean steady-state distribution of the team across the various tasks/sites [11, 13, 1]. In both formulations, the k_{ij}'s translate into a set of stochastic guard conditions for the single robot controllers. The result is a set of decentralized agent-level control policies that allow the team to maintain the steady-state mean of the ensemble distribution. Different from the mean-field approach, the formulation of the ensemble dynamics in terms of the moments of the robot population enables us to synthesize distributed control strategies to enable the team to maintain *both the mean and the variance* of the robot team distribution across the various tasks/sites. We describe the approaches in the following sections.

3 Methodology

Regardless of the approach, both (1) (or (2)) and (5) are macroscopic models of the ensemble activity. In this section, we show how these models can be used to analyze the effects of uncertainty on the performance of the team of robots servicing a collection of spatially distributed tasks. Furthermore, we will show how these models can be used to design distributed coordination strategies to improve the team's performance in the presence of the these uncertainties.

3.1 Characterizing and Filtering the Ensemble Noise

Berman *et al.* showed that when the task execution times are stochastic, the ensemble dynamics given by (2) can be approximated using an equivalent expanded linear system [1], *i.e.*, a Multi-Pole approximation. This is achieved by introducing additional dummy transitions between states to approximate the effects of the stochastic delay times. When delay times are deterministic or near deterministic, Mather *et al.* showed that Padé approximants employed in the frequency domain do a better job of capturing the effects of the delay in (2) [22].

To determine the effects of the deterministic (or near deterministic) delays on the overall system performance, we can analyze (2) in the frequency domain [22]. In the frequency domain, the time delay is modeled as an exponential variable. As the frequency increases, the output signal is delayed by more and more periods, in effect, worsening the phase error. The advantage of the Padé approximation is the ability to more accurately capture these effects while retaining the algebraic structure of the differential equations, *i.e.*, (2), in the frequency domain.

Consider the deployment of an ensemble of 10 robots moving in the plane to 2 distinct locations/sites. Initially, robots are randomly assigned to each of the two sites. To simulate the execution of a task at a given site, each robot is tasked to circle the site in a clockwise direction for a fixed time $\tau_i = \tau$. Once the task has been executed, the robot moves to the next site and performs the same task at the new site. The variability in each robot's site-to-site navigation times depends on the amount of traffic on the road, which is affected by the number of collision avoidance maneuvers each robot must execute. Fig. 2(a) shows the Fast Fourier Transform (FFT) of the average output, *e.g.*, x_i in (2), of 54 agent-based simulations performed in USARSim [24]. The frequency response of the agent-based simulations was obtained by logging the population fractions at each site over time and applying the FFT to these variables for each run. The FFT results were then averaged over all 54 runs. The agent-based system exhibits a maximum gain at approximately 7.5 mHz while both the Padé and Multi-Pole macroscopic models exhibit peaks at approximately the same frequency. However, the Padé model shows larger gain [22].

The spurious frequency components shown in Fig. 2(a) manifest as oscillations in the ensemble states in the time domain and is an effect of robots clustering together as they travel from one site to another. These clusters form because too many robots are traveling between tasks resulting in more collision avoidance maneuvers and thus further delaying the arrival of robots at their next task. This leads to degraded

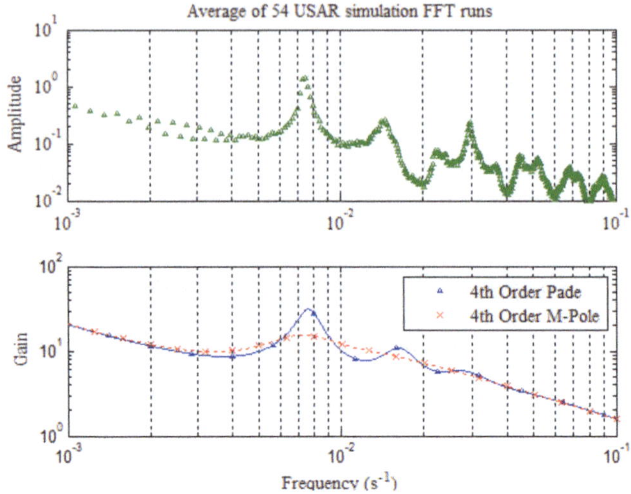

(a) Ensemble Response

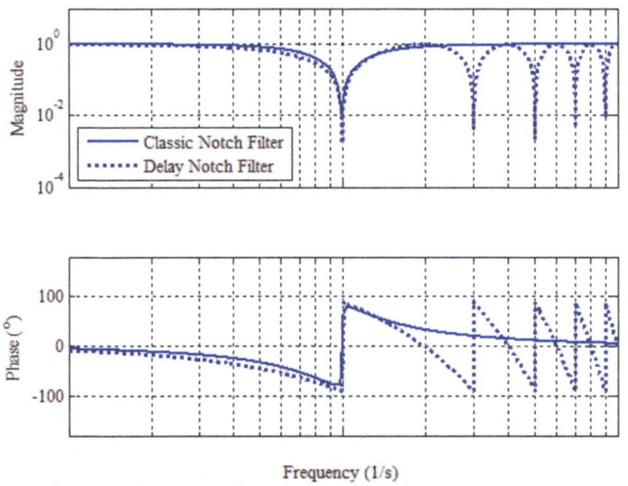

(b) Notch Filter Response

Fig. 2 (a) Top: Average of the FFT of the population fraction at site 2 obtained from 54 micro-discrete simulations. Bottom: Magnitude portion of the Bode plots relating to the number of Robots at building 2. For the 4th order Padé, and a 4th order Multi-Pole macro-continuous systems. (b) Frequency response of the classical 2nd order notch filter and the delayed notch filter given by $H_{\tau}(s) = \frac{1}{2}(1+e^{-s\omega_{\tau}})$.

performance as the average transit time between sites will increase due to these traffic concerns. The Padé approximated macro-continuous model has the ability to better predict the spurious frequency component that is present in the agent-based simulations and can provide insight into the synthesis of agent-level controllers to filter out these spurious frequencies [22].

In general, it is difficult to directly compare the macroscopic results with the microscopic results. This is because the FFT of the system states only considers the outputs of the system. The magnitude portion of the Bode plots, on the other hand, gives the response of the ratio of the output to input of the system for all frequencies. In other words, the macroscopic frequency response is based on a unity gain input at all frequencies. The difference between the two plots is dependent on the form of the noise input to the system and is related by the shape of the frequency spectrum of the noise input to the system.

3.1.1 Distributed Filtering

To smooth the response of the system, a common approach is to implement a notch filter to get rid of the spurious behavior. A notch selectively filters out a specific frequency while leaving other frequency components unchanged, effectively reducing the gain of the single spurious frequency component. A typical 2nd order notch controller has the transfer function $H_1(s)$ given by

$$H_1(s) = \frac{s^2 + 2\zeta_1\omega_N + \omega_N^2}{s^2 + 2\zeta_2\omega_N + \omega_N^2}$$

where ω_N, ζ_1, and ζ_2 set the location and magnitude of the notch. However, careful inspection of the closed-loop time domain equations suggest implementation of the filter will require individual robots to estimate the higher order derivatives of the populations at the various sites. Instead, we propose an approximate solution, where the spurious frequency response can be removed without extra knowledge of the system states by the individual robots.

This can be achieved by splitting the team into two sub-teams where one team purposely adds an additional delay at a given site for each cycle path in $\mathscr{G}$. This approach can eliminate a frequency by adding a signal to a copy of itself, 180° degrees out of phase. The transfer function for the proposed notch filter is $H_\tau(s) = \frac{1}{2}(1 + e^{-s\omega_\tau})$. The advantage of this approach is that it can be implemented in a completely distributed way without requiring any inter-agent communication. The frequency response plot for this distributed delay notch filter and the classical 2nd order filter are shown in Fig. 2(b). The difference is that the delay filter cancels every odd harmonic of the primary notched frequency [21].

The addition of a single notch filter will suppress a single spurious population behavior. If the task precedence graph has multiple cycles with spurious loops, multiple notches are required to eliminate spurious behavior. While introduction of a delay into a system with feedback can be dangerous since it can lead to enough phase lag to turn negative feedback into positive feedback resulting in unstable oscillations, the systems discussed here fall into a family of systems that are stable independent of delay [3]. Thus, no amount of extra phase delay can drive the system unstable.

Fig. 3(a) shows the frequency response of the average of 50 agent-based simulations for a team of 10 with two tasks with and without the delay notch filter. Our results show that the distributed notch filter suppressed the undesired frequency component by 70%. Fig. 3(b) shows the frequency response of the average of three experimental trials for a team of 10 robots with and without the delay notch filter. Each experiment ran for roughly 45 to 50 minutes with the robots executing 750 state transitions.

3.2 Controlling the Ensemble Moment Dynamics

While the mean-field approach enables us to design decentralized coordination strategies that can be implemented without any inter-agent wireless communication, it is limited since it can only affect the mean of the distribution. On the other hand, the explicit modeling of the moment dynamics of the robot population distribution using (5) gives the ability to devise ensemble feedback strategies that enables the team to affect the mean, the variances, and any higher order moments of the ensemble distribution.

As shown with equation (3), the rate in which agents in state X_i transition to X_j depends on the population in state X_i. As such, the more agents in state X_i, the faster they transition to X_j. However, Klavins recently showed that if we allow for both positive and negative transition rates, it is possible to shape both the mean and the variance of the ensemble distribution [17]. In other words, by introducing a negative feedback rate, it is possible to slow the population growth at a given state and thus affect the population variance in that state.

For the M state system described by (5), consider the following ensemble feedback controller

$$\mathbf{u} = -\mathbf{K}_\beta \mathrm{E}[X] \quad \mathbf{K}_\beta^{ij} = \begin{cases} \beta_{ji} & \forall (i,j) \in \mathscr{E} \\ -\sum_{(i,j)\in\mathscr{E}} \beta_{ji} & \forall i = j \\ 0 & \text{otherwise} \end{cases} . \tag{6}$$

Expression (6) can be seen as a form of linearizing feedback control that inhibits transitions from X_i to X_j as X_j becomes larger than X_i. This results with the following closed-loop moment dynamics

$$\begin{aligned} \tfrac{d}{dt}\mathrm{E}[X] &= (\mathbf{K}_\alpha + \mathbf{K}_\beta)\mathrm{E}[X] \\ \tfrac{d}{dt}\mathrm{E}[XX^T] &= (\mathbf{K}_\alpha + \mathbf{K}_\beta)\mathrm{E}[XX^T] + \mathrm{E}[XX^T](\mathbf{K}_\alpha + \mathbf{K}_\beta)^T \\ &\quad + \Gamma(\alpha, \beta, \mathrm{E}[X]). \end{aligned} \tag{7}$$

The steady-state values of $\mathrm{E}[X_i]$ and $\mathrm{E}[X_iX_j]$ can be independently set by adjusting parameters α and β. The above equations are obtained by simply substituting

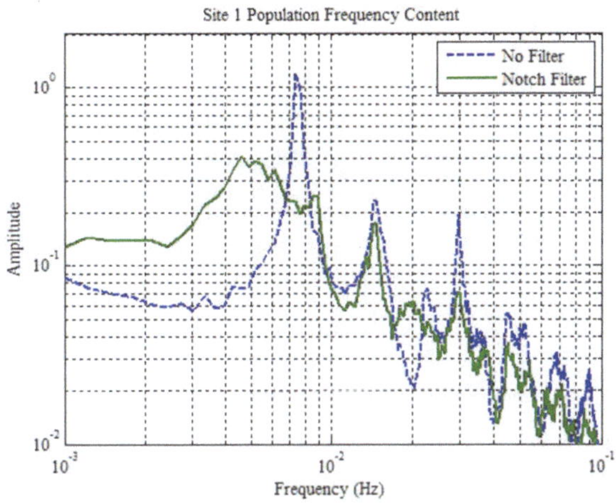

(a)

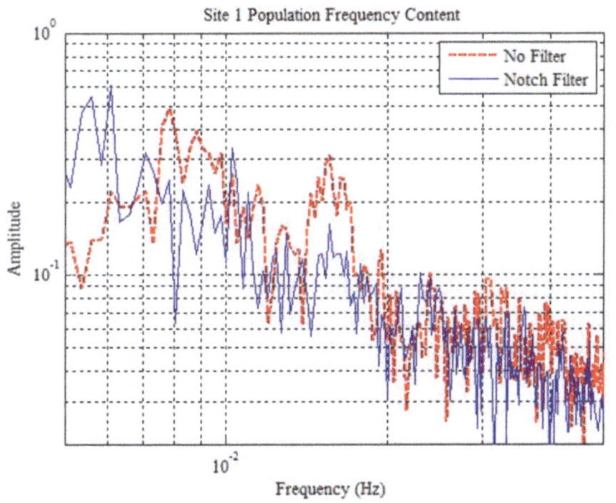

(b)

Fig. 3 (a) Frequency response results for the simulated system. This plot shows the average response of 50 simulations of the unfiltered (no notch) system and 50 simulations of the system with the delay notch filter. The system with the distributed filter shows a depression in the frequency response at the active notching point. (b) Frequency response of the initial experimental trials. The plots shows the frequency content of 3 averaged response for the unfiltered and notched system. The peak, though small, is properly located according to the transition times. The high peak at 6.1mHz in the notched response is due to the round trip time if all agents went along the delay route that includes the τ_{NOTCH} delay.

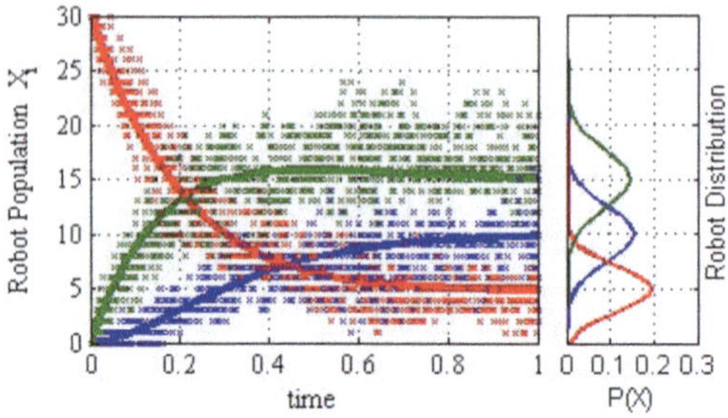

(a)

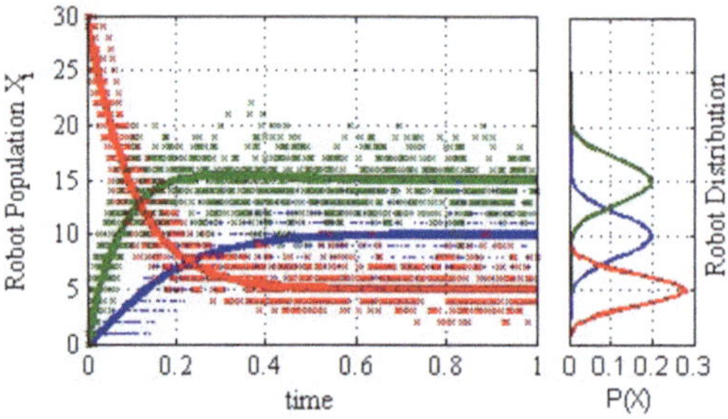

(b)

Fig. 4 These plots compare the steady state distributions and the convergence rate of the system with and without ensemble feedback. Each left side plot shows the transient behavior from an initial condition of $X = [0,0,30]$. The solid lines denote the numerical solutions of the first moment dynamics and the data points are 10 representative stochastic simulation runs. The right side plots are the steady state distributions represented as Gaussians.

$k_{ij} = \alpha_{ij} - \beta_{ij}\frac{X_j}{X_i}$ in the reactions given by (3) and applying the extended generator to $\psi(X_i) = X_i$ [23].

The advantage of the proposed ensemble feedback strategy, over any other negative feedback strategy, is that it maintains the moment closure property for the closed-loop system. This enables us to show that the close-loop moment dynamics remain stable when $\beta_{ij}X_j \leq \alpha_{ij}X_i$ [23]. When $\beta_{ij}X_j > \alpha_{ij}X_i$, the system experiences a backwards flow. As such, in practice, we restrict this rate to be greater than or equal to zero. The addition of these saturation effects will slightly complicate the stability analysis [23].

While the feedback strategy (6) gives robots in state X_i the ability to set their own state transition rates to be independent from the number of robots X_i, it requires robots at task i to know how many robots are at adjacent sites, *i.e.*, robots in X_j where $e_{ij} \in \mathscr{E}$. This differs from the delayed notch filter presented in Section 3.1 which can be implemented with no communication. As such, the implementation of (6) will depend on the available communication infrastructure.

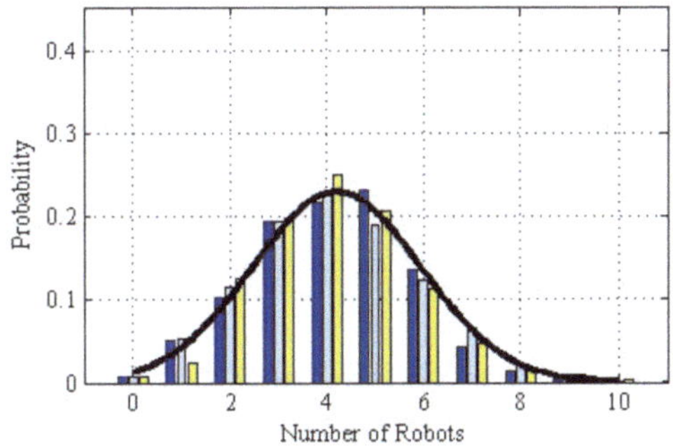

(a)

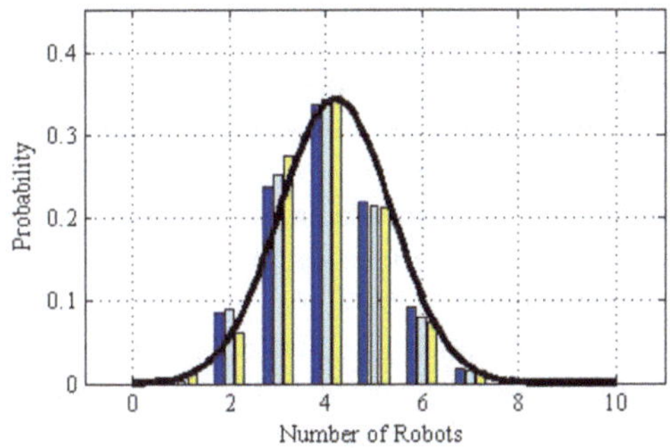

(b)

Fig. 5 Probability distribution of the robot ensemble at each site, $\{X_1, X_2, X_3\}$. (a) Without ensemble feedback and (b) with ensemble feedback using local communication.

If we endow each task location with the computational capability to track the number of robots at the site and the ability to communicate with adjacent task sites, then the estimation of the ensemble states would be similar to having a single global estimator. When individual robots arrive at a location, the information is updated and broadcasted to all adjacent task sites. Fig. 4 shows the first moments of a three state system ($M = 3$ tasks) over time with and without the ensemble feedback strategy. Note how the system with ensemble feedback has both faster convergence and smaller variance on its populations [23].

In practice, not only is it unreasonable to assume full and perfect communication among the robots, it is often unreasonable to assume full and perfect communication between the task sites. This is especially true when sites are distributed across vast geographic regions or in situations where long-range communication is difficult/impossible, *e.g.*, underground/underwater environments. A distributed implemention of (6) can be achieved through local inter-robot communication alone. We assume robots have finite communication ranges and can only communicate with other robots that are co-located at the same site and/or within each other's communication range. As robots move from one site to another, they can exchange information with other robots they encounter and construct their own estimates of the population levels at the various sites. Fig. 5 shows the allocation of the 15 robots to three task sites with and without the distributed implementation of the ensemble feedback strategy.

4 Discussion and Outlook

In this chapter, we presented a method for synthesizing distributed ensemble feedback control strategies through the development and analysis of an appropriate macroscopic description of the ensemble dynamics. In one case, mean-field models allowed for the identification of the spurious interactions between robots as they moved within a workspace executing a collection of spatially distributed tasks. The macroscopic analysis lead to the development of a distributed filtering strategy that could be implemented without requiring any inter-robot wireless communication nor estimation of population variables. In a second case, moment closure techniques where used to model the dynamics of a team of robots servicing a collection of spatially distributed tasks. The analysis provided a linearizing ensemble feedback strategy to enable the team to maintain the mean and the variance of the robot population distribution across the various tasks.

The key advantage of this approach is a lower dimensional parameterization of the ensemble dynamics that retains the salient features of the underlying agent-based system. These techniques are particularly well-suited for analyzing the effects of uncertain interactions on overall system performance in multi-agent robotic systems. Specifically, these techniques enable the analysis of highly redundant systems in a lower dimensional space while simultaneously retaining the systems-level view of the dynamics. Furthermore, since interaction uncertainties can be explicitly encoded in these models, the feedback strategies developed using these techniques would be robust to any changes in population sizes.

Despite these advantages, further investigation is needed to determine the classes of multi-agent coordination problems that are amenable to these macroscopic modeling and controller synthesis techniques. Specifically, we are interested in investigating the viability of these techniques in modeling and controlling multi-agent robotic systems executing highly coupled tasks. For any ensemble derived feedback strategy, there is also the added challenge of determining the appropriate distributed implementation. However, this presents an opportunity for network resource aware synthesis of distributed coordination and control strategies for multi-agent systems. We are interested in investigating ensemble controller synthesis techniques that can take into account the trade-off between more precise control and the need for estimating ensemble states.

Acknowledgements. We gratefully acknowledge the support of NSF grant DGE-0947936 and CNS-1143941.

References

[1] Berman, S., Halasz, A., Hsieh, M.A., Kumar, V.: Navigation-based optimization of stochastic deployment strategies for a robot swarm to multiple sites. In: Proc. of the 47th IEEE Conference on Decision and Control, Cancun, Mexico (2008)

[2] Berman, S., Halasz, A., Hsieh, M.A., Kumar, V.: Optimized stochastic policies for task allocation in swarms of robots. IEEE Transactions on Robotics 25(4), 927–937 (2009)

[3] Chen, J., Latchman, H.: Frequency sweeping tests for stability independent of delay. IEEE Transactions on Automatic Control 40(9), 1640–1645 (1995), doi:10.1109/9.412637

[4] Dahl, T.S., Matariè, M.J., Sukhatme, G.S.: A machine learning method for improving task allocation in distributed multi-robot transportation. In: Braha, D., Minai, A., Bar-Yam, Y. (eds.) Understanding Complex Systems: Science Meets Technology, pp. 307–337. Springer, Berlin (2006)

[5] Dias, M.B.: Traderbots: A new paradigm for robust and efficient multirobot coordination in dynamic environments. PhD thesis, Robotics Institute, Carnegie Mellon University, Pittsburgh, PA (2004)

[6] Dias, M.B., Zlot, R.M., Kalra, N., Stentz, A.T.: Market-based multirobot coordination: a survey and analysis. Proceedings of the IEEE 94(7), 1257–1270 (2006)

[7] Gerkey, B.P., Matarić, M.J.: Sold!: Auction methods for multi-robot control. IEEE Transactions on Robotics & Automation 18(5), 758–768 (2002)

[8] Gerkey, B.P., Matarić, M.J.: A formal framework for the study of task allocation in multi-robot systems. International Journal of Robotics Research 23(9), 939–954 (2004)

[9] Golfarelli, M., Maio, D., Rizzi, S.: Multi-agent path planning based on task-swap negotiation. In: Proceedings of the 16th UK Planning and Scheduling SIG Workshop, PlanSIG, Durham, England (1997)

[10] Guerrero, J., Oliver, G.: Multi-robot task allocation strategies using auction-like mechanisms. In: Artificial Research and Development, Frontiers in Artificial Intelligence and Applications, vol. 100, pp. 111–122. IOS Press (2003)

[11] Halasz, A., Hsieh, M.A., Berman, S., Kumar, V.: Dynamic redistribution of a swarm of robots among multiple sites. In: Proceedings of the Conference on Intelligent Robot Systems (IROS 2007), San Diego, CA, pp. 2320–2325 (2007)

[12] Hespanha, J.P.: Moment closure for biochemical networks. In: Proc. of the Third Int. Symp. on Control, Communications and Signal Processing (2008)

[13] Hsieh, M.A., Halasz, A., Berman, S., Kumar, V.: Biologically inspired redistribution of a swarm of robots among multiple sites. Swarm Intelligence (2008)

[14] Hsieh, M.A., Halasz, A., Cubuk, E.D., Schoenholz, S., Martinoli, A.: Specialization as an optimal strategy under varying external conditions. In: Accepted the International Conference on Robotics and Automation, ICRA 2007, Kobe-Japan (2009)

[15] Jones, E.G., Browning, B., Dias, M.B., Argall, B., Veloso, M., Stentz, A.T.: Dynamically formed heterogeneous robot teams performing tightly-coordinated tasks. In: Proceedings of the 2006 IEEE International Conference on Robotics and Automation (ICRA 2006), pp. 570–575. IEEE, Los Alamitos (2006)

[16] Jones, E.G., Dias, M.B., Stentz, A.: Learning-enhanced market-based task allocation for oversubscribed domains. In: Proceedings of the Conference on Intelligent Robot Systems (IROS 2007), pp. 2308–2313. IEEE, Los Alamitos (2007)

[17] Klavins, E.: Proportional-integral control of stochastic gene regulatory networks. In: Proc. of the 2010 IEEE Conf. on Decision and Control (CDC 2010), Atlanta, GA, USA (2010)

[18] Lerman, K., Jones, C., Galstyan, A., Matarić, M.J.: Analysis of dynamic task allocation in multi-robot systems. International Journal of Robotics Research (2006)
[19] Lin, L., Zheng, Z.: Combinatorial bids based multi-robot task allocation method. In: Proceedings of the 2005 IEEE International Conference on Robotics and Automation (ICRA 2005), pp. 1145–1150. IEEE, Los Alamitos (2005)
[20] Martinoli, A., Easton, K., Agassounon, W.: Modeling of swarm robotic systems: a case study in collaborative distributed manipulation. International Journal of Robotics Research: Special Issue on Experimental Robotics 23(4-5), 415–436 (2004)
[21] Mather, T.W., Hsieh, M.A.: Distributed filtering for time-delayed deployment to multiple sites (best paper award winner). In: 10th International Symposium on Distributed Autonomous Robotics Systems (DARS 2010), Lausanne, Switzerland (November 2010)
[22] Mather, T.W., Hsieh, M.A.: Analysis of stochastic deployment policies with time delays for robot ensembles. International Journal of Robotics Research: Special Issue on Stochasticity in Robotics & Biological Systems, Part 1 30(5) (2011)
[23] Mather, T.W., Hsieh, M.A.: Distributed robot ensemble control for deployment to multiple sites. In: 2011 Robotics: Science and Systems, Los Angeles, CA USA (2011)
[24] USARSim, Unified system for automation and robot simulation (2005), `http://usarsim.sourceforge.net`
[25] Vail, D., Veloso, M.: Multi-robot dynamic role assignment and coordination through shared potential fields. In: Schultz, A., Parker, L., Schneider, F. (eds.) Multi-Robot Systems, pp. 87–98. Kluwer (2003)

Cooperating Mobile Cable Robots: Screw Theoretic Analysis

Xiaobo Zhou, Chin Pei Tang, and Venkat Krovi

Abstract. Cable robots form a class of parallel architecture robots with significant benefits including simplicity of construction, large workspace, significant payload capacity and end effector stiffness. While conventional cable robots have fixed bases, we seek to explore inclusion of mobility into the bases (in the form of gantries, and/or vehicle bases) which can significantly further enhance the capabilities of cable robots. However, this also introduces redundancy and complexity into the system which needs to be carefully analyzed and resolved. To this end, we propose a generalized modeling framework for systematic design and analysis of cooperative mobile cable robots, building upon knowledge base of multi-fingered grasping, and illustrate it with a case study of four cooperating gantry mounted cable robots transporting a planar payload. We show its wrench closure workspace and reconfiguration to extend the workspace, as well as redundancy resolution by optimally repositioning the bases to maximize tension factor along a given trajectory.

1 Introduction

Cable driven parallel manipulators, also called cable-driven robots or cable robots, are formed by attaching multiple cables (instead of articulated links) to an end-effector/platform. They have significantly improved workspace as compared to conventional rigid-link architectures, while possessing many of the desirable features such as high payload-to-weight ratios, low inertial properties, low energy consumption, ease of assembly/disassembly and reconfiguration. Overall low cost and

X. Zhou · V. Krovi
Department of Mechanical and Aerospace Engineering,
State University of New York at Buffalo, Buffalo, NY 14260 USA
e-mail: {xzhou9,vkrovi}@buffalo.edu

C.P. Tang
Caterpillar Global Mining Division, Denison TX 75020 USA
e-mail: nonholonomic@gmail.com

D. Milutinović & J. Rosen (Eds.): Redundancy in Robot Manipulators, LNEE 57, pp. 109–123.
DOI: 10.1007/978-3-642-33971-4_7

reliability contribute to their deployment in many real-world applications, such as heavy payload handling for manufacturing [1], extraterrestrial exploration [2], haptics [3, 4], large scale radio telescopes [5], and load transport [6].

Cooperative payload manipulation using cables comes in two flavors: one class of approaches focuses on fixed bases and varying cable lengths [1, 3, 7, 8] (i.e. conventional cable robots); the other class is with fixed cable lengths and moving bases for manipulating of objects [9] and payload manipulation and transportation on land [10], sea [11], and in the air [12] (i.e. cable towing). In this work, we explore merging the two, i.e. coupling mobile bases with articulated-cable-arms together to create composite mobile-cable collectives for the combined payload transportation and reconfiguration tasks (such as shown in Fig. 1). We call this type of cable robots with moving bases cooperating mobile cable robots. While this combination potentially could greatly increase the capability of cable robots, it also introduces redundancy and complexity into the system. Hence, we will focus on developing a systematic framework for design, analysis and control of such mobile cable robot collectives.

There are many challenges to the development of such a framework. Cable robot systems can function only when the cables are in tension, which creates unilateral constraints on the controlled-input rendering conventional control schemes developed for typical parallel robots incompatible. Workspace determination in the presence of these unilateral constraints creates challenges that will be reviewed in Section 2. Further, despite many parallels exist between the unilateral tension requirements and unidirectional normal-force constraints arising in multi-fingered hands and multi-legged walkers efforts to relate this wealth of literature to cable robots have been very limited [13].

In modularly composed systems, both the nature of the individual modules as well as their interactions can affect the overall system performance. Hence, a systematic (and preferably quantitative) framework for evaluation of the individual

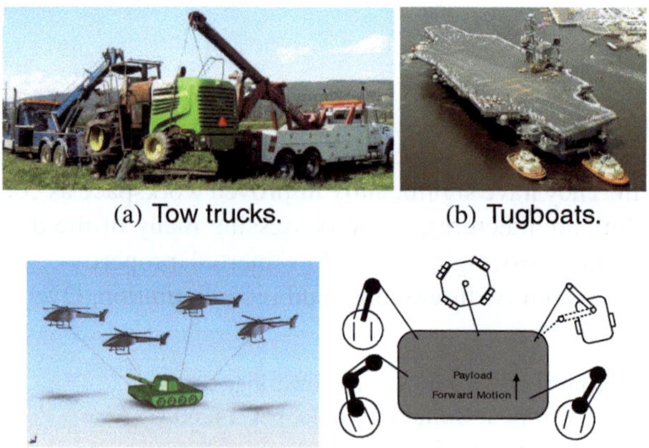

(a) Tow trucks. (b) Tugboats.

(c) Aerial towing. (d) Composite system.

Fig. 1 Illustrative examples of mobile-agent teams tethered to a payload module- and system-level characteristics is desirable. This is an aspect that we examine in the context of cooperative payload transport by mobile cable robot collectives in this chapter. To this end, we leverage the rich history and background of analysis methods for constrained articulated mechanical systems. In particular, a twist- and wrench-based analysis of in-parallel systems [14, 15] provides the underlying framework for examining the performance of the cooperative system here. *The unique contributions of this chapter come from: (i) the constructive modeling of the individual- and group-capabilities of the cooperating mobile cable robots; (ii) systematic design evaluation of options such as attachment points choices and mobile base positioning; (iii) redundancy resolution by optimal reconfiguration to maximize tension factor along a desired trajectory.*

2 Background

Besides the recent interest in multiple mobile agents, other forms of cooperative multi-robot systems including multi-fingered hands and multi-legged vehicles have been extensively studied in many contexts, as reviewed in [16]. Traditionally, such systems have been modeled as articulated mechanical systems, with the characteristic formation of closed kinematic chains. Apart from the structural classification of Type I (legged) and Type II (multi-arm/multi-finger hand) systems, an alternate functional classification into under-actuated, fully-actuated and redundantly-actuated systems is also possible [16]. It is meaningful to exploit the redundancy in the system to optimize secondary criteria, such as the contact/internal force distribution, in addition to the performance of the motion tasks in multi-legged walkers [17], multi-fingered hands [18] and multi-arm systems [19]. It is important to note that most of these efforts have been addressed in a centralized control context - with the notable exception of [20].

A cable robot with an n DOF end effector requires at least $n+1$ cables to fully constrain the end-effector, leading to minimum limits of four cables for planar robots and seven cables for the spatial case [3]. This also leads to a natural classification into fully-, over- and under-constrained cable-robot systems. In the fully- and over-constrained cable-robot systems, the posture (position/orientation) of the end-effector can be completely determined by the given lengths of the cables and force closure can be achieved [13]. The workspace determination [21, 22, 23, 24] poses challenges - while the potentially-reachable workspace is a function of the geometric configuration (cable lengths, motor mounting position, cable attachment location, etc.), not all postures may be feasible under positive-tension constraints. Hence an additional functional workspace classification becomes possible [24, 7, 25]. In [26], a generic method for determining wrench closure for a fully constrained cable robot is presented. A measure of workspace quality named tension factor is presented in [27], which we will adopt as our optimization criteria. The similarities between cable robots and other parallel architecture manipulators lead to systematic formulation of system performance from individual agent contribution, and we explore this perspective next.

3 Formulation

Using the matrix Lie Group representation of $SE(3)$ based on the notation introduced in [14], let $\{s\}$ and $\{b\}$ be the spatial and body fixed frames, the relative pose of a rigid body may be expressed as $g_{sb} = \begin{bmatrix} R_{sb} & p_{sb} \\ O & 1 \end{bmatrix}$, where $R \in SO(3)$ is a rotation matrix and $p \in \mathbb{R}^3$ is a displacement vector. The body twists can be computed as $\underset{\sim}{t}^b_{sb} = g^{-1}_{sb}\dot{g}_{sb}$. The body twist vector corresponding to this twist matrix can be interpreted in terms of linear and angular velocities in body fixed coordinates $\underset{\sim}{t}^b_{sb} = {}^b[v_x, v_y, v_z, \omega_x, \omega_y, \omega_z]^T$. Wrenches $\underset{\sim}{w}^b_{sb} = {}^b[f_x, f_y, f_z, \tau_x, \tau_y, \tau_z]^T$ correspond to co-vector fields and satisfy the virtual work relationship $\underset{\sim}{w} \cdot \underset{\sim}{t} = 0$. The adjoint transformation $Ad_g = \begin{bmatrix} R & \hat{p}R \\ O & R \end{bmatrix}$ and the co-adjoint transformation $Ad^T_{g^{-1}} = \begin{bmatrix} R & O \\ \hat{p}^T R & R \end{bmatrix}$ serves to transform twists and wrenches between various frames of references.

3.1 Agent Twists

For each cooperating robot (they do not have to be identical), we treat the cable as an articulated prismatic joint extending from its end-effector. Then it is a straight forward process to derive its spatial twist. We can assign the preferred frames and find its body twists of successive joints and then transform to a common frame (for which we choose fixed world frame $\{F\}$ here) to compose the agent's spatial Jacobian $J^s_i(\mathbf{q}_i)$:

$$^F\left[{}^0\underset{\sim}{t}_n\right]_i = J^s_i(\mathbf{q}_i)\dot{\mathbf{q}}_i = \left[{}^F\left[{}^0\underset{\sim}{t}_1\right]_i \cdots {}^F\left[{}^{n-1}\underset{\sim}{t}_n\right]_i\right], \tag{1}$$

where $\mathbf{q}_i$ are the joint space coordinates. This way, we can incorporate heterogeneous mobile agent collectives (such as shown in Fig. 1(d)) to perform cooperative manipulation.

3.2 Payload Attachment Statics

Since the cables are firmly attached to the payload, there is no slipping. We note that unlike finger pushing, cable pulling does not depend upon object's shape/contact normal direction, rather, it is the cable attachment point's relative position with respect to payload center of mass (COM) that matters. Thus, we define the cable attachment contact frame $\{c_i\}$ to have the same orientation as the payload object COM frame $\{o\}$, as shown in Fig. 2. The transformation from cable attachment frame $\{c_i\}$ to payload frame $\{o\}$ is given by $g_{oc_i} = \begin{bmatrix} R_{oc_i} & p_{oc_i} \\ O & 1 \end{bmatrix}$, where $R_{oc_i} = I$ since we choose attachment frame to have the same orientation as payload object frame, and p_{oc_i}, which are all fixed, once the attachment locations are chosen.

The basis direction of the cable tension in contact frame $\{c_i\}$ is given by B_{c_i}, then the cable wrench can be expressed in the payload object frame $\{o\}$ via co-adjoint

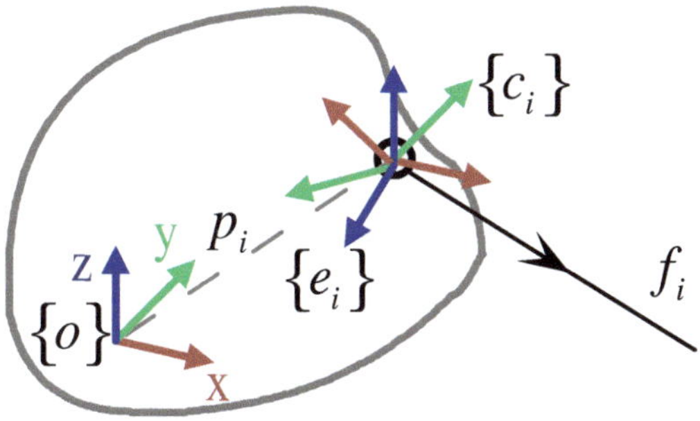

Fig. 2 Cable attachment contact model

transformation as ${}^o\underset{\sim}{w}_i = Ad^T_{g^{-1}_{oc_i}} B_{c_i} f_i$. Thus, in the payload object frame, the mapping P from the space of the m cable tensions $\mathbf{f}$ to the object wrench $\underset{\sim}{w}_o$ can be expressed as

$$\underset{\sim}{w}_o = P\,\mathbf{f}_{m\times 1} = \begin{bmatrix} | & | & & | \\ P_1 & P_2 & \cdots & P_m \\ | & | & & | \end{bmatrix}_{6\times m} \begin{bmatrix} f_1 \\ f_2 \\ \vdots \\ f_m \end{bmatrix}, \tag{2}$$

where $P_i = Ad^T_{g^{-1}_{oc_i}} B_{c_i}$. Analogous to the grasp map, we call P the pulling map.

3.3 *The Attachment Pulling Constraint*

Now that we have both the agent Jacobian and payload attachment statics model, we can write the pulling constraint in terms of relative velocity between attachment contact frame $\{c_i\}$ and cable end frame $\{e_i\}$. The constrained motion direction is the cable pulling direction, which means:

$$B^T_{c_i} \underset{\sim}{t}^b_{e_i c_i} = 0, \tag{3}$$

where $\underset{\sim}{t}^b_{e_i c_i}$ is the body twist between the payload attachment contact frame $\{c_i\}$ and mobile agent cable end frame $\{e_i\}$, expressed in $\{c_i\}$.

Now we seek to rewrite the constraint (3) in known quantities, i.e. we wish to relate payload velocity and agent velocity. We expand $\underset{\sim}{t}^b_{e_i c_i}$ as:

$$\underset{\sim}{t}^b_{e_i c_i} = Ad_{g_{c_i F}} \underset{\sim}{t}^b_{e_i F} + \underset{\sim}{t}^b_{F c_i} = -Ad_{g^{-1}_{F c_i}} \underset{\sim}{t}^s_{F e_i} + Ad_{g^{-1}_{oc_i}} \underset{\sim}{t}^b_{F o}, \tag{4}$$

where the agent's spatial twist $\underset{\sim}{t}^s_{F e_i} = J^s_{F e_i} \dot{\mathbf{q}}_i$ as derived earlier, and $\underset{\sim}{t}^b_{F o} = \dot{\mathbf{x}}_o$ is the payload body twist.

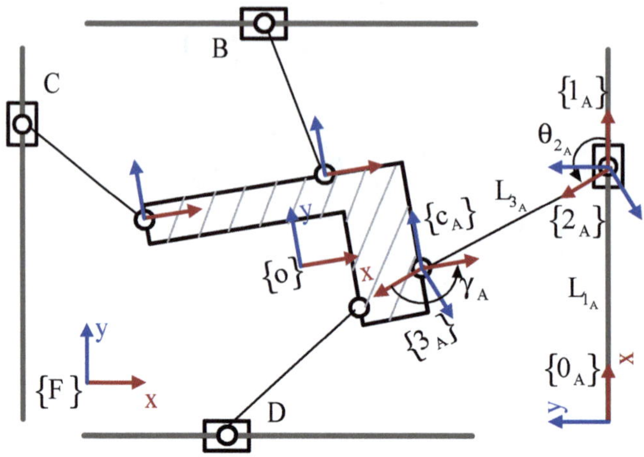

Fig. 3 Payload with 4 Agents (A,B,C,D)

Substituting (4) into (3), we get:

$$J_T(\mathbf{q_T}, \mathbf{x}_o)\dot{\mathbf{q_T}} = P_o^T \dot{\mathbf{x}}_o, \tag{5}$$

where the team Jacobian and the pulling map:

$$J_T = \begin{bmatrix} J_{t1} & & O \\ & \ddots & \\ O & & J_{tm} \end{bmatrix}, \quad P_o^T = \begin{bmatrix} P_1^T \\ \vdots \\ P_m^T \end{bmatrix} = \begin{bmatrix} B_{c_1}^T Ad_{g_{oc_1}}^{-1} \\ \vdots \\ B_{c_m}^T Ad_{g_{ocm}}^{-1} \end{bmatrix}, \tag{6}$$

where $J_{ti} = B_{c_i}^T Ad_{g_{Fc_i}}^{-1} J_{Fe_i}^s$. We can rewrite the pulling constraint (5) into the form $A_p(q)\dot{q} = 0$, where $A_p = [J_T, \ -P^T]$ and $q = [q_T^T, \ x_o^T]^T$.

4 Planar Gantry Cable Robot Example

As an illustrating example of the process, for simplicity, we consider a planar payload being manipulated by four cooperating gantry-type mobile-crane modules. Each mobile cable robot agent consists of a linear gantry that can translate along one axis with a mounted winch to control the cable length. The distal end of the cable is assumed to be attached to the payload using a pin joint (for simplicity, although a variety of other attachments are possible). Four such mobile cable robot agents are assumed to be attached to a common payload as shown in Fig. 3.

4.1 Formulation

4.1.1 Kinematics

Following the framework presented above, we show how to systematically derive the equations. For this planar case:

$$Ad_g = \begin{bmatrix} R & \begin{bmatrix} p_y \\ -p_x \end{bmatrix} \\ O & 1 \end{bmatrix}, \; Ad^T_{g^{-1}} = \begin{bmatrix} R & 0 \\ -[p_y \;\; -p_x]R & 1 \end{bmatrix}. \tag{7}$$

The joints of each gantry cable agent are $\mathbf{q}_i = [\, l_{1i},\ \theta_{2i},\ l_{3i}\,]^T, \forall i \in \{A,B,C,D\}$. The example reference frames for agent A are shown in Fig. 3, where the $\{3_A\}$ frame is the cable end frame $\{e_A\}$ noted in Section 3. In each successive joint frame, the body twists can be easily found as:

$$\underset{\sim}{t}^b_{01_i} = \begin{bmatrix} \dot{l}_{1_i} \\ 0 \\ 0 \end{bmatrix}, \; \underset{\sim}{t}^b_{12_i} = \begin{bmatrix} 0 \\ 0 \\ \dot{\theta}_{2_i} \end{bmatrix}, \; \underset{\sim}{t}^b_{23_i} = \begin{bmatrix} \dot{l}_{3_i} \\ 0 \\ 0 \end{bmatrix}. \tag{8}$$

Then the spatial Jacobian of each agent can be found as:

$$J^s_{Fei} = \begin{bmatrix} \cos\phi_{0i} & y_{0i} + l_{1i}\sin\phi_{0i} & \cos(\phi_{0i}+\theta_{2i}) \\ \sin\phi_{0i} & -x_{0i} - \cos\phi_{0i} l_{1i} & \sin(\phi_{0i}+\theta_{2i}) \\ 0 & 1 & 0 \end{bmatrix}, \tag{9}$$

where $x_{0i},\ y_{0i},\ \phi_{0i}$ is the position and orientation of the gantry starting frame $\{0_i\}$ in world fixed frame $\{F\}$.

4.1.2 Statics

As shown in Fig. 2, the transformation from cable attachment contact frame $\{c_i\}$ to payload COM frame $\{o\}$ is given by $g_{oc_i} = \begin{bmatrix} I & p_{oc_i} \\ O & 1 \end{bmatrix}$. The basis direction of the cable tension in attachment contact frame $\{c_i\}$ is given by $B_{c_i} = [-\cos\gamma_i,\ \sin\gamma_i,\ 0]^T$, where γ_i is the angle from $\{e_i\}$ to $\{c_i\}$, then the cable wrench can be expressed in the object frame $\{o\}$ via co-adjoint transformation as:

$${}^o\underset{\sim}{w}_i = Ad^T_{g^{-1}_{oc_i}} B_{c_i} f_i = \begin{bmatrix} -\cos\gamma_i \\ \sin\gamma_i \\ y_{oc_i}\cos\gamma_i + x_{oc_i}\sin\gamma_i \end{bmatrix} f_i\,. \tag{10}$$

We can concatenate the four cable wrenches into:

$$\begin{bmatrix} F_x \\ F_y \\ M_z \end{bmatrix} = \begin{bmatrix} | & | & | & | \\ P_A & P_B & P_C & P_D \\ | & | & | & | \end{bmatrix} \begin{bmatrix} f_A \\ f_B \\ f_C \\ f_D \end{bmatrix}, \tag{11}$$

where $P_i = Ad^T_{g_{oc_i}^{-1}} B_{c_i}$, $\forall i = A,\ldots,D$. This is the pulling map (2) that maps the cable tension forces $\mathbf{f}$ to the object wrench $\underset{\sim}{w}_o$ in the payload frame.

4.1.3 Pulling Constraint

The velocity level constraints can be developed by projecting the relative velocity difference of the cable tip and the payload along the line of action of the cable (which is the pulling direction). This relative velocity is now expected to be equal to zero in order to avoid cable slack.

As derived in Section 3.3, for the pulling constraint (5) in body frame, we have:

$$P_{oi} = \begin{bmatrix} -\cos(\phi_o - \phi_{0i} - \theta_{2i}) \\ \sin(\phi_o - \phi_{0i} - \theta_{2i}) \\ y_{c_i}\cos(\phi_o - \phi_{0i} - \theta_{2i}) + x_{c_i}\sin(\phi_o - \phi_{0i} - \theta_{2i}) \end{bmatrix}, \tag{12}$$

and the team Jacobian is:

$$J_T = \begin{bmatrix} J_{t1} & & O \\ & \ddots & \\ O & & J_{tm} \end{bmatrix}, \tag{13}$$

where $J_{ti} = [-\cos\theta_{2_i},\ \cos(\phi_{0_i}+\theta_{2_i})y_o + \sin(\phi_{0_i}+\theta_{2_i})(x_{0_i}+\cos\phi_{0_i} l_{1_i}) - \cos(\phi_{0_i} + \theta_{2_i})(y_{0_i}+l_{1_i}\sin\phi_{0_i}) - \sin(\phi_{0_i}+\theta_{2_i})x_o + y_{c_i}\cos(\phi_o - \phi_{0_i} - \theta_{2_i}) + x_{c_i}\sin(\phi_o - \phi_{0_i} - \theta_{2_i}),\ -1]$.

4.2 *Case Study 1: Wrench Closure Workspace and Its Quality*

We now consider the ability of the system to both generate and resist arbitrary payload wrenches (i.e. wrench closure). We use the simple planar example of an "L" shaped payload manipulated by four gantry cable robots to showcase the benefits of the systematic formulation. In particular, we will focus attention on using quantitative metrics derived from this formulation to determine the wrench closure workspace and its quality.

4.2.1 Wrench Closure Condition

For planar cases, it is possible to analytically determine the workspace such as shown in [24]. Considering applicability to spatial cases, the numerical algorithm presented in [26] can be used to determine wrench closure for general $m > n$ cases and it is relatively fast in computation. It basically says a necessary and sufficient condition for wrench closure is a test vector such as $p_t = -\sum_{i=1}^{n} p_i$ can be positively spanned by another set of basis of the pulling map P. The algorithm itself is straight forward, whose details can be found in [26].

We show a sample calculation of the workspace when the gantries are positioned in the middle of their stroke and the attachment points are at the four tips of the payload. We check for wrench closure condition across the workspace with the payload orientation angle ϕ varying from -40 to 40 degrees. Fig. 4 shows the resulting workspace.

4.2.2 Workspace Quality

Apart from simple wrench closure, it would be useful to know the quality of the workspace. One measure is the tension factor(TF) as defined in [27]:

$$TF = \frac{min(\mathbf{f})}{max(\mathbf{f})}. \tag{14}$$

Since cable tensions are positive, then $0 < TF \leq 1$. A larger TF means a more even distribution of tensions in cables. It is shown that maximizing TF is equivalent to the following linear optimization problem:

$$\begin{aligned} \text{minimize} \quad & \sum_{i=1}^{m} f_i \\ \text{subject to} \quad & P\mathbf{f} = 0 \\ & f_i \geq f_{i_{min}} > 0, \ \ (i = 1, 2, \ldots, m) \end{aligned} \tag{15}$$

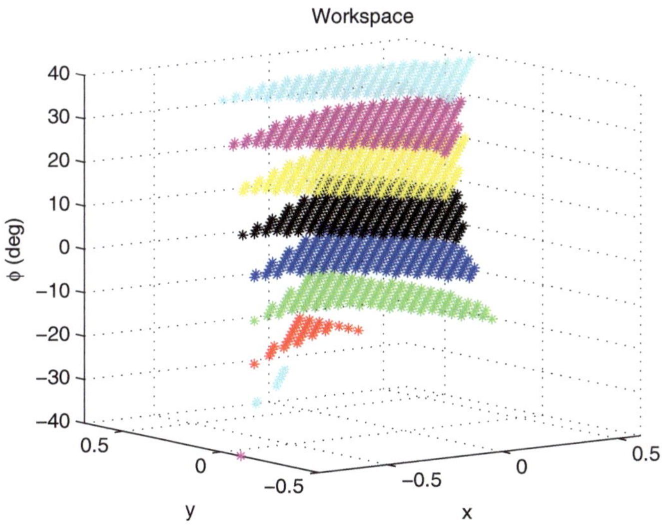

Fig. 4 Wrench closure workspace

Figure 5 shows the quality of the workspace when the gantries are positioned in the middle of their stroke and the attachment points are at the four tips of the payload. The tension factors are represented by the relative size of the square markers. We can see that due to the asymmetric payload shape, the workspace is irregular and its quality in the sense of tension factors is even more limited. The high redundancy in the base positioning allows us to optimize the design.

4.3 Case Study 2: Design Optimization

We first show a simple illustration of the idea. As can be easily seen, Fig. 6(a) is not wrench closure. Intuitively, we have two design choices to make it wrench closure: one is by changing attachment point location (as shown in Fig. 6(b)); and the other is by moving the base location (as shown in Fig. 6(c)). While this is simply done by inspection, in general, the selection of cable attachment point position/base location for asymmetric payloads tends to be non-intuitive.

Next we perform design optimization based on our systematic formulation to the example. The "L" shaped payload is assumed to be general with non uniform mass density and thus its COM does not coincide with its geometric area center. It is in circumstances such as this that the systematic formulation coupled with quantitative analysis can be very useful for design and analysis.

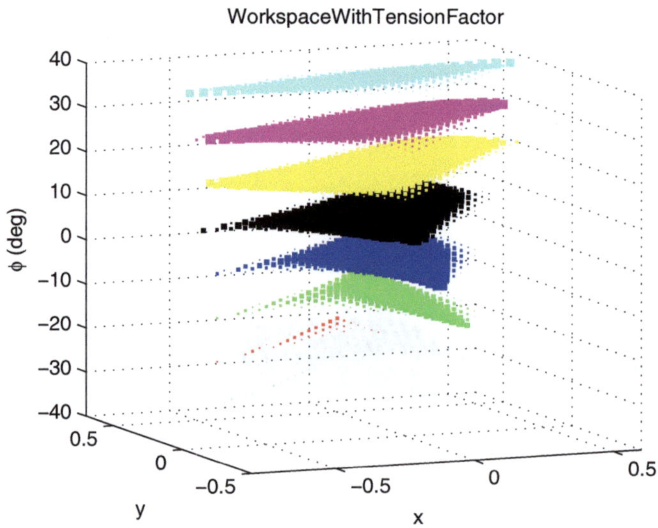

Fig. 5 Workspace tension factor (size of the cubes proportional to TF value)

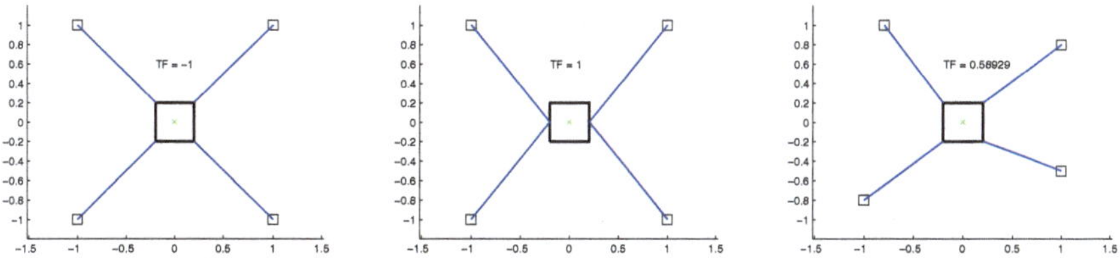

(a) Non closure. (b) Move attachment. (c) Move base.

Fig. 6 Simple illustration of reconfiguration

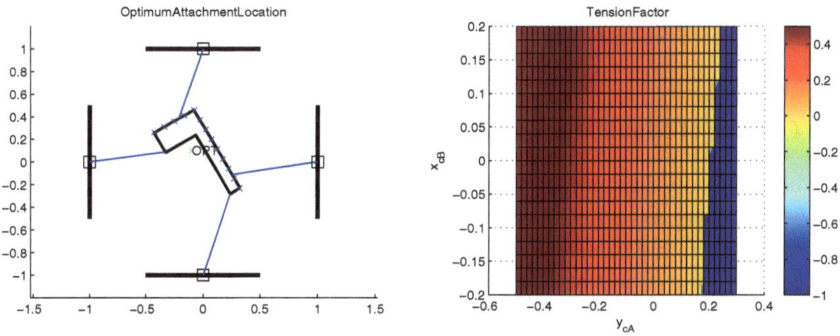

(a) Optimal attachment. (b) TF vs attachment.

Fig. 7 Effect of cable attachment choice

4.3.1 Cable Attachment Choices

We first consider optimizing the cable attachment point locations on the payload. The gantries are fixed at the center of their stroke, which essentially reduces our model to the conventional fixed-base cable robot case.

The base gantries are immobilized at the mid point ($l_{1_i} = 0.5$, $\forall i = A, \ldots, D$) of their full stroke. The payload COM is at $x_o = -0.1$, $y_o = 0.1$, $\phi_o = 30°$. We perform a parametric sweep to study the role of cable attachment positioning to payload on the wrench closure condition of the pulling map, as shown in Fig. 7. Two of the cable attachment points on the payload are held fixed at the tips while the other two attachments can be repositioned anywhere along the corresponding sides (y_{c_A} and x_{c_B}).

The tension factors for each configuration is shown in Fig. 7(b). We set the tension factor to -1 to clearly represent wrench singular (not wrench closure) configuration. The attachment locations corresponding to optimal tension factor is shown in Fig. 7(a).

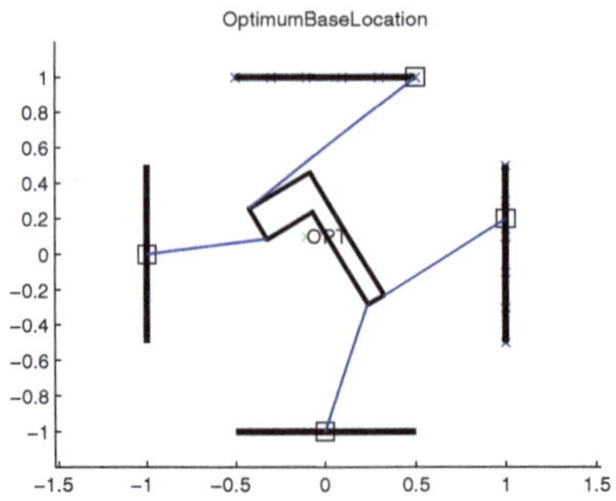

(a) Optimal base.

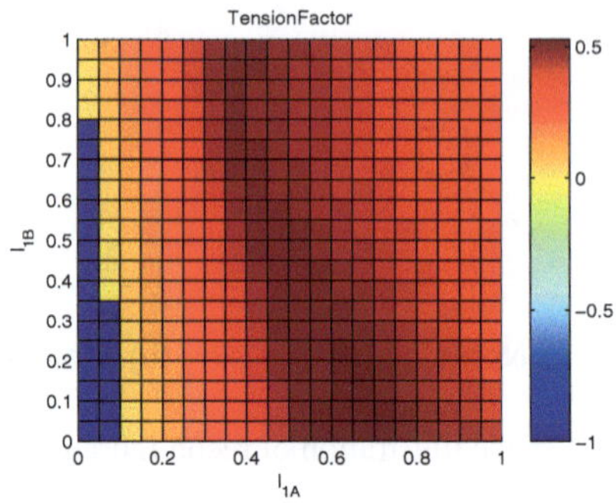

(b) TF vs base.

Fig. 8 Effect of base repositioning

4.3.2 Choices of Mobile Bases

It may be relatively inconvenient to change attachment points on the fly (i.e. re-grasping); reconfiguring the mobile bases is more useful in practical situations. Again, we perform a parametric sweep to study the role of base positioning on the wrench closure of the pulling map. The cable attachment points on the payload are held fixed at the four corner. Two of the base gantries are immobilized at the mid point ($l_{1_C} = 0.5$, $l_{1_D} = 0.5$) of their full stroke, while the other two base gantries can be repositioned anywhere along their full stroke (l_{1_A} and l_{1_B}).

Fig. 8(b) showcases the tension factor plotted against the gantry positions (l_{1_A} and l_{1_B}). As a result, the pose that has the largest tension factor is shown in Fig. 8(a). While the above results were shown in the form of parametric sweep results for two design variables at a time, this was done solely for visual illustration. Various optimization methods can now be systematically applied to a full fledged multi-variable case which is shown next.

4.4 Case Study 3: Maintaining Optimal Tension Factor Along Trajectory

In addition to the previous "static" design optimization of either attachment location or base position, a more useful way would be to "dynamically" resolve the redundancy by optimally reconfiguring the base gantry location along a desired trajectory (such as the case in Fig. 9). This way, tension factor can be maintained the highest possible all the time. This problem can be solved using "cascaded" optimization. Basically we wrap an optimization of the four gantry positions on top of the optimization of tension factor.

Different from conventional parallel robot singularities, there is no analytical solution to wrench closure. Therefore conventional singularity avoidance techniques using redundancy to optimize the smallest singular value are infeasible for the mobile cable robots here. There is no analytical gradient, and numerical approximation

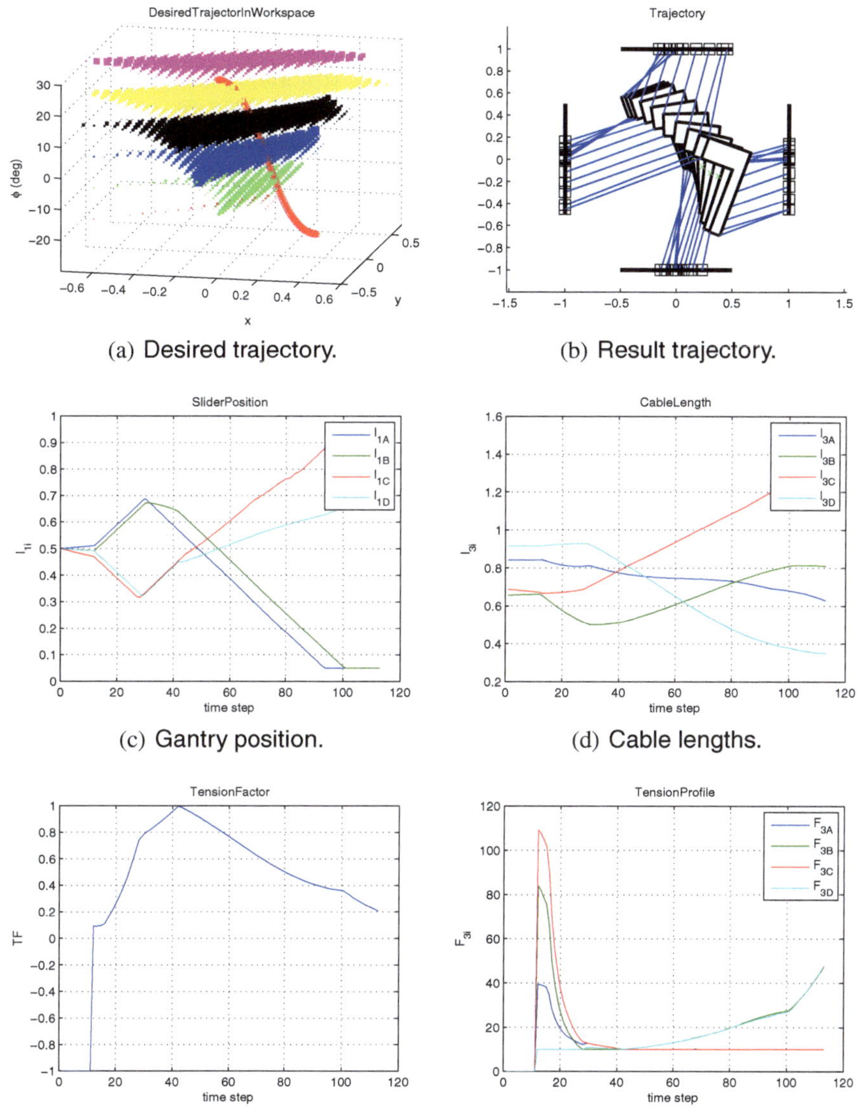

(a) Desired trajectory. (b) Result trajectory.

(c) Gantry position. (d) Cable lengths.

(e) Tension factor. (f) Cable tensions.

Fig. 9 Reconfiguration to optimize tension factor along trajectory

is not a good solution due to high nonlinearity of the pulling map with respect to configuration change. Either a pre-calculation/planning of trajectories to avoid singularity (which is computationally expensive) or reconfiguration along the way is needed.

In case of wrench singular (not wrench closure) configurations of the mobile cable robots during the trajectory, two simple approaches can be used to resolve it: one is to perform a global optimization to find a feasible bases configuration, then interpolate in the internal joint space to reach the feasible configuration and initialize from there; the other is to perform a local search by exploratory moves of the mobile bases which should be faster and possible if it is not far off. Also this might lead to a local optimal tension factor, but it is acceptable in the case of trajectory tracking. To ensure continuity, we also impose maximum velocity of gantry as an additional constraint to be practical, otherwise optimization results may drive the gantries all over the place and thus causing discontinuity just to get the best tension factor. Higher order of continuity can be achieved by imposing acceleration/jerk level constraints to get a smoother result.

We note here in our case with only one redundant cable, the null space of the pulling map has only one dimension. So the wrench closure condition in [26] and tension factor in [27] can be simplified. It is straight forward to show wrench closure is equivalent to requiring components of the null space vector to have same sign, and tension factor is equivalent to the ratio of minimum and maximum of the absolute values of the null space vector. This way, the lower level optimization is reduced and thus saving us some computation time.

Here we show an example of transporting the payload along a desired trajectory as shown in Fig. 9(a). The starting point is out of the wrench closure workspace for the initial base configuration, which is also evident in Fig. 9(e), as the starting tension factor is -1 indicating non wrench closure. It can be seen from Fig. 9(b), after a few exploratory steps, the gantry bases move to a feasible configuration, and then afterwards the redundant gantry positions are optimized (using MATLAB Optimization Toolbox) to maintain a configuration that yields the best possible tension factor as shown in Fig. 9(e). We note that instead of performing optimization, planning algorithms such as RRT may also be used for reaching an initial feasible configuration. The resulting gantry position is shown in Fig. 9(c), cable length profile in Fig. 9(d), and corresponding positive cable tensions in Fig. 9(f). We also note that with high reconfigurability of mobile bases, wrench closure condition can actually be relaxed (i.e. three mobile cable robots transporting payload) if it is not required to exert/resist arbitrary wrench, as along as the configuration is able to exert certain required dynamics forces/moments along a given trajectory. This aspect is currently being pursued.

5 Discussion

The addition of base mobility provides cable robots greater flexibility, yet it requires careful investigation. In this chapter we extend a systematic screw theoretic formulation approach to the general cooperating cable robots on mobile bases. In particular, creating a formalism for studying system level configuration by composing the contributions of individual agents and thereby creating a parametric model is attractive. Various parametric analysis including parameter sweeps, optimization and sensitivity analysis, can now be brought to aid design and analysis. Another benefit is the ready extensibility of framework to full fledged spatial cases using this formalism. Using this formulation also permits close linkage between grasping, walking and in general parallel robots, allowing for cross-pollination of results. Physical system validation for cooperative ground mobile robots is currently underway.

References

1. Albus, J., Bostelman, R., Dagalakis, N.: Journal of Robotic Systems 10(5), 709 (1993)
2. Huntsberger, T., Stroupe, A., Aghazarian, H., Garrett, M., Younse, P., Powell, M.: Journal of Field Robotics 24(11), 1015 (2007)
3. Kawamura, S., Ito, K.: Proceedings of the IEEE/RSJ International Conference on Intelligent Robots and Systems, Yokohama, Japan (1993)
4. Williams II, R.L.: International Journal of Virtual Reality 3(3), 13 (1998)
5. Nan, R.: Science in China Series G: Physics Mechanics and Astronomy 49(2), 129 (2006), doi:10.1007/s11433-006-0129-9
6. Oh, S.R., Ryu, J.C., Agrawal, S.K.: ASME Journal of Mechanical Design 128(5), 1113 (2006)
7. Fattah, A., Agrawal, S.K.: ASME Journal of Mechanical Design 127(5), 1021 (2005)
8. Meunier, G., Boulet, B., Nahon, M.: IEEE Transactions on Control Systems Technology 17(5), 1043 (2009)
9. Donald, B., Gariepy, L., Rus, D.: Proceedings of IEEE International Conference on Robotics and Automation, ICRA 2000, vol. 1, pp. 450–457 (2000)
10. Cheng, P., Fink, J., Kumar, V., Pang, J.S.: ASME Journal of Mechanisms and Robotics 1(1), 1 (2009)
11. Esposito, J., Feemster, M., Smith, E.: IEEE International Conference on Robotics and Automation, ICRA 2008, pp. 1501–1506 (2008)
12. Michael, N., Fink, J., Kumar, V.: Robotics: Science and Systems V (2009)
13. Voglewede, P.A., Ebert-Uphoff, I.: IEEE Transactions on Robotics 21(4), 713 (2005)
14. Murray, R.M., Li, Z., Sastry, S.S.: A Mathematical Introduction to Robotic Manipulation. CRC Press, Boca Raton (1994)
15. Waldron, K.J., Hunt, K.H.: The International Journal of Robotics Research 10(5), 473 (1991), http://ijr.sagepub.com/content/10/5/473.abstract, doi:10.1177/027836499101000503
16. Tang, C.P., Krovi, V.: Robotica 25(1), 29 (2007)
17. Kumar, V., Waldron, K.J.: ASME Journal of Mechanisms, Transmissions, and Automation in Design 112(1), 90 (1990)
18. Salisbury, J.K., Roth, B.: ASME Journal of Mechanisms, Transmissions and Automation in Design 105(1), 35 (1983)
19. Paljug, E., Yun, X., Kumar, V.: IEEE Transactions on Robotics and Automation 10(4), 441 (1994)
20. Liu, Y.H., Arimoto, S.: IEEE Transactions on Automatic Control 41(8), 1193 (1996)
21. Bosscher, P., Riechel, A.T., Ebert-Uphoff, I.: IEEE Transactions on Robotics 22(5), 890 (2006)
22. Ghasemi, A., Eghtesad, M., Farid, M.: ASME Journal of Mechanisms and Robotics 1(4), 044502 (2009)
23. Stump, E., Kumar, V.: ASME Journal of Mechanical Design 128(1), 159 (2006)
24. Gouttefarde, M., Gosselin, C.M.: IEEE Transactions on Robotics 22(3), 434 (2006)
25. Barrette, G., Gosselin, C.M.: Journal of Mechanical Design 127(2), 242 (2005)
26. Lim, W.B., Yang, G., Yeo, S.H., Mustafa, S.K., Chen, I.M.: IEEE International Conference on Robotics and Automation, ICRA 2009, pp. 2187–2192 (2009)
27. Pham, C.B., Yeo, S.H., Yang, G., Chen, I.M.: Robotics and Autonomous Systems 57(9), 901 (2009)

Deployment Algorithms for Dynamically Constrained Mobile Robots

Sonia Martínez

Abstract. The use of unmanned vehicles in exploration and surveillance operations has become evermore pervasive in today's world. The development of cooperative motion strategies has been fueled by this increasing demand. However, many dynamical models for these autonomous vehicles remain simple and are not accurate representations of a vehicle where such cooperative motion strategies may be physically implemented. This paper reviews complementary solutions to the problem of cooperative deployment of autonomous vehicles using multi-center functions. In particular, vehicles are subject to three types of dynamic constraints, such as those due to remaining power supplies, nonholonomic dynamics, and constraints due to external environmental forces. Simulations illustrate the convergence properties of the algorithms when applicable.

1 Introduction

The study of coordination mechanisms in multi-agent systems is relevant for both the understanding of scientific phenomena and the development of new technologies. A main class of examples from nature is given by swarming in animal species such as ant colonies, termites, flocking birds, and schooling fish. Emergence and self-organization is also a characteristic of human politics, societies, and economic groups. In these groups, each member makes decisions repeatedly based on local information signals sent or left by other members of the group and the environment. This decentralized process with no leaders yields complex emerging behaviors that translate into robust and efficient global structures, patterns, or organizations.

Complex systems in biology and society can help us understand, model, and design large-scale engineered systems composed of autonomous and semi-autonomous agents. Their potential advantages are those found in their biological counterparts—robustness to failure thanks to system redundancy, and increased efficiency in the

Sonia Martínez
MAE - UC San Diego, 9500 Gilman Dr, La Jolla, CA 92093-0411
e-mail: soniamd@ucsd.edu

D. Milutinović & J. Rosen (Eds.): Redundancy in Robot Manipulators, LNEE 57, pp. 125–143.
DOI: 10.1007/978-3-642-33971-4_8

number and quality of the global tasks that can be accomplished. However, the realization of multi-agent systems poses new challenges induced by scalability problems, agent heterogeneity, and intermittent interactions in uncertain, dynamically changing environments.

Research in mobile robotics has helped enormously in the understanding of these challenges through the study of several important benchmark problems including *rendezvous, formation control, deployment, and task assignment.* In particular, the problem of robot deployment to provide better coverage or task servicing in an environment lends itself to geometric optimization formulations, which have been extensively studied. An incomplete list of references on coverage includes [1, 2], based on potential field methods, [3] using the theory of coverage point processes, [4] making use of non-smooth analysis techniques, and [5, 6, 7] based on behavioral control approaches. More recently, the robotics community is developing new non-model based algorithms for coordinated deployment and map building [8].

An alternative, popular approach makes use of multi-center performance metrics and Voronoi-based control algorithms to stabilize multi-robot systems to locally optimal positions [9, 10]. This approach has been adapted to deal with non-convex environments with obstacles [11, 12, 13], distributed environment learning [14, 15], and equitable partitions [16, 17]. The resulting algorithms can be adapted so that limited-range, distributed interactions are possible as well [18].

More recently, different researchers have started to reconsider the difficulties introduced by vehicle dynamics. Coordination algorithms typically assume simple dynamics for vehicles. The idea is to implement this strategy as a high-level plan, together with low-level local motion plans that each vehicle uses to reconfigure to the prescribed upper-level positions. If synchronization is possible, each vehicle can wait for others to reach their positions before moving forward. However, in asynchronous regimes this strategy can be just infeasible. On the other hand, dynamic constraints may require a re-definition of the deployment objectives in order to produce more meaningful solutions. However, the inclusion of constraints in the coordination objectives can impose additional computational and control challenges.

In this chapter, we review and summarize several extensions of Voronoi-based deployment to account for different types of dynamical constraints. These include power constraints, and vehicle controllable and uncontrollable dynamics. The paper is organized as follows. In Section 2, we review the basic Lloyd's algorithm to the Locational Optimization problem for coverage control. In Section 3 we address the problem of deployment under power limitation to vehicle motion. Section 4 adapts Lloyd's algorithm to deal with Dubin's type of unicycles via a hybrid coordinated motion law. Finally, Section 5 introduces a heuristic to deal with underactuated and uncontrollable vehicles in river environments.

2 Benchmark Problem: Coverage Control

In this section, we present a basic coverage control problem formulation together with gradient-based algorithmic solutions for unconstrained vehicle dynamics

[9, 10]. Some of these will be extended in the following sections to account for different types of dynamical constraints.

Basic coverage and task-assignment objectives can be formulated by a meaningful class of Locational Optimization or multicenter performance metrics. Let $Q \subseteq \mathbb{R}^2$ be a convex, bounded environment, and $\phi : Q \to \mathbb{R}_{\geq 0}$ be a scalar field with bounded support Q. Here, ϕ represents an *a priori* measure of information on Q—the higher the value of $\phi(q)$, $q \in Q$, the more attention that should be afforded to $q \in Q$. Let $P = (p_1, \dots, p_n)$ denote the agent positions in Q. In the following, we interchangeably refer to the elements of the network as sensors, agents, vehicles, or robots. Let $f : \mathbb{R} \to \mathbb{R}$ be a non-decreasing and piecewise differentiable function relating the Euclidean distance from p_i to $q \in Q$, $\|p_i - q\|$, to coverage performance from p_i on q, for all $i \in \{1, \dots, n\}$. For example, the function f can encode the signal-to-noise ratio between a source with location q and the sensor located at p_i. Or it can define the cost of servicing a location q by an agent placed at p_i; e.g. the traveling time from p_i to q when moving on a straight line with constant velocity. With these elements, a coverage metric can be defined as:

$$\mathscr{H}(P) = \int_Q \min_{p_i} f(\|p_i - q\|) dq, \tag{1}$$

where $\min_{p_i} f(\|p_i - q\|)$ has the interpretation of the best coverage of q provided by the multi-robot system. The minimization of this metric results into a minimum average cost to cover Q using the multi-robot group.

For the purpose of defining a distributed algorithm that optimizes this metric, it is helpful to restate (1) in terms of the individual contribution that each agent in the network adds to $\mathscr{H}$. For example, assume that $f(x) = x^2$ for all $i \in \{1, \dots, n\}$ and denote the associated $\mathscr{H}$ by $\mathscr{H}_{\text{centr}}$. Let $\mathscr{V}(P) = (V_1, \dots, V_n)$ be the so-called *Voronoi partition* of Q, where

$$V_i = \{q \in Q \mid \|q - p_i\| \leq \|q - p_i\|, \forall j \neq i\}, \qquad \forall i \in \{1, \dots, n\}, \tag{2}$$

satisfy $\cup_{i=1}^n V_i = Q$. Then $\mathscr{H}$ may be rewritten as

$$\mathscr{H}(P) \equiv \mathscr{H}(P, \mathscr{V}(P)) = \sum_{i=1}^{N} \int_{V_i} \|q - p_i\|^2 \phi(q) dq. \tag{3}$$

Given a region $W \subseteq Q$, one can define its *mass* and *centroid*, as follows:

$$M_W = \int_W \phi(q) dq, \qquad \text{CM}_W = \int_W q\phi(q) dq. \tag{4}$$

It can be shown, see [19, 9], that if agents are in a *centroidal Voronoi configuration*; that is, $p_i = \text{CM}_{V_i}$ for all i, then the cost function $\mathscr{H}$ is at a local minimum.

An alternative metric that considers flat and limited sensor footprints (resp. travel ranges) R can be obtained by taking $f(\|p - q\|) = -1_{[0,R)}(\|p - q\|)$, which leads to:

$$\mathscr{H}_{\text{area}}(P) = -\int_{\bigcup_{i=1}^{n} B(p_i,R)} \phi(q)dq,.$$

The minimization of this metric results into a maximization of the area covered by the group of agents. This objective can be combined with the previous one, leading to a mixed metric of the form $\mathscr{H}_{\text{mixed}}(P) = \alpha \mathscr{H}_{\text{centr}} + \beta \mathscr{H}_{\text{area}}(P)$, for $\alpha, \beta \geq 0$.

Once a metric is chosen, a gradient-based distributed control algorithm can be implemented by each agent to asymptotically reach the corresponding set of local minima. The following law is a continuous-time version of the algorithm in [9]:

$$\dot{p}_i = -\text{sat}\left(\frac{\partial \mathscr{H}_{\text{centr}}}{\partial p_i}\right) = -M_{V_i}(\text{CM}_{V_i} - p_i), \qquad i \in \{1,\ldots,n\},$$

where the function $\text{sat}(v) = v$, if $\|v\| \leq 1$, otherwise $\text{sat}(v) = \frac{v}{\|v\|}$. Essentially, agents need to be able to compute the regions V_i, and follow the corresponding centroid. If Q is compact, agents will converge to centroidal locations. The corresponding Voronoi regions may be computed by agents using information of a limited set of other agents, the Delaunay neighbors. Thus, the algorithm is distributed in the sense of the Delaunay graph. Correspondingly, the metrics $\mathscr{H}_{\text{area}}$ and $\mathscr{H}_{\text{mixed}}$ give rise to algorithms that are distributed in the sense of the $2R$-disk graph (for appropriate α and β). Discrete-time versions of these algorithms can be seen to be convergent even if partial asynchronous behavior is permitted [9, 10].

A simulation of the above gradient algorithm is provided below.

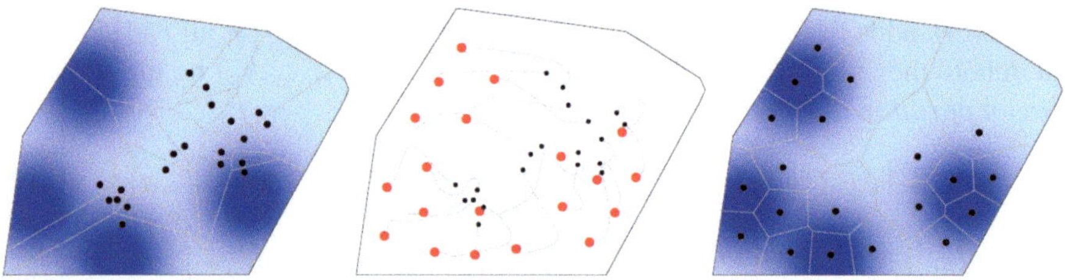

Fig. 1 A simulation run of the gradient-based algorithm associated with $\mathscr{H}_{\text{centr}}$. The figure on the left depicts robots' initial positions. The figure on the right represents robots' final positions and correspond to centroidal Voronoi configurations.

3 Power-Limited Deployment

Power-aware algorithms have been the subject of extensive research in static sensor networks and mobile middleware, see [20, 21]. However, there is limited work on power constraints, and how these may affect cooperative control algorithms. For instance, the final agent configurations provided by Figure 1 for a task assignment objective do not seem reasonable when agents have different motion restrictions.

Here we describe a first approach [22] to deal with this problem in the context of the Locational Optimization or multicenter type of metrics of Section 2. We assume that there is enough group redundancy so that the loss of a particular agent can be afforded. This allows to account for energy limitations by means of a modified performance metric of the form of (1) and associated generalized Voronoi partitions.

Let $P = (p_1, \dots, p_n)$ be the positions of n robots in Q. The sensors have an associated energy content E_i such that $0 \leq E_i \leq E_{\max}$, for all $i \in \{1, \dots, n\}$. As agents move, their energy reserve will decrease. We propose the following simple agent dynamics in the augmented state $(p_i, E_i) \in Q \times \mathbb{R}_{\geq 0}$:

$$\dot{p}_i = u_i, \qquad \dot{E}_i = -g_i(\|\dot{p}_i\|), \tag{5}$$

where $\dot{p}_i$ denotes the velocity of agent i such that $\|\dot{p}_i\| \in [0, v_{\max}]$, u_i is the control input, and $g_i : [0, v_{\max}] \to \mathbb{R}_{\geq 0}$ is any increasing function such that $g_i(x) = 0$ only at $x = 0$. Intuitively, $g_i(x)$ captures the fact that energy expenditure increases as velocity increases. This modeling assumption is based on the consideration that power is consumed to change absolute speed and counteract drag forces; the latter being the predominant force [22]. In the following, we will take $g_i(x) = g(x) = x^2$, for all $i \in \{1, \dots, n\}$, for simplicity.

Energy expenditure will affect the travel range that a sensor can cover with maximum velocity before running out of batteries. Suppose that agent i travels with a maximum velocity $\dot{p}_i(t) = (v_{\max}, 0)^T \in \mathbb{R}^2$. Then, the vehicle runs out of energy at time $T(v_{\max}) = E_i(0)/g(v_{\max})$. The associated travel range is the distance $R_* = v_{\max} T(v_{\max})$. This motivates the use of a mixed type of performance metric as in Section 2 that accounts for travel-range limits. Thus, we consider:

$$\mathscr{H}(P, E) = \int_Q \min_{i \in \{1, \dots, n\}} f_i(d_{E_i}(q, p_i)) \phi(q) dq, \tag{6}$$

where $E = (E_1, \dots, E_n)$ are current energy levels of agents, the maps $f_i : \mathbb{R} \to \mathbb{R}$ are non-decreasing functions associated with the travel cost of each agent i, and $d_{E_i} : Q \times Q \to \mathbb{R}$ is a weighted (quasi) pseudo-metric function such as the following:

1. The power-weighted metric, $d_{E_i, \text{pow}}(q, p_i) = \|q - p_i\|^2 - (E_i)^2$,
2. The multiplicatively-weighted metric, $d_{E_i, \text{mult}}(q, p_i) = \frac{1}{E_i^2} \|q - p_i\|^2$,
3. The additively-weighted metric, $d_{E_i, \text{aw}}(q, p_i) = \|q - p_i\| - (E_i)^2$.

All these metrics lead to generalized Voronoi regions [19] whose size depends on the relative energy content of neighboring robots:

$$V_i^{\text{gen}} = \{q \in Q \mid d_{E_i}(q, p_i) \leq d_{E_i}(q, p_j), \forall j \neq i\}, \text{ where } d_{E_i} \text{ is a pseudometric.}$$

Boundaries of these Voronoi regions are (1) straight lines for the power metric, (2) circles of radii E_i, $i \in \{1, \dots, n\}$, for the multiplicatively-weighted metric, and (3) hyperbolic boundaries for the additively-weighted metric. Due to the difficulty of representing and intersecting hyperbolic boundaries, we focus on the first two types. As opposed to standard Euclidean Voronoi regions, these generalized regions can be non-convex and their generators may lie outside them. If vehicles only have a limited amount of energy to move using a maximum velocity, E_i, $i \in \{1, \dots, n\}$, we propose that in order for an agent i to be able to cover a point $q \in Q$, agent i must be able to reach q with its current energy level. This leads to new assignment regions for agents given by the intersection of Voronoi regions with circles of radii E_i, $i \in \{1,\dots,n\}$.

Let $B_i(E_i)$ be a closed ball centered at p_i with radius E_i. Then, the space that can be covered by the robots to $\cup_{i=1}^n B_i(E_i) \subseteq Q$. The new limited-Voronoi regions assigned to each agent are defined as follows:

$$V^{\text{gen}}_{\text{LD},\,i} = \{q \in Q \,|\, d_{E_i}(q,p_i) \le d_{E_i}(q,p_j), \forall\, j \neq i \text{ and } \|q-p_i\| \le E_i\},$$

where d_{E_i} is one of the pseudometrics above. Figure 2 compares limited-Voronoi regions.

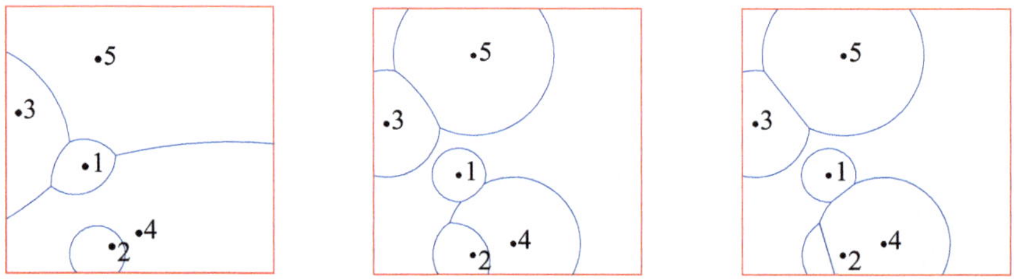

Fig. 2 Figures from left to right: (i) Voronoi partition associated with the multiplicatively weighted pseudometric, (ii) limited-range cells associated with the multiplicatively-weighted pseudometric, (iii) limited-range cells associated with the power-weighted pseudometric

The computation of limited types of regions are spatially distributed over the Delaunay graph and the third one over the $2E_{\max}$-disk graph. The new regions emphasize different energy levels of agents. The corresponding metrics for centroidal, area, and mixed coverage are then given respectively as follows:

$$\mathscr{H}_{\text{centr}}(P,E) = \int_{\cup_{i=1}^n B_i(E_i)} \min_{i\in\{1,\dots,n\}} \{d_{E_i}(q,p_i)\}\phi(q)dq\,,$$

$$\mathscr{H}_{\text{area}}(P,E) = \int_{Q} \min_{i\in\{1,\dots,N\}} (-1_{[0,E_i]}(\|q-p_i\|))\phi(q)dq = -\int_{\cup_{i=1}^n B_i(E_i)} \phi(q)dq\,,$$

$$\mathscr{H}_{\text{mixed}}(P,E) = \kappa_{\text{area}}\mathscr{H}_{\text{area}}(P,E) + \kappa_{\text{cent}}\mathscr{H}_{\text{cent}}(P,E)\,.$$

In particular, it is still possible to rewrite $\mathscr{H}_{\text{centr}}(P,E)$ as follows:

$$\mathscr{H}_{\text{centr}}(P,E) = \sum_{i=1}^n \int_{V^{\text{gen}}_{\text{LD},\,i}} d_{E_i}(q,p_i)\phi(q)dq\,.$$

For any of these functions $\mathscr{H}$, we can define a gradient descent control algorithm for agents as follows:

$$\dot{p}_i = -k^*(p_i, E_i)\text{sat}\left(\frac{\partial \mathscr{H}}{\partial p_i}\right), \qquad k^*(p_i, E_i) = \frac{\text{sat}\left(\frac{\partial \mathscr{H}}{\partial p_i}\right) \cdot \left(\frac{\partial \mathscr{H}}{\partial p_i}\right)}{2\left|\text{sat}\left(\frac{\partial \mathscr{H}}{\partial p_i}\right)\right|^2 \frac{\partial \mathscr{H}}{\partial E_i}}$$

$$\dot{E}_i = -\|\dot{p}_i\|^2, \qquad i \in \{1, \ldots, n\}. \tag{7}$$

That is, assuming that energy decreases according to $\|\dot{p}_i\|^2$, we modulate the velocity of agents via the gain k_i^* while collectively decreasing the cost function $\mathscr{H}$. In this way, vehicles with lower energy will spend less in moving toward their goal positions, while vehicles with larger energy will spend more. With this strategy, controllability to critical positions is possible for those agents that have enough energy to move. For the particular case of $\mathscr{H}_{\text{centr}}(P, E)$, the algorithm makes agents follow the centroids of the corresponding generalized Voronoi regions if they do not run out of energy. Convergence is stated in the following theorem.

Theorem 1 (Critical configurations for centroidal coverage and MWVD, [23]). *The critical points of a gradient descent flow characterized by* (7) *using an objective function* $\mathscr{H}_{\text{centr}}$ *are configurations where each agent is either:*

1. *located at the centroid,* $p_i = \text{CM}_{V_{\text{LD},i}^{gen}}$,
2. *has no energy,* $E_i = 0$.

Agents approach these critical configurations as $t \to \infty$.

A simulation run of the energy-aware gradient-descent algorithm for a mixed metric $\mathscr{H}_{\text{mixed}}$ is provided in Figure 3. Agents that need to travel further away, will eventually have smaller assigned regions. In this case vehicles end up at the centroids of their regions. Similar convergence results can be established for the other cost functions such as $\mathscr{H}_{\text{area}}$. The number of agents that can run out of energy depends on the initial agents' positions, the density $\phi(q)$, and size of the environment. The extension of these results to deal with asynchronous implementations can be done in a similar way to [9].

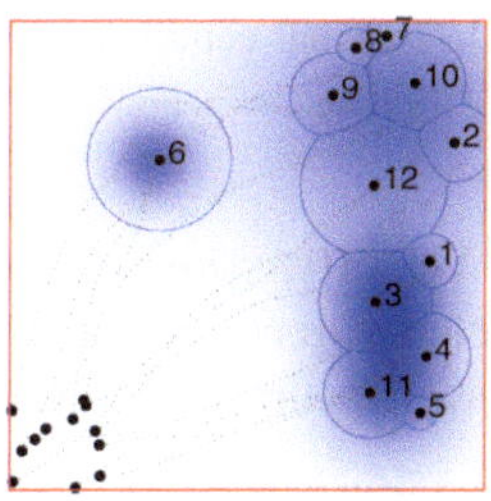

Fig. 3 Multi-robot system trajectories evolving under (7) for $\mathscr{H}_{\text{centr}}$. The final energy regions, and final agent energy levels, are also shown here.

4 Unicycle Vehicle Dynamics

Vehicles' actuation can severely be affected by dynamical constraints on their positions and velocities. For example, the control laws of the previous section force individual agents to move directly towards the centroid of their Voronoi regions; however, this is not always possible. When vehicles are controllable, one can consider vehicle dynamics into the design of coordination algorithms from the start. This can help avoid severe performance degradation due to lack of synchronization.

In the following we introduce control algorithms that propel a class of nonholonomic vehicles to centroidal Voronoi configurations while the minimization of $\mathscr{H}_{\text{centr}}$ is satisfied in certain sense. This results into a hybrid system that can be analyzed via the novel theory of [24, 26]. Suppose that vehicles are deployed in a convex, compact environment Q. Referencing Figure 4, each vehicle has configuration variables $(p_i, \theta_i) \in \mathrm{SE}(2)_Q$, and a body coordinate frame with basis $e_{i,1} = (\cos\theta_i, \sin\theta_i)$ and $e_{i,2} = (-\sin\theta_i, \cos\theta_i)$. We denote $d_i = \mathrm{CM}_{V_i} - p_i$ as in Figure 4 and define the angle $\Omega_i \in [-\pi, \pi]$ to be the angle between $e_{i,1}$ and d_i. We assume the vehicles have bounded velocity and turning rate, $|v_i| \leq v_{\max}$ and $|\omega_i| \leq \omega_{\max}$ respectively. Next, we introduce a Dubin's type of nonholonomic vehicles that we shall consider.

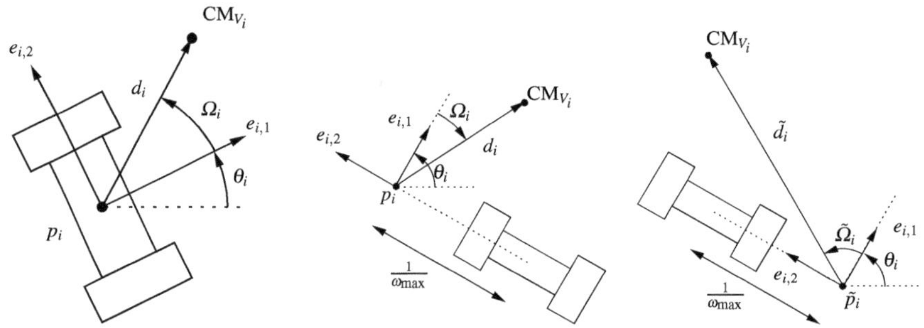

Fig. 4 Vehicle with wheeled mobile dynamics (left). The variables are redefined for a vehicle with fixed forward velocity and a left virtual center (middle). The non-active virtual center quantities are shown with a tilde, ˜ (right).

Vehicles with Fixed Forward Velocity. Simple models for UAVs and underwater gliders can be modeled as vehicles with constant forward velocity, constant altitude and a minimum turning radius such as the following [25]:

$$\dot{p}_i = (\cos\theta_i, \sin\theta_i)^T, \qquad \dot{\theta}_i = \omega_i,$$

where ω_i is the only input. Define the vehicle *virtual center* as its center of rotation when the turning input is $\pm\omega_{\max}$. These centers can be on either side of the vehicle, and a strategy to switch virtual center locations will be introduced later. Our coverage objective will be formulated in terms of the virtual center of each vehicle to a desired centroid target. This target will be the centroid of the vehicle's

Voronoi region calculated using the virtual centers of all vehicles in the network. Once the virtual center has arrived at the centroid, the vehicle will *hover* about it by maintaining the maximum steering input $\pm\omega_{\max}$.

The virtual centers' coordinates in the global frame are

$$p_i' = p_i \pm \frac{1}{\omega_{\max}}(-\sin\theta_i, \cos\theta_i)^T . \tag{8}$$

with time derivative:

$$\dot{p}_i' = \dot{p}_i \pm \frac{1}{\omega_{\max}}(-(\cos\theta_i)\dot{\theta}_i, -(\sin\theta_i)\dot{\theta}_i)^T = \left(1 \mp \frac{\omega_i}{\omega_{\max}}\right)(\cos\theta_i, \sin\theta_i)^T . \tag{9}$$

Indeed, with $\omega_i = \pm\omega_{\max}$, the vehicle is hovering since the virtual center remains fixed, $\dot{p}_i' = 0$. At any point in time, the current virtual center is chosen by a vehicle to be located on either side of the direction of travel. To simplify notation, let a current virtual center be p_i', and the opposite virtual center be $\tilde{p}_i' = p_i' \pm \frac{2}{\omega_{\max}} e_{i,2}$. Let $\tilde{d}_i = \mathrm{CM}_i(\mathscr{V}(P)) - \tilde{p}_i$ and let $\tilde{\Omega}_i$ denote the angle between $e_{i,1}$ and $\tilde{d}_i$, see Figure 4.

Each vehicle will either be in forward motion or hovering motion about one of the centers. This will result into four possible modes of operation for each vehicle depending on the center location: forward-left, hover-left, forward-right, and hover-right. We enumerate each mode with the state $l_i \in \{1,2,3,4\}$, thus we describe each agent by a state variable, $x_i \in \mathrm{SE}(2)_Q \times \{1,2,3,4\}$, and the multi-agent system state by $x = (x_1, \ldots, x_N) \in \mathbb{R}^{4N}$.

The choice of the center for each vehicle is based on the following observation. Starting arbitrarily with a center position p_i', we propose that each vehicle can switch to the other center, $\tilde{p}_i$, only if the actual improvement in cost satisfies $\mathscr{H}(P, \mathscr{V}(P)) - \mathscr{H}(\tilde{P}, \mathscr{V}(\tilde{P})) \geq \beta$, where $\beta \geq 0$ is a fixed constant, $P = (p_1', \ldots, p_i', \ldots, p_n')$ and $\tilde{P} = (p_1', \ldots, \tilde{p}_i, \ldots p_n')$ considers the new virtual center position $\tilde{p}_i$. This improvement can be evaluated locally by each vehicle, by knowing V_i.

We now describe more precisely the hybrid system that formalizes the cooperative algorithm for the multi-UAV group. The system state-space is $\mathrm{SE}(2)_Q \times \{1,2,3,4\} \subseteq O = \mathbb{R}^{4N}$. First, the sets $A_{i,1}, \ldots, A_{i,4}$ define the states where each vehicle i can flow continuously in each of the four modes, and are given as follows:

(1) An individual vehicle can be in $A_{i,1}$ (resp. $A_{i,3}$) if the centroid is in front of the left (resp. right) virtual center at p_i', and if p_i' is not sufficiently close to $\mathrm{CM}_i(\mathscr{V})$. Additionally, the improvement from switching between forward-left to forward-right (resp. vice-versa) must be better than β. However, if the opposite virtual center $\tilde{p}_i'$ is not in Q, then the vehicle may maintain its current virtual center despite violating the improvement threshold β:

$$\begin{aligned}
A_{i,1} = &\{x \in O \mid x_i \in \mathrm{SE}(2)_Q \times \{1\},\ e_{i,1} \cdot d_i \geq \underline{\varepsilon},\ M_i\|d_i\|^2 - M_i\|\tilde{d}_i\|^2 \leq \beta,\ \|d_i\| \geq \varepsilon\} \\
&\cup \{x \in O \mid x_i \in \mathrm{SE}(2)_Q \times \{1\},\ e_{i,1} \cdot d_i \geq \underline{\varepsilon},\ \tilde{p}_i' \in \overline{Q^c},\ \|d_i\| \geq \varepsilon\}, \\
A_{i,3} = &\{x \in O \mid x_i \in \mathrm{SE}(2)_Q \times \{3\},\ e_{i,1} \cdot d_i \geq \varepsilon,\ M_i\|d_i\|^2 - M_i\|\tilde{d}_i\|^2 \leq \beta,\ \|d_i\| \geq \varepsilon\} \\
&\cup \{x \in O \mid x_i \in \mathrm{SE}(2)_Q \times \{3\},\ e_{i,1} \cdot d_i \geq \underline{\varepsilon},\ \tilde{p}_i' \in \overline{Q^c},\ \|d_i\| \geq \varepsilon\},
\end{aligned}$$

(2) A vehicle can be in $A_{i,2}$ (resp. $A_{i,4}$) if $\mathrm{CM}_i(\mathscr{V})$ is behind the left (resp. right) virtual center p_i', or if p_i' is on the boundary Q and heading outwards, or if p_i' is sufficiently close to $\mathrm{CM}_i(\mathscr{V})$:

$$A_{i,2} = \{x \in O \mid x_i \in \mathrm{SE}(2)_Q \times \{2\}, e_{i,1} \cdot d_i \leq \varepsilon, \|d_i\| \geq \varepsilon\} \cup \\ \{x \in O \mid x_i \in \mathrm{SE}(2)_Q \times \{2\}, e_{i,1} \cdot \hat{\mathbf{n}}_{\mathrm{in}} \leq 0\} \cup \{x \in O \mid x_i \in \mathrm{SE}(2)_Q \times \{2\}, \|d_i\| \leq \overline{\varepsilon}\}, \\ A_{i,4} = \{x \in O \mid x_i \in \mathrm{SE}(2)_Q \times \{4\}, e_{i,1} \cdot d_i \leq \varepsilon, \|d_i\| \geq \varepsilon\} \cup \\ \{x \in O \mid x_i \in \mathrm{SE}(2)_{\partial Q} \times \{4\}, e_{i,1} \cdot \hat{\mathbf{n}}_{\mathrm{in}} \leq 0\} \cup \{x \in O \mid x_i \in \mathrm{SE}(2)_Q \times \{4\}, \|d_i\| \leq \overline{\varepsilon}\}.$$

The hysteresis variables $0 < \underline{\varepsilon} < \varepsilon < \overline{\varepsilon}$ serve to insure that Zeno effects do not occur. Combining these sets together, the entire hybrid system flow set is $A = \bigcap_{i=1}^{N} (A_{i,1} \cup A_{i,2} \cup A_{i,3} \cup A_{i,4})$. When the system configuration $x \in A$, the state evolves under the $\dot{x} = F(x)$, where $F(x)$ is defined as follows. First, let $F_i(x)$ with:

$$F_{i,1}(x) = (\cos\theta_i, \sin\theta_i, \frac{2\Omega_i\omega_{\max}}{\pi}, 0)^T, \qquad F_{i,2}(x) = (\cos\theta_i, \sin\theta_i, \omega_{\max}, 0)^T, \\ F_{i,3}(x) = (\cos\theta_i, \sin\theta_i, \frac{2\Omega_i\omega_{\max}}{\pi}, 0)^T, \qquad F_{i,4}(x) = (\cos\theta_i, \sin\theta_i, -\omega_{\max}, 0)^T.$$

Then, $F(x) = (F_1(x), \cdots, F_N(x))^T$, $F_i(x) = F_{i,k}(x)$ if and only if $l_i = k \in \{1,2,3,4\}$.

We now describe the set where discrete jumps can occur. We will consider:

1. Switching from forward-left to forward-right:
$$B_{i,1} = \{x \in O \mid x_i \in \mathrm{SE}(2)_Q \times \{1\}, e_{i,1} \cdot d_i \geq \underline{\varepsilon}, M_i(\|d_i\|^2 - \|\tilde{d}_i\|^2) \geq \beta, \tilde{p}_i \in Q\},$$
2. Switching from forward-right to forward-left:
$$B_{i,2} = \{x \in O \mid x_i \in \mathrm{SE}(2)_Q \times \{3\}, e_{i,1} \cdot d_i \geq \underline{\varepsilon}, M_i(\|d_i\|^2 - \|\tilde{d}_i\|^2) \geq \beta, \tilde{p}_i \in Q\},$$
3. Switching from forward-left to hover-left:
$$B_{i,3} = \{x \in O \mid x_i \in \mathrm{SE}(2)_Q \times \{1\}, e_{i,1} \cdot d_i \leq \underline{\varepsilon}\} \cup \\ \{x \in O \mid x_i \in \mathrm{SE}(2)_{\partial Q} \times \{1\}, e_{i,1} \cdot \hat{\mathbf{n}}_{\mathrm{in}} \leq -\varepsilon\} \cup \{x \in O \mid x_i \in \mathrm{SE}(2)_Q \times \{1\}, \|d_i\| \leq \varepsilon\},$$
4. Switching from hover-left to forward-left:
$$B_{i,4} = \{x \in O \mid x_i \in SE2_Q \times \{2\}, e_{i,1} \cdot d_i \geq \varepsilon, e_{i,1} \cdot \hat{\mathbf{n}}_{\mathrm{in}} \geq 0, \|d_i\| \geq \overline{\varepsilon}\}$$
5. Switching from forward-right to hover-right:
$$B_{i,5} = \{x \in O \mid x_i \in \mathrm{SE}(2)_Q \times \{3\}, e_{i,1} \cdot d_i \leq \underline{\varepsilon}\} \cup \\ \{x \in O \mid x_i \in \mathrm{SE}(2)_{\partial Q} \times \{3\}, e_{i,1} \cdot \hat{\mathbf{n}}_{\mathrm{in}} \leq -\varepsilon\} \cup \{x \in O \mid x_i \in \mathrm{SE}(2)_Q \times \{3\}, \|d_i\| \leq \varepsilon\},$$
6. Switching from hover-right to forward-right:
$$B_{i,6} = \{x \in O \mid x_i \in SE2_Q \times \{4\}, e_{i,1} \cdot d_i \geq \varepsilon, e_{i,1} \cdot \hat{\mathbf{n}}_{\mathrm{in}} \geq 0, \|d_i\| \geq \overline{\varepsilon}\}.$$

The switching domain is the union $B = \bigcup_{i=1}^{N} \bigcup_{k=1}^{6} B_{i,k}$. The jump map G is then defined as follows. First, let $g_{i,1}(x), \ldots, g_{i,6}(x)$ be the maps for an individual vehicle i. These maps are:

$$g_{i,1}(x) = (3, p_i - \frac{2}{\omega_{\max}} e_{i,2}, \theta_i), \qquad g_{i,2}(x) = (1, p_i + \frac{2}{\omega_{\max}} e_{i,2}, \theta_i),$$
$$g_{i,3}(x) = (2, p_i, \theta_i), \qquad g_{i,4}(x) = (1, p_i, \theta_i),$$
$$g_{i,5}(x) = (4, p_i, \theta_i), \qquad g_{i,6}(x) = (3, p_i, \theta_i).$$

We combine the above functions for each vehicle and obtain

$$G_i(x) = \{(x_1, \ldots, g_{i,k}(x), \ldots, x_N) \mid x \in B_{i,k}, \text{ for } k \in \{1, \ldots, 6\}\}.$$

The complete set-valued jump map is then $G(x) = \bigcup_{i=1}^{N} G_i(x)$.

Concisely, the hybrid system of unicycles is described as

$$\dot{x} = F(x), \qquad x \in A,$$
$$x^+ \in G(x), \qquad x \in B.$$

It can be seen that the system satisfies the Basic Conditions of [26], Section VI. This allows us to apply the hybrid LaSalle invariance theorem derived therein:

Theorem 2 (Goebel, Sanfelice, Teel [27]). *Given a hybrid system (F, G, A, B) on a state space $O \subseteq \mathbb{R}^M$ which satisfies the Basic Conditions, suppose that:*

1. *there is a $V : O \to \mathbb{R}$, Lyapunov function continuous on O and Locally Lipschitz on a neighborhood of A,*
2. *$\mathscr{U} \subseteq O$ is non-empty,*
3. *$u_A(x) = \max_{f \in F(x)} \mathscr{L}_f V(x) \leq 0$, for all $x \in A$,*
4. *$u_B(x) = \max_{x^+ \in G(x)} (V(x^+) - V(x)) \leq 0$, for all $x \in B$.*

Let x be precompact with $\overline{\text{range}}(x) \subseteq \mathscr{U}$. Then for some constant $r \in V(\mathscr{U})$, x approaches the largest weakly invariant set in $V^{-1}(r) \cap \mathscr{U} \cap \left(\overline{u_A^{-1}(0)} \cup u_B^{-1}(0)\right)$.

A direct application of the above result leads to:

Theorem 3 (Kwok, Martínez [28]). *Let $\mathscr{U} = O$. Given the hybrid system for fixed forward velocity vehicles defined above and with virtual center dynamics* (9)*, any precompact trajectory $x(t, j)$ with* rge$x \in \mathscr{U}$*, will approach the set of points*

$$\mathscr{M} = \{x \in O \mid \|\text{CM}_{V_i} - p_i'\| \leq \overline{\varepsilon}, \forall i \in \{1, \ldots, n\}\}. \tag{10}$$

The proof makes use of $\mathscr{H}_{\text{centr}}(x_1, \ldots, x_N) = \mathscr{H}(p_1', \ldots, p_N')$ as a locally Lipschitz Lyapunov function. It can be seen that (i) $\dot{\mathscr{H}}_{\text{centr}}(x) \leq 0$ for all $x \in A$ and, that (ii) $\mathscr{H}_{\text{centr}}(x^+) = \mathscr{H}_{\text{centr}}(x)$, for all $x \in B$. The proof follows from the analysis that the only possible set that can contain the largest invariant set is $\mathscr{M}$.

We present a simulation case below where vehicles have a fixed forward velocity in Figure 5. All vehicles begin with random positions and orientations in the lower left corner. They start with a left virtual center, but agent 5 switches to a right virtual center early in the simulation. It can be seen how the vehicles navigate their virtual centers to the centroids of their Voronoi cells. The plot in the right shows a plot of the cost function minimization to a critical value.

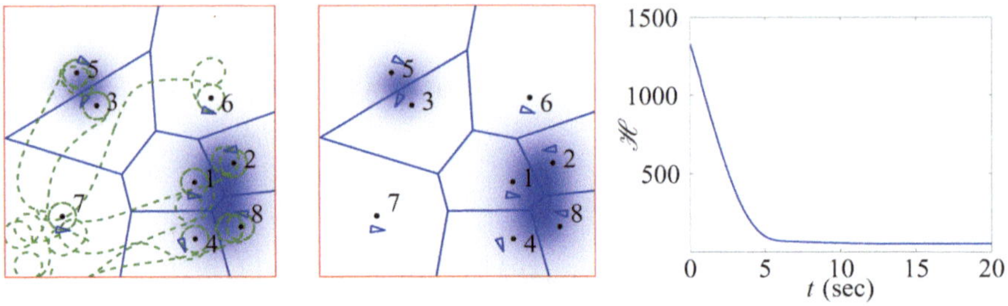

Fig. 5 Fixed forward velocity deployment simulation. The agents start in the lower left corner and path lines are shown in the left figure with final positions and orientations shown in the right figure. Virtual center locations are denoted by a dot.

5 Uncontrollable Vehicles in River Environments

It is generally assumed that vehicles have fully actuated, or at least controllable, dynamics. However, potential applications may involve the deployment of vehicles in hazardous environments where agents lack the actuation to counteract external forces. Example applications include the deployment of micro-UAVs in wind or gliders in a swift current. One can still aim to factor such significant environmental dynamics into the cooperative control algorithms.

In this section we summarize the results found in [29, 30] for the deployment of vehicles in fast flow environments. Assume the following kinematic model for each of the agents:

$$\dot{p}_i = u_i + V(p_i), \tag{11}$$

where $u_i(t)$ is piecewise smooth, $\|u_i\| \leq 1$, and $\|V\| > 1$. Time-optimal trajectories can only be obtained with maximum velocity, thus $u_i = (\cos\theta_i, \sin\theta_i)^T$, see [31].

Our notion of coverage will be associated with the set of points that an agent can travel to faster than other agents. First, let us recall the definition of reachable set:

Definition 1 (Reachable set). We define the *reachable set*, $\mathcal{R}(p_i)$, of an agent at position p_i to be the set of points $x \in X$ that an agent can reach in finite time starting from the initial position p_i and using a piecewise smooth control input $u_i(t)$ with $\|u_i\| \leq 1$. The *T-limited reachable set*, $\mathcal{R}_T(p_i)$, of an agent at position p_i, is the set of points that an agent can reach within time T using a piecewise smooth control input $u_i(t)$ with $\|u_i\| \leq 1$.

Figure 6 shows two examples of reachable sets in an affine and constant flows.

A distributed algorithm for the deployment of agents in a flow environment can be now based on the maximization of the following area coverage metric:

$$\mathcal{H}_{\text{area}}(p_1,\ldots,p_n) = \int_{\bigcup \mathcal{R}_T(p_i)} 1 dx. \tag{12}$$

This must be done while taking into account the flow environment and how it affects the dynamics of each agent. The consideration of other metrics is still possible,

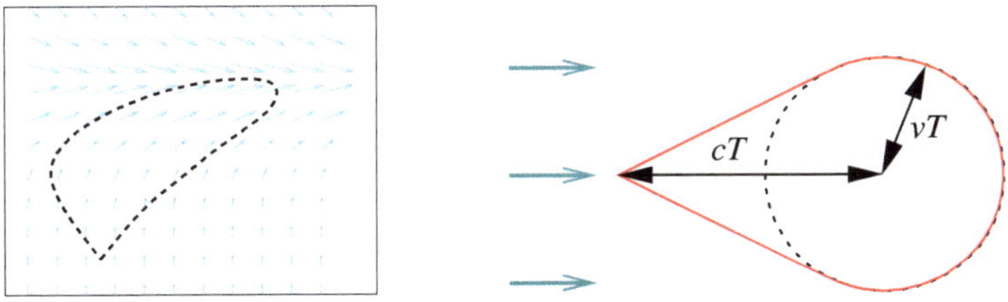

Fig. 6 Reachable sets in affine environments. The figure on the right corresponds to the T reachable set of a vehicle moving with velocity v in a constant flow of magnitude $V = (c, 0)$.

however, external drifts give rise to generalized Voronoi regions with complex boundaries; see [32, 33] for some initial work in this regard. In order to maximize $\mathscr{H}_{\text{area}}$, one can follow the next steps: (a) determine minimum time trajectories in the flow environment, and (b) use knowledge of the properties of these optimal trajectories to compute a gradient direction.

In order to find $\mathscr{R}_T(p_i)$, one must solve the following optimal control problem:

$$\begin{aligned} \text{minimize:} \qquad & J = \int_0^{t_f} 1 dt\,, \\ \text{subject to:} \qquad & \dot{p}_i = u_i + V(p_i)\,, \|u_i\| \leq 1\,, \\ & p_i(0) \,\text{and}\, p_i(t_f) \,\text{given}\,. \end{aligned} \tag{13}$$

For a smooth flow field V, this is known as Zermelo's problem, and a solution can be found in [31]. The optimal solution is to consider a control input of the form

$$\begin{aligned} u_i &= (\cos\theta_i, \sin\theta_i)\,, \\ \dot{\theta}_i &= \sin^2\theta_i \frac{\partial V_2}{\partial x_1} + \sin\theta_i\cos\theta_i\left(\frac{\partial V_1}{\partial x_1} - \frac{\partial V_2}{\partial x_2}\right) - \cos^2\theta_i \frac{\partial V_1}{\partial x_2}\,. \end{aligned} \tag{14}$$

The minimum-time trajectories are obtained by using this input in combination with (11). Note that a constant V produces straight-line optimal trajectories.

To obtain the T-limited boundary of $\mathscr{R}_T(p_i)$, one could integrate (11) using (14) to time T starting at the agent location p_i and initial heading $\theta_i(0) \in [\alpha - \beta - \frac{\pi}{2}, \alpha + \beta + \frac{\pi}{2}]$, where $\alpha = \arctan(V_2(x), V_1(x))$ and $\beta = \arcsin\left(\frac{1}{\|V(x)\|}\right)$. The solutions for various initial headings at time T could then be recorded and combined with the solutions $\gamma(t, -1)$ and $\gamma(t, 1)$ for $t \in [0, T]$. Note that this procedure works well for affine flows, for which optimal trajectories are well behaved. That is, the trajectories do not intersect and they fill up the cone between the extreme optimal trajectories $\gamma(t, -1)$ and $\gamma(t, 1)$.

The consideration of piecewise constant flows changes the nature of optimal solutions and reachable sets. We summarize some of their properties under the following assumption.

Assumption 1. The flow environment X may have obstacles and:

1. The flow V is piecewise constant. That is, $X = \bigcup_{k=1}^{m} X_k$ such that $V_{|X_k}$ is constant and satisfies $\|V_{|X_k}\| > 1$ for all k.
2. The regions X_k, $k \in \{0,\ldots,m\}$, are separated by piecewise differentiable curves. Let $\psi_{k,\ell} : X_k \to \mathbb{R}$ be piecewise differentiable common boundary of X_k and X_ℓ.
3. Along the interface between two flows k and ℓ, we consider any $V(x) \in \text{co}\{V_{|X_k}, V_{|X_\ell}\}$ for $\{x \mid \psi_{k,\ell}(x) = 0\}$.

Thus, the optimal paths in the interior of each X_k will be straight lines. As a path reaches X_k, several situations may arise. We briefly describe these in the following.

Catalog of Optimal Trajectories. For simplicity, in this chapter we assume the boundary of the environment X to be parallel to the flow in the inner region X, and there will not be obstacles present.

For the case that a trajectory intersects a boundary between two flows, defined by $\psi_{k,\ell}(x) = 0$, the intersection can again occur either transversely or tangentially. Based on this, we classify trajectories into *simple (transversal)* or *non-simple (tangential) trajectories*; see Figure 7. The transversal simple trajectories are non-pathological and undergo a direction change at the interface, following an analogous rule to that of the Snell's law in physics:

Proposition 1 (Kwok, Martínez [34]). *Let $V_- = (c_1, c_2)^T$ and $V_+ = (d_1, d_2)^T$ be the flows in two neighboring regions, and α_1, α_2 be their respective flow orientations. Let ξ be the orientation of the normal vector of the smooth curve $\psi(x) = 0$ at the point where the optimal trajectory crosses into the second flow region. A necessary condition for an optimal trajectory across the interface of the two flow regions requires that:*

$$\frac{1 + \|V_-\| \cos(\theta_- - \alpha_1)}{\sin(\theta_- - \xi)} = \frac{1 + \|V_+\| \cos(\theta_+ - \alpha_2)}{\sin(\theta_+ - \xi)}. \tag{15}$$

Given (15)*, and a fixed heading θ_-, the final heading satisfies*

$$\sin\theta_+ = \frac{B \pm C\sqrt{B^2 + C^2 - 1}}{B^2 + C^2}, \tag{16}$$

where $B = \frac{1+\|V_-\|\cos(\theta_- - \alpha_1)}{\sin(\theta_- - \xi)} \cos\xi - d_2$ and $C = \frac{1+\|V_-\|\cos(\theta_- - \alpha_1)}{\sin(\theta_- - \xi)} \sin\xi + d_1$.

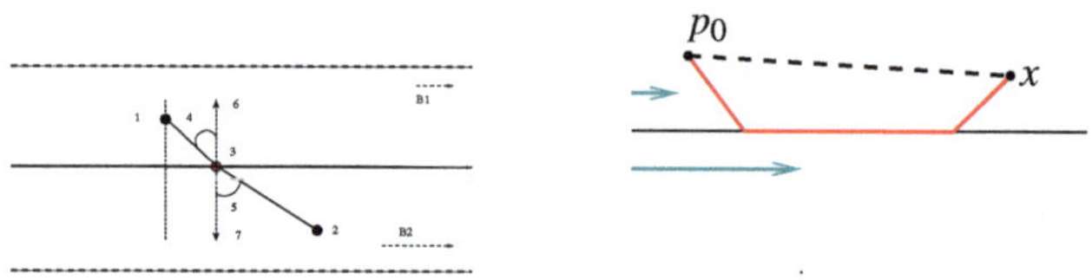

Fig. 7 Example of simple trajectories (left figure) and of non-simple trajectories (right figure)

However, the application of (15) can also result in a trajectory (non-simple trajectory) that travels along the boundary between two different flows; see Figure 7, right picture. The use of the same result can also give a way to compute a heading back into the first region. When an agent is moving along a flow boundary, and it is possible to switch back into the first region, the agent may choose to switch back at any time, making this process indeterminate. However, the result above dictates that there is only one possible outgoing heading back into the first flow region. Furthermore, it is possible for these trajectories that flow along boundaries and later return to intersect other trajectories that remained in the original flow region. For completeness, the following result summarizes necessary conditions for the incoming and outgoing angles for these cases.

Proposition 2 (Kwok, Martínez [34]). *Assume two flow regions defined by the parameters $\|V_-\|, \alpha_1$ and $\|V_+\|, \alpha_2$, respectively, separated by an interface whose normal angle is ξ. If it is possible for an agent to flow along the boundary under the second flow, then θ_+ satisfies*

$$\theta_+ \in \left\{\xi \pm \arccos\left[-\|V_+\| \sin(\alpha_2+\xi)\right], -\xi \pm \arccos\left[\|V_+\| \sin(\alpha_2+\xi)\right]\right\}. \quad (17)$$

Let $D = \frac{1+\|V_+\|\cos(\theta_+-\alpha_2)}{\sin(\theta_+-\xi)}$. Then, the incoming heading resulting in flow along the boundary, if it exists, satisfies

$$\begin{aligned}\theta_- = &\arctan\left[\frac{\|V_-\| \sin\alpha_1 - D\cos\xi}{\|V_-\| \cos\alpha_1 + D\sin\xi}\right] \\ &\pm \arccos\left(\frac{-1}{\sqrt{(\|V_-\| \sin\alpha_1 - D\cos\xi)^2 + (D\sin\xi + \|V_-\| \cos\alpha_1)^2}}\right).\end{aligned} \quad (18)$$

Knowledge about these trajectories can be used to derive a gradient-ascent algorithm that aims to maximize $\mathscr{H}_{\text{area}}$. We begin by taking the gradient of $\mathscr{H}_{\text{area}}$ with respect to p_i in order to obtain a set directions each agent must travel in.

Proposition 3 (Kwok, Martínez [34]). *Given the area objective* (12), *let*

$$\mathring{A}_i = \partial\mathscr{R}_T(p_i) \cap \left(\bigcup_{j\in\mathscr{N}_{i,\text{flow}}} \mathscr{R}_T(p_j)\right)^c \cap X, \quad (19)$$

the set of points in $\partial\mathscr{R}_T(p_i)$ is not in the interior of neighboring reachable sets. Then the gradient with respect to p_i is:

$$\frac{\partial\mathscr{H}}{\partial p_i} = \int_{\mathring{A}_i} \hat{\mathbf{n}}_{\text{out}}^T(\zeta_i)\frac{\partial\zeta_i}{\partial p_i} d\zeta_i, \quad (20)$$

where $\zeta_i : \mathbb{S} \to \mathbb{R}^2$ is a parametrization of $\partial\mathscr{R}_T(p_i)$, and $\hat{\mathbf{n}}_{\text{out}} : \mathbb{R}^2 \to \mathbb{R}^2$ is the unit outward-pointing normal vector at ζ_i.

For piecewise constant flows, $\frac{\partial \zeta_i}{\partial p_i}$ can analytically be computed using the previous analysis of the course changes of optimal trajectories in flows; see [34]. We can further analyze the algorithm

$$\dot{p}_i = \frac{\partial \mathscr{H}_{\text{area}}}{\partial p_i}, \qquad i \in \{1, \dots, n\}, \tag{21}$$

above for the special case of a single constant flow field. For a constant flow the gradient according to (20) becomes $\frac{\partial \mathscr{H}_{\text{area}}}{\partial p_i} = \int_{\mathring{A}_i} \hat{\mathbf{n}}_{\text{out}}^T(\zeta_i) d\zeta_i$. This result has an intuitive interpretation. In order to maximize area covered, agents move towards locations that are not occupied by other agents' reachable sets. Now the following can be proven:

Proposition 4 (Kwok, Martínez [34])

1. *For a constant flow field, $V = c$, if no regions intersect the boundaries ∂X, then $\mathscr{H}_{\text{area}}$ is non-decreasing if agents use the control law* (21).
2. *For constant flows, if the flow boundaries are parallel with the flow direction and X is unbounded (the flow domain is an infinitely long strip), then $\mathscr{H}_{\text{area}}$ is maximized by* (21).

Other flow cases make difficult the analysis of the evolution of $\mathscr{H}_{\text{area}}$, similarly to what happens with time-dependent coverage functions. The current strategy makes agents follow the direction of maximum ascent of $\mathscr{H}_{\text{area}}$. However, one can imagine

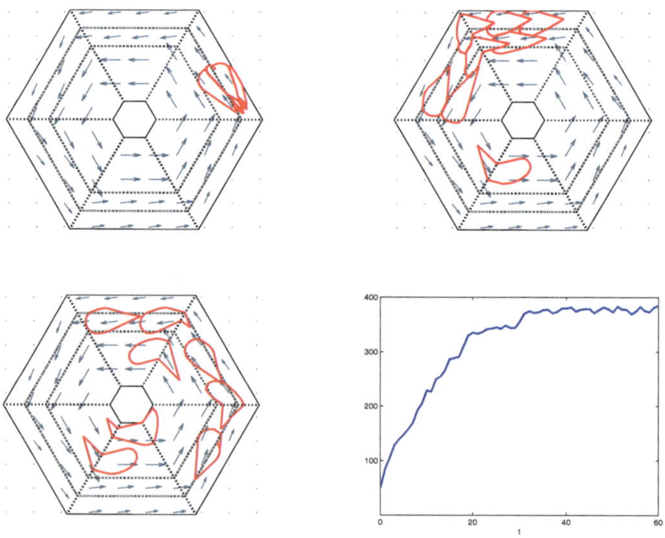

Fig. 8 The central "eye" of the storm is treated as an obstacle, or equivalently a "no-fly zone." The simulation snapshots occur for $t = 0$ (top left), $t = 20$ (top right), and $t = 60$ (bottom left). A plot of the total reachable area is shown in the bottom right.

situations in which the value of $\mathscr{H}_{\text{area}}$ decreases despite of this if, for example, boundaries of X become closer and closer. The following is a simulation showing how $\mathscr{H}_{\text{area}}$ oscillates around a given value when the flow regions force vehicles into a bounded region; see Figure 8.

6 Conclusions

This chapter summarized several results concerning the deployment of vehicles subject to dynamic constraints. In general, the algorithms are distributed over the associated Delaunay graphs or, in r -disk graphs with r sufficiently large. The algorithms provide convergence guarantees to the set of local minima of different classes of Locational Optimization or multicenter metrics. Dynamic constraints were dealt with in essentially three ways: (i) in a soft manner, by modifying the Locational Optimization metric and working with easy-to-compute generalized Voronoi regions, and (ii) by resorting to controllability properties of the vehicles, and (iii) by using the dynamic constraints in the definition of generalized regions assigned to each vehicle. In general, dynamic constraints will lead to involved generalized Voronoi regions, whose boundary is hard to compute and represent, as it reduces to the solution of an optimal control problem. We are currently investigating how this can be alleviated by considering upper and lower approximations of Voronoi regions, which can be refined to any degree at a higher computational expense. By defining an algorithm that allows each agent follow the direction of an approximated gradient using the lower Voronoi region approximation, it can be seen how local minima can still be reached. We are exploring this in the context of constant river environments in the manuscript [35] with Voronoi regions given by hyperbolas, but we believe the approach can be extended to general cost functions.

Acknowledgements. This work surveys partially the thesis work of the graduate student Andrew N. Kwok. Andrew defended his thesis in January 2011 and was funded by grants NSF CNS-0930946, NSF CAREER CMMI-0643679.

References

1. Payton, D., Daily, M., Estowski, R., Howard, M., Lee, C.: Pheromone robotics 11, 319–324 (2001)
2. Howard, A., Matarić, M.J., Sukhatme, G.S.: Mobile sensor network deployment using potential fields: A distributed scalable solution to the area coverage problem. In: International Conference on Distributed Autonomous Robotic Systems (DARS 2002), Fukuoka, Japan, pp. 299–308 (2002)
3. Poduri, S., Sukhatme, G.S.: Constrained coverage for mobile sensor networks, New Orleans, LA, pp. 165–172 (2004)
4. Choset, H.: Coverage for robotics - a survey of recent results. Annals of Mathematics and Artificial Intelligence 31, 113–126 (2001)
5. Arkin, R.C.: Behavior-Based Robotics (1998)

6. Schultz, A.C., Parker, L.E. (eds.): Multi-Robot Systems: From Swarms to Intelligent Automata (2002); Proceedings from the 2002 NRL Workshop on Multi-Robot Systems
7. Balch, T., Parker, L.E. (eds.): Robot Teams: From Diversity to Polymorphism. A K Peters Ltd., Natick (2002)
8. Bulusu, N., Heidemann, J., Estrin, D.: Adaptive beacon placement. In: Int. Conference on Distributed Computing Systems, Mesa, AZ, pp. 489–498 (2001)
9. Cortés, J., Martínez, S., Karatas, T., Bullo, F.: Coverage control for mobile sensing networks. IEEE Trans. Rob. 20, 243–255 (2004)
10. Cortés, J., Martínez, S., Bullo, F.: Spatially-distributed coverage optimization and control with limited-range interactions. ESAIM: Control Optim. Calc. Var. 11, 691–719 (2005)
11. Zhong, M., Cassandras, C.G.: Distributed coverage control in sensor network environments with polygonal obstacles, Seoul, Korea, pp. 4162–4167 (2008)
12. Pimenta, L.C.A., Kumar, V., Mesquita, R.C., Pereira, G.A.S.: Sensing and coverage for a network of heterogeneous robots, pp. 3947–3952 (2008)
13. Caicedo-Nùñez, C.H., Žefran, M.: Performing coverage on nonconvex domains, San Antonio, TX, pp. 1019–1024 (2008)
14. Schwager, M., Rus, D., Slotine, J.J.: Decentralized, adaptive coverage control for networked robots. Int. J. Rob. Res. 28, 357–375 (2009)
15. Martínez, S.: Distributed interpolation schemes for field estimation by mobile sensor networks. IEEE Trans. Control Syst. Technol. (to appear, 2009)
16. Cortés, J.: Area-constrained coverage optimization by robotic sensor networks, pp. 1018–1023 (2008)
17. Pavone, M., Frazzoli, E., Bullo, F.: Distributed policies for equitable partitioning: Theory and applications, pp. 4191–4197 (2008)
18. Bullo, F., Cortés, J., Martínez, S.: Distributed Control of Robotic Networks (2009), `http://www.coordinationbook.info`
19. Okabe, A., Boots, B., Sugihara, K., Chiu, S.N.: Spatial Tessellations: Concepts and Applications of Voronoi Diagrams (2000)
20. Raghunathan, V., Pereira, C., Srivastava, M., Gupta, R.: Energy-aware wireless systems with adaptive power-fidelity tradeoffs. IEEE Trans. Very Large Scale Integration Systems 13, 211–225 (2005)
21. Mohapatra, S., Dutt, N., Nicolau, A., Venkatasubramanian, N.: DYNAMO: A cross-layer framework for end-to-end QoS and Energy Optimization in Mobile Handheld Devices. IEEE J. Selected Areas in Communication (2007)
22. Kwok, A., Martínez, S.: Deployment algorithms for a power-constrained mobile sensor network (to appear, 2009)
23. Kwok, A., Martínez, S.: Deployment algorithms for a power-constrained mobile sensor network. International Journal of Robust and Nonlinear Control 20, 725–842 (2010)
24. Sanfelice, R.G., Goebel, R., Teel, A.R.: Results on convergence in hybrid systems via detectability and an invariance principle. In: American Control Conference, pp. 551–556 (2005)
25. Dubins, L.E.: On curves of minimal length with a constraint on average curvature and with prescribed initial and terminal positions and tangents. American Journal of Mathematics 79, 497–516 (1957)
26. Goebel, R., Hespanha, J.P., Teel, A.R., Cai, C., Sanfelice, R.G.: Hybrid systems: generalized solutions and robust stability, Stuttgart, Germany, pp. 1–12 (2004)
27. Goebel, R., Sanfelice, R., Teel, A.: Hybrid dynamical systems. IEEE Control Systems Magazine 29, 28–93 (2009)
28. Kwok, A., Martínez, S.: Unicycle coverage control via hybrid modeling. IEEE Transactions on Automatic Control 55, 528–532 (2010)

29. Kwok, A., Martínez, S.: A coverage algorithm for drifters in a river environment. In: 2010 American Control Conference, Baltimore, MD, USA, pp. 6436–6441 (2010)
30. Kwok, A., Martínez, S.: Deployment of drifters in a piecewise-constant flow environment. In: IEEE International Conference on Decision and Control, Atlanta, GA, USA, pp. 6584–6589 (2010)
31. Bryson, A.E., Ho, Y.: Applied Optimal Control. Hemisphere Publishing Corporation, New York (1969)
32. Bakolas, E., Tsiotras, P.: The Zermelo–Voronoi diagram: A dynamic partition problem. Automatica 46, 2059–2067 (2010)
33. Ru, Y., Martínez, S.: Energy-based Voronoi partition in constant-flow environments. IEEE Transactions on Automation Science and Engineering (2011) (submitted)
34. Kwok, A., Martínez, S.: A coverage algorithm for drifters in a river environment. In: American Control Conference, Baltimore, MD (to apear, 2010)
35. Ru, Y., Martínez, S.: Coverage control in constant flow environments using a mixed energy-time metric. In: IEEE International Conference on Decision and Control (Submitted 2012)

Kalman Smoothing for Distributed Optimal Feedback Control of Unicycle Formations

Ross P. Anderson and Dejan Milutinović

Abstract. In many multi-agent control problems, the ability to compute an optimal feedback control is severely limited by the dimension of the state space. In this work, deterministic, nonholonomic agents are tasked with creating and maintaining a formation based on observations of their neighbors, and each agent in the formation independently computes its feedback control from a Hamilton-Jacobi-Bellman (HJB) equation. Since an agent does not have knowledge of its neighbors' future motion, we assume that the unknown control to be applied by neighbors can be modeled as Brownian motion. The resulting probability distribution of its neighbors' future trajectory allows the HJB equation to be written as a path integral over the distribution of optimal trajectories. We describe how the path integral approach to stochastic optimal control allows the distributed control problems to be written as independent Kalman smoothing problems over the probability distribution of the connected agents' future trajectories. Simulations show five unicycles achieving the formation of a regular pentagon.

1 Introduction

The focus of this work is on formation control, in which each agent, a robotic nonholonomic vehicle, in a team is tasked with attaining and maintaining pre-specified distances from the agents in its neighborhood. Problems of this type are beginning to demonstate their significance and potential impact in a variety of applications in both the public and private sector [3, 22, 29, 35]. Nonholonomic vehicle formations, in particular, have attracted much attention [1, 7, 8, 9, 31, 32, 33], but these studies have typically relied on stability analyses, or on ad-hoc artificial potential functions or navigation functions.

Ross P. Anderson · Dejan Milutinović
University of California, Santa Cruz, 1156 High Street, Santa Cruz, CA 95060
e-mail: {anderson, dejan}@soe.ucsc.edu

D. Milutinović & J. Rosen (Eds.): Redundancy in Robot Manipulators, LNEE 57, pp. 145–166.
DOI: 10.1007/978-3-642-33971-4_9

Alternatively, the formation control problem may be defined as an optimal feedback control problem. To compute an optimal feedback control, one must solve the Hamilton-Jacobi-Bellman (HJB) equation, which is a nonlinear partial differential equation (PDE). However, the computational complexity of solving the PDE grows exponentially with the size of the team, and this severely limits the effectiveness of conventional techniques of stochastic optimal control. A number of promising approaches to address this so-called "curse of dimensionality" have been proposed, including reinforcement learning [36], neurodynamic programming [4], and approximate dynamic programming [30], just to name a few.

In this work, we approach the problem of formation control based on the path integral formulation of stochastic optimal control [17]. We explicitely take into account the fact that although agents may be capable of observing or receiving the current state of their neighbors, the future trajectories of these neighbors will seldom be known exactly, since they individually compute their control based on their own available information and observations. From this point of view, the distributed formation control problem is inherently stochastic, and not only due to unpredictable neighbors. In addition, the control is a function of an agent's noisy observations, and it must also deal with agent model uncertainties and environmental uncertainties (e.g., wind). Along these lines, this work considers the problem of controlling one agent based on observations of its neighbors *and the probability of their future motion*. This probability distribution arises from an assumption that the unknown control of an agent can be modeled as Brownian motion [15], so that based on the system kinematics, we can infer the probability of finding the relative state $\mathbf{x}$ to all neighbors in an interval $(\mathbf{x}, \mathbf{x}+d\mathbf{x})$ at a particular future time [40].

Perhaps more importantly, this probability distribution over future system trajectories can be used to statistically infer the probability distribution of the *control*, and, hence, the optimal control. When an agent's neighbors are treated as nondeterministic, one can consider that agent's optimal control to be the action that minimizes the expected value of the accumulated cost with respect to the distribution of neighbors' future trajectories (see [20] and references therein for a more precise interpretation in terms of minimization of Kullback-Leibler divergence). Along these lines, the path integral (PI) formulation of stochastic optimal control [16, 19, 18] transforms the problem of solving the HJB equation into an estimation problem on the distribution of optimal trajectories in continuous state space [39].

The path integral approach is made possible by the relation between the solutions to optimal control PDEs and the probability distribution of stochastic differential equations [27, 44] (see [24, 25, 26, 28] for an analogous approach in the open-loop control case), and it has shown great potential for systems with large state space. For example, Theodorou et al. [37, 38] have examined the the link between reinforcement learning and the path integral method for motor control and robotics, while van den Broek et al. [5, 6] and Wiegerinck et al. [41, 42] apply the path integral framework to multi-agent systems. In the latter, the agents exhibit explicitly-stochastic kinematics and cooperatively compute their control from a marginalization of the joint probability distribution of the group's system trajectory.

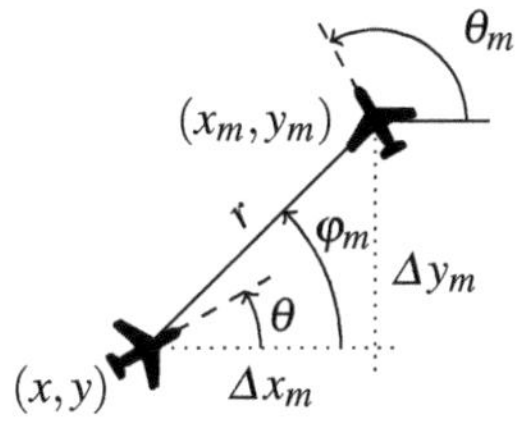

Fig. 1 Diagram of the AiF moving at direction θ and at a distance r to the neighbor m. The turning rate and acceleration of the neighbor are unknown.

In this chapter, we develop a method to apply the path integral approach to fully-distributed multi-agent systems by constructing a fast-switching process that randomly select a single neighboring agent from which the optimal control may be inferred. We also establish a connection between the optimal feedback control problem for multi-agent systems and nonlinear Kalman smoothing algorithms, which allows each agent to independently compute its control in real-time.

This chapter is organized as follows. Section 2 introduces the formation control problem as viewed by a single agent in the group and describes a way to parameterize the kinematic model so that the resulting HJB equation solution can be represented as a path integral. Section 3 reviews the path integral formulation of stochastic optimal control. Section 4 presents a duality between stochastic optimal feedback control and Kalman smoothing algorithms. Section 5 illustrates our methods with a simulated five-agent formation, and conclusions are in Section 6.

2 Control Problem Formulation

In this section we formulate the optimal feedback control problem for unicycle formations and describe a way to manipulate the model kinematics into a form that allows the HJB equation solution to be described by a path integral.

2.1 Preliminary Kinematic Model

In the problem formulation, each agent is modeled as a unicycle, which moves in the direction if its heading angle θ at a speed v:

$$dx(t) = v\cos\theta dt \tag{1}$$
$$dy(t) = v\sin\theta dt \tag{2}$$
$$d\theta(t) = \omega dt \tag{3}$$
$$dv(t) = udt, \tag{4}$$

where u is the feedback acceleration control and ω is the feedback turning rate control.

Each agent independently computes its respective control based on observations of its M neighbors, labeled $m = 1, \ldots, M$. To this end, we focus on the system state as viewed by one agent, which we call the agent-in-focus, or AiF for short. Define $\Delta x_m = x_m - x$ and $\Delta y_m = y_m - y$ as the Cartesian components of the distance from the AiF to the neighbor m (Fig. 1). These states evolve as:

$$d\Delta x_m(t) = -v\cos(\theta)\,dt + v_m\cos(\theta_m)\,dt \tag{5}$$
$$d\Delta y_m(t) = -v\sin(\theta)\,dt + v_m\sin(\theta_m)\,dt, \tag{6}$$

where v_m is the speed of neighbor m. Although the kinematics of the neighboring vehicles are identical, their turning rate control and acceleration control are unknown. Based on the motivation of the previous section, we assume that the turning rate and acceleration controls of neighbor m can be modeled as Wiener processes with mutually independent increments $dw_{\theta,m}$ and $dw_{v,m}$, and intensities σ_θ and σ_v, respectively:

$$d\theta_m = \sigma_\theta dw_{\theta,m} \tag{7}$$
$$dv_m = \sigma_v dw_{v,m}. \tag{8}$$

Finally, introducing the distance from the AiF to the neighbor m as $r_m = \sqrt{\Delta x_m^2 + \Delta y_m^2}$ and the angle to the neighbor m as $\varphi_m = \tan^{-1}(\Delta y_m/\Delta x_m)$, we arrive at a preliminary model for the AiF and a single neighbor m:

$$dr_m(t) = -v\cos(\varphi_m - \theta)\,dt + v_m\cos(\varphi_m - \theta_m)\,dt \tag{9}$$
$$d\varphi_m(t) = \frac{v}{r_m}\sin(\varphi_m - \theta)\,dt - \frac{v_m}{r_m}\sin(\varphi_m - \theta_m)\,dt \tag{10}$$
$$d\theta(t) = \omega dt \tag{11}$$
$$d\theta_m(t) = \sigma_\theta dw_{\theta,m} \tag{12}$$
$$dv(t) = udt \tag{13}$$
$$dv_m(t) = \sigma_v dw_{v,m}. \tag{14}$$

Note that this system, when augmented to account for M neighbors, would have a two-dimensional control $\mathbf{u} = [\omega, u]^T$, but that the stochastic process $\mathbf{w}(t) = [w_{\theta,1}, \ldots, w_{v,M}]^T$ would be of dimension M.

2.2 Switching Kinematic Model

Since the AiF will be drawing from the random processes describing its neighbors kinematics as a source from which to compute its control, we wish to devise a way to connect the random motion in (12) and (14) to the controls in (11) and (13). In particular, for reasons that will become more clear in Section 3, any controlled state should be affected by just one Wiener process, and vice versa. In the model developed in the previous section, this is not the case since there is a two dimensional control $\mathbf{u}$ and M dimensional stochastic process. In order to manipulate the model into such a form, we introduce a second, faster time scale t/ε, for a small $\varepsilon > 0$, and we assume that in each infinitesimal time increment in this faster time scale, the AiF is using the relative state to just one neighbor m to compute its control, and that the choice of neighbor m switches randomly among the M neighbors at a fast rate. It will turn out [43] that under a sufficiently fast secondary time scale, the randomly-switching model to be developed in this section is equivalent to (9)-(14).

Let us define the difference in heading angle $\gamma = \theta - \theta_m$ and difference in speed $\kappa = v - v_m$. Then we can obtain (Appendix 1) the following model, one for each of M neighbors.

$$
\begin{aligned}
dr_m(t) = &-\left(\frac{1}{M}\sum_{j=1}^{M}\left(\mathbb{E}\left(\kappa_j\right)\right) + \overline{v_m}(0)\right)\cos\left(\varphi_m - \frac{1}{M}\sum_{j=1}^{M}\left(\mathbb{E}\left(\gamma_j\right)\right) - \overline{\theta_m}(0)\right)dt \\
&+ \left(-\left(\kappa_m - \mathbb{E}\left(\kappa_m\right)\right) + v_m(0)\right)\cos\left(\varphi_m + \left(\gamma_m - \mathbb{E}\left(\gamma_m\right)\right) - \theta_m(0)\right)dt \qquad (15)\\
d\varphi_m(t) = &\left(\frac{1}{M}\sum_{j=1}^{M}\left(\mathbb{E}\left(\kappa_j\right)\right) + \overline{v_m}(0)\right)\sin\left(\varphi_m - \frac{1}{M}\sum_{j=1}^{M}\left(\mathbb{E}\left(\gamma_j\right)\right) - \overline{\theta_m}(0)\right)dt \\
&- \frac{1}{r_m}\left(-\left(\kappa_m - \mathbb{E}\left(\kappa_m\right)\right) + v_m(0)\right)\sin\left(\varphi_m + \left(\gamma_m - \mathbb{E}\left(\gamma_m\right)\right) - \theta_m(0)\right)dt \\
&\qquad (16)\\
d\gamma_m(t) = &\left(M\omega dt - \sqrt{M}\sigma_\theta dw_{\theta,m}\right)\delta_{\xi(t/\varepsilon),m} \qquad (17)\\
d\kappa_m(t) = &\left(Mudt - \sqrt{M}\sigma_v dw_m\right)\delta_{\xi(t/\varepsilon),m}. \qquad (18)
\end{aligned}
$$

When taking into account all M neighbors of the AiF, the system state is defined through a concatenation of the model (15)-(18), one for each neighbor $m = 1, \ldots, M$.

The fast switching behavior [11] is captured by an ergodic Markov chain $\xi(t/\varepsilon)$ with a fast time scale $\varepsilon > 0$, taking on values in $\{1, 2, \ldots, M\}$. We assume that this chain is independent of the Wiener process $\mathbf{w}(t)$ affecting the neighbors' heading angles and speeds in (7)-(8). The Kronecker deltas δ in (17)-(18), therefore, select the pair of "actively evolving" states among the M states in $[\gamma_1, \ldots, \gamma_M, \kappa_1, \ldots, \kappa_M]^T$. In other words, if $\xi(t/\varepsilon) = M-1$, for example, only the states γ_{M-1} and κ_{M-1} evolve as in (17) and (18), while all other relative heading angle and relative speed states have zero increment ($d\gamma_m = d\kappa_m = 0$). In this case, the AiF is using the random motion of neighbor $M-1$ to compute its control.

The evolution equation for M neighbors may now be written in a more general stochastic differential equation for the state vector $\mathbf{x}(t)$:

$$d\mathbf{x}(t) = f(\mathbf{x})dt + B_i\mathbf{u}dt + \Gamma_i d\mathbf{w}, \qquad (19)$$

where $f(\mathbf{x})$ describes the deterministic motion in states (15)-(16), and the matrices B_i and Γ_i in the state $\xi(t/\varepsilon) = i$ and state vector $\mathbf{x}(t)$ are constructed as:

$$B_i = M \begin{bmatrix} 0 & 0 \\ \vdots & \vdots \\ 0 & 0 \\ \delta_{i1} & 0 \\ \vdots & \vdots \\ \delta_{i,M} & 0 \\ 0 & \delta_{i1} \\ \vdots & \vdots \\ 0 & \delta_{i,M} \end{bmatrix} \begin{matrix} \left.\vphantom{\begin{matrix}0\\ \vdots\\0\end{matrix}}\right\} r_1,\dots,r_M,\varphi_1,\dots,\varphi_M \\ \left.\vphantom{\begin{matrix}0\\ \vdots\\0\end{matrix}}\right\} \gamma_1,\dots,\gamma_M \\ \left.\vphantom{\begin{matrix}0\\ \vdots\\0\end{matrix}}\right\} \kappa_1,\dots,\kappa_M \end{matrix}$$

$$\Gamma_i = -\sqrt{M} \begin{bmatrix} 0 & \dots & 0 \\ \vdots & \ddots & \vdots \\ 0 & \dots & 0 \\ \sigma_\theta \mathbb{1}_i & & 0 \\ 0 & & \sigma_v \mathbb{1}_i \end{bmatrix} \begin{matrix} \left.\vphantom{\begin{matrix}0\\ \vdots\\0\end{matrix}}\right\} r_1,\dots,r_M,\varphi_1,\dots,\varphi_M \\ \} \gamma_1,\dots,\gamma_M \\ \} \kappa_1,\dots,\kappa_M \end{matrix} \qquad \mathbf{x}(t) = \begin{bmatrix} r_1 \\ \vdots \\ r_m \\ \hdashline \varphi_1 \\ \vdots \\ \varphi_M \\ \hdashline \gamma_1 \\ \vdots \\ \gamma_M \\ \hdashline \kappa_1 \\ \vdots \\ \kappa_M \end{bmatrix}, \tag{20}$$

where $\mathbb{1}_i = diag\,([\delta_{i,1}, \dots, \delta_{i,M}])$.

The transition probabilities from state $\xi(t/\varepsilon) = i$ to $\xi(t/\varepsilon) = j$ are defined in terms of a $M \times M$ generator matrix $Q(t)/\varepsilon$

$$P(\xi(t+\Delta t) = j \mid \xi(t) = i) = q_{ij} + o(\Delta t), \;\; j \neq i, \tag{21}$$

and we choose

$$q_{ij}(t) = 1, \qquad\qquad j \neq i \tag{22}$$
$$q_{ii}(t) = -(M-1), \tag{23}$$

so that the chain has an equal probability of transitioning into any of the M states. It is well known that in the limit of $\varepsilon \to 0$, the evolution of a switching model like (19) under fast Markov switching $\xi(t/\varepsilon)$ converges weakly to an average, or homogenized, model [11, 21]. Because of the symmetry of the transition probabilities (21), the switching model (15)-(18) then converges weakly the same kinematics as the original model (9)-(12). However, unlike the original model, the switching model has the advantage that $B_i B_i^T \propto \Gamma_i \Gamma_i^T \; \forall i$, which will become important in Section 3. Therefore, we seek to control the state of $\mathbf{x}(t)$ as described by the switching kinematic model (19).

2.3 Cost Functional

In our control approach, the agents create and maintain a formation described by a vector of nominal distances. From the perspective of the AiF, these distances are $\mu = [\mu_1, \ldots, \mu_M]^T$, and it computes its respective feedback control $\mathbf{u}$ from a cost functional of the form:

$$J(\mathbf{x},t,\xi) = \min_u \mathbb{E}\left\{\int_t^{t_f} \left(\frac{1}{2}(h(\mathbf{x}(s)) - \mu)^T A(h(\mathbf{x}(s)) - \mu) + \frac{1}{2}\mathbf{u}^T R\mathbf{u}\right) ds\right\},$$

where $h(\mathbf{x}(t)) = [r_1(t), \ldots, r_M(t)]^T$ is the vector of distances to each neighbor of the AiF, and R is the penalty on control, i.e., turning rate and acceleration control. The nominal distances μ are assumed to be constant over the planning horizon, but any change in μ could reflect the inclusion of dynamic formation changes. We define the general state cost $k(\mathbf{x})$,

$$k(\mathbf{x}) = (h(\mathbf{x}) - \mu)^T A(h(\mathbf{x}) - \mu), \tag{24}$$

to yield the cost-to-go function for the AiF

$$J_i(\mathbf{x},t) \equiv J(\mathbf{x},t,\xi = i) = \min_{\mathbf{u}} \mathbb{E}\left\{\int_t^{t_f} \frac{1}{2}\left(k(\mathbf{x}(s)) + \mathbf{u}(\mathbf{x}(s))^T R\mathbf{u}(\mathbf{x}(s))\right) ds\right\}. \tag{25}$$

3 Path Integral Construction

In this section the path integral representation of the switching kinematics model is derived. We begin with the (stochastic) Hamilton-Jacobi-Bellman equation for the model (19) and cost functional (25), which, for the state of the fast-switching Markov chain is $\xi(t/\varepsilon) = i$, is

$$\begin{aligned} 0 = \partial_t J_i + \min_u &\left\{ (f + B_i\mathbf{u})^T \partial_{\mathbf{x}} J_i + \frac{1}{2}\mathrm{Tr}\left(\Gamma_i \Gamma_i^T \partial_{\mathbf{x}}^2 J_i\right)\right. \\ &\left. + \frac{1}{2}k(\mathbf{x}) + \frac{1}{2}\mathbf{u}^T R\mathbf{u} + \frac{Q(t)}{\varepsilon} J(\mathbf{x},t)(i)\right\}, \qquad i = 1, \ldots, M, \end{aligned} \tag{26}$$

where

$$\frac{Q(t)}{\varepsilon} J(\mathbf{x},t)(i) = \frac{1}{\varepsilon}\sum_{j \neq i} q_{ij}(t)\left(J_j(\mathbf{x},t) - J_i(\mathbf{x},t)\right). \tag{27}$$

We have chosen zero terminal cost (at $t = t_f$) for this system of PDEs,

$$J_i(\mathbf{x},t_f) = \phi(\mathbf{x}) = 0, \qquad \forall i,\ \forall \mathbf{x}, \tag{28}$$

and added reflective boundary conditions to constrain the speed of agents to remain between $v_{LB} \leq v \leq v_{UB}$:

$$\partial_{\mathbf{x}} J_i(\mathbf{x},t) \cdot \hat{n} = 0, \qquad \forall i,\ \forall t, \quad \mathbf{x} \in \mathscr{V} \tag{29}$$

$$\begin{aligned}\mathscr{V} = \Big\{\mathbf{x} :\ & \frac{1}{M}\sum_{j=1}^{M}(\mathbb{E}(\kappa_j)) + \overline{v_m}(0) = v_{LB} \bigcup \frac{1}{M}\sum_{j=1}^{M}(\mathbb{E}(\kappa_j)) + \overline{v_m}(0) = v_{UB} \\ & \bigcup -(\kappa_m - \mathbb{E}(\kappa_m)) + v_m(0) = v_{LB} \\ & \bigcup -(\kappa_m - \mathbb{E}(\kappa_m)) + v_m(0) = v_{UB}\Big\},\end{aligned} \tag{30}$$

at the domain normals $\hat{n}$.

The HJB equation is typically solved numerically (see [23], for example), which is impossible for a problem of this size. However, we can exploit the structure of the formation control problem to formulate the HJB PDE solution as a solution to an equivalent estimation problem through a path integral representation.

The optimal control $\mathbf{u}(\mathbf{x},t,i)$ that minimizes (26) is

$$\mathbf{u}(\mathbf{x},t,i) = -R^{-1}B_i^T \partial_{\mathbf{x}} J_i(\mathbf{x},t), \tag{31}$$

which, when substituted back into the HJB equation, yields:

$$\begin{aligned}0 = \partial_t J_i + f^T \partial_{\mathbf{x}} J_i - \frac{1}{2}(\partial_{\mathbf{x}} J_i)^T B_i R^{-1} B_i^T \partial_{\mathbf{x}} J_i \\ + \frac{1}{2}\mathrm{Tr}\left(\Gamma_i \Gamma_i^T \partial_{\mathbf{x}}^2 J_i\right) + \frac{1}{2}k(\mathbf{x}(t)) + \frac{Q(t)}{\varepsilon} J(\mathbf{x},t)(i).\end{aligned} \tag{32}$$

A logarithmic transformation [10] is applied for each state i:

$$J_i(\mathbf{x},t) = -\lambda \log \Psi_i(\mathbf{x},t), \tag{33}$$

yielding a new PDE

$$\begin{aligned}\frac{1}{\Psi_i}\partial_t \Psi_i = \frac{1}{2\lambda}k(\mathbf{x}) - \frac{f^T}{\Psi_i}\partial_{\mathbf{x}}\Psi_i - \frac{1}{2\Psi_i}\mathrm{Tr}\left(\Gamma_i \Gamma_i^T \partial_{\mathbf{x}}^2 \Psi_i\right) - \frac{Q(t)}{\varepsilon}\log \Psi(\mathbf{x},t)(i) \\ - \frac{\lambda}{2\Psi_i^2}(\partial_{\mathbf{x}}\Psi_i)^T B_i R^{-1} B_i^T \partial_{\mathbf{x}}\Psi_i + \frac{1}{2}\frac{1}{\Psi_i^2}(\partial_{\mathbf{x}}\Psi_i)^T \Gamma_i \Gamma_i^T \partial_{\mathbf{x}}\Psi_i.\end{aligned} \tag{34}$$

In the relative model (17)-(18), it can be seen that the states $\gamma_m(t)$ and $\kappa_m(t)$ collectively describe the evolution of the difference between the AiF control and the unknown control of a neighbor. This suggests that we might be able to compute a control that can, in some sense, compensate for the uncertainty associated with a neighbor's control. Moreover, this implies that a large disturbance in the relative states (i.e., (17)-(18)) likely requires a greater control input, and conversely, that non-actuated states must be noiseless. Because of this, we assume that the noise in the controlled components is inversely proportional to the control penalty, or

$$\Gamma_i \Gamma_i^T = \lambda B_i R^{-1} B_i^T, \qquad \forall i. \tag{35}$$

This selects the value of the control penalty that we shall use in the sequel as

$$R = diag\left(\lambda \sigma_\theta^{-2}, \lambda \sigma_v^{-2}\right). \tag{36}$$

From (36), the quadratic terms on the second line of (34) cancel, and the remaining PDE for Ψ_i is

$$\partial_t \Psi_i = \frac{1}{2\lambda} k(x) \Psi_i - f^T \partial_{\mathbf{x}} \Psi_i - \frac{1}{2} \mathrm{Tr}\left(\Gamma_i \Gamma_i^T \partial_{\mathbf{x}}^2 \Psi_i\right) - \Psi_i \frac{Q(t)}{\varepsilon} \log \Psi(\mathbf{x},t)(i). \tag{37}$$

Note that this cancellation is only possible in the switching model.

Next, it is shown in Appendix 2 that a first order asymptotic approximation to (37), denoted $\Psi_0(\mathbf{x},t)$, is independent of the state i of the chain $\xi(t/\varepsilon)$, and that this approximation satisfies the following linear PDE:

$$\partial_t \Psi_0(\mathbf{x},t) = -f^T \partial_x \Psi_0(\mathbf{x},t) - \frac{1}{2} \mathrm{Tr}\left(\Sigma \partial_{\mathbf{x}}^2 \Psi_0\right) + \frac{1}{2\lambda} k(\mathbf{x}) \Psi_0(\mathbf{x},t) \tag{38}$$

$$= -\left(f^T \partial_{\mathbf{x}} + \frac{1}{2} \mathrm{Tr}\left(\Sigma \partial_{\mathbf{x}}^2\right) - \frac{1}{2\lambda} k(\mathbf{x})\right) \Psi_0 \tag{39}$$

$$= -H \Psi_0(\mathbf{x},t), \tag{40}$$

where

$$\Sigma = \frac{1}{M} \sum_{j=1}^{M} \Gamma_j \Gamma_j^T \tag{41}$$

is the average of the covariance of the stochastic disturbances in the switching model (15)-(18). The boundary conditions become

$$\Psi_0(\mathbf{x},t_f) = \exp(0) = 1, \qquad \forall \mathbf{x} \tag{42}$$

$$\partial_{\mathbf{x}} \Psi_0(\mathbf{x},t) \cdot \hat{n} = 0, \qquad \forall t, \ \mathbf{x} \in \mathscr{V}. \tag{43}$$

This could be numerically solved backward in time from the terminal condition. However, the Feynman-Kac equations [27, 44] connect certain linear differential operators, H included, to adjoint operators that describe the evolution of a *forward* diffusion process beginning from the current state $\widetilde{\mathbf{x}}(t_0) = \widetilde{\mathbf{x}}_0 = \mathbf{x}$ and ending at $\widetilde{\mathbf{x}}_N = \widetilde{\mathbf{x}}(t_N) = \widetilde{\mathbf{x}}(t_f)$.

In expected value, the solution to (40) is

$$\Psi_0(\widetilde{\mathbf{x}}_0, t_0) = \mathbb{E}_{p(\chi|\widetilde{\mathbf{x}}_0)} \left\{ \exp\left(-\frac{1}{2\lambda} \int_{t_0}^{t_N} k(\mathbf{x}(s)) ds \right) \right\}, \tag{44}$$

where $\widetilde{\mathbf{x}}(t)$ satisfies the path integral-associated, uncontrolled dynamics (cf. (19))

$$d\widetilde{\mathbf{x}}(t) = f(\widetilde{\mathbf{x}}(t))dt + \sqrt{\Sigma}d\mathbf{w} \tag{45}$$

$$\widetilde{\mathbf{x}}(t_0) = \mathbf{x}. \tag{46}$$

Note that since the process $\widetilde{\mathbf{x}}(t)$ is uncontrolled, the expected values of the relative states γ_m and κ_m reduce to $\theta(0) - \theta_m(0)$ and $v(0) - v_m(0)$, respectively. This simplifies the passive components of the uncontrolled model, i.e., $f(\cdot)$ in (15)-(16), to the following form, one for each neighbor m:

$$dr_m(t) = -\kappa(0)\cos(\varphi_m - \theta(0))\,dt - (\kappa_m - v_m(0))\cos(\varphi_m + \gamma_m - \theta(0))\,dt \tag{47}$$

$$d\varphi_m(t) = \kappa(0)\sin(\varphi_m - \theta(0))\,dt + \frac{1}{r_m}(\kappa_m - v_m(0))\sin(\varphi_m + \gamma_m - \theta(0))\,dt \tag{48}$$

The expectation in (44) is taken with respect to the distribution $p(\chi|\widetilde{\mathbf{x}}_0)$ of sample paths χ that begin at $\widetilde{\mathbf{x}}_0 = \mathbf{x}$ and evolve as (45). By discretizing the interval $[t_0, t_N]$ into N intervals of equal length Δt, $t_0 < t_1 < \ldots < t_N$, we can write a sample of the discretized trajectory χ_N^i as

$$\chi_N^i = \left(\widetilde{\mathbf{x}}_1^i, \ldots, \widetilde{\mathbf{x}}_N^i\right),$$

which is sampled from

$$\chi_N^i \sim p(\chi_N|\widetilde{\mathbf{x}}_0) = p(\widetilde{\mathbf{x}}_1, \ldots, \widetilde{\mathbf{x}}_N|\widetilde{\mathbf{x}}_0).$$

Under this discretization in time, the solution (44) can be written as

$$\Psi_0(\widetilde{\mathbf{x}}_0, t_0) = \lim_{\Delta t \to 0} \int d\chi_N p(\chi_N|\widetilde{\mathbf{x}}_0) \exp\left[-\frac{\Delta t}{2\lambda} \sum_{k=1}^{N} k(\widetilde{\mathbf{x}}_k)\right], \tag{49}$$

where $d\chi_N = \prod_{k=1}^{N} d\widetilde{\mathbf{x}}_k$ and where $p(\chi_N|\widetilde{\mathbf{x}}_0)$ is the probability of a discretized sample path, conditioned on the starting state $\widetilde{\mathbf{x}}_0$, given by

$$p(\chi_N|\widetilde{\mathbf{x}}_0) = \prod_{k=0}^{N-1} p(\widetilde{\mathbf{x}}_{k+1}|\widetilde{\mathbf{x}}_k). \tag{50}$$

Since, in the uncontrolled process (45), the noise is Gaussian with zero mean and covariance Σ, the transition probabilities may be written as

$$p(\widetilde{\mathbf{x}}_{k+1}|\widetilde{\mathbf{x}}_k) = \frac{1}{\sqrt{2\pi|\Sigma|\Delta t}} \exp\left(-\frac{1}{2\Delta t}(\widetilde{\mathbf{x}}_{k+1} - \widetilde{\mathbf{x}}_k - f(\widetilde{\mathbf{x}}_k)\Delta t)^T \Sigma^{-1} (\widetilde{\mathbf{x}}_{k+1} - \widetilde{\mathbf{x}}_k - f(\widetilde{\mathbf{x}}_k)\Delta t)\right). \tag{51}$$

We can then write the probability of a complete trajectory as

$$p(\chi_N|\widetilde{\mathbf{x}}_0) \propto \exp\left(-\frac{\Delta t}{2}\sum_{k=0}^{N-1}\left(\frac{\widetilde{\mathbf{x}}_{k+1}-\widetilde{\mathbf{x}}_k}{\Delta t}-f(\widetilde{\mathbf{x}}_k)\right)^T \Sigma^{-1}\left(\frac{\widetilde{\mathbf{x}}_{k+1}-\widetilde{\mathbf{x}}_k}{\Delta t}-f(\widetilde{\mathbf{x}}_k)\right)\right). \tag{52}$$

The path integral representation of $\Psi_0(\widetilde{\mathbf{x}}_0,t_0)$ is obtained from equations (49-52), and can be written as an exponential of an "action" [14] $S(\chi_N|\widetilde{\mathbf{x}}_0)$ along the time-discretized sample trajectory $(\widetilde{\mathbf{x}}_1,\ldots,\widetilde{\mathbf{x}}_N)$:

$$\Psi_0(\widetilde{\mathbf{x}}_0,t_0) = \frac{1}{|2\pi\Sigma\Delta t|^{N/2}}\lim_{\Delta t\to 0}\int d\chi_N \exp\left(-S(\chi_N|\widetilde{\mathbf{x}}_0)\right) \tag{53}$$

$$\begin{aligned} S(\widetilde{\mathbf{x}}_1,\ldots,\widetilde{\mathbf{x}}_N|\widetilde{\mathbf{x}}_0) &= \sum_{k=1}^{N}\frac{\Delta t}{2\lambda}k(\widetilde{\mathbf{x}}_k) \\ &+\sum_{k=0}^{N-1}\frac{1}{2\Delta t}\left(\widetilde{\mathbf{x}}_{k+1}-\widetilde{\mathbf{x}}_k-\Delta t f(\widetilde{\mathbf{x}}_k)\right)^T \\ &\times\Sigma^{-1}\left(\widetilde{\mathbf{x}}_{k+1}-\widetilde{\mathbf{x}}_k-\Delta t f(\widetilde{\mathbf{x}}_k)\right). \end{aligned} \tag{54}$$

From (31), (33), the optimal control is given by

$$\begin{aligned} \mathbf{u}(\widetilde{\mathbf{x}}_0,t_0,i) &= \lim_{\Delta t\to 0}\lambda R^{-1}B_i^T\partial_{\widetilde{\mathbf{x}}_0}\log\Psi_0 \\ &= \lim_{\Delta t\to 0}\int d\chi_N P(\chi_N|\widetilde{\mathbf{x}}_0)\mathbf{u}_L(\chi_N|\widetilde{\mathbf{x}}_0,i) \\ &= \lim_{\Delta t\to 0}\mathbb{E}_{P(\chi_N|\widetilde{\mathbf{x}}_0)}\left\{\mathbf{u}_L(\chi_N|\widetilde{\mathbf{x}}_0,i)\right\} \\ &= \mathbb{E}_{P(\chi|\widetilde{\mathbf{x}}_0)}\left\{\mathbf{u}_L(\chi|\widetilde{\mathbf{x}}_0,i)\right\} \end{aligned} \tag{55}$$

where $\lim_{\Delta t\to 0}P(\chi_N|\widetilde{\mathbf{x}}_0) = P(\chi|\widetilde{\mathbf{x}}_0)$ is the probability of an *optimal* trajectory:

$$P(\chi_N|\widetilde{\mathbf{x}}_0) \propto e^{-S(\chi_N|\widetilde{\mathbf{x}}_0)}, \tag{56}$$

and the local controls $\mathbf{u}_L(\chi_N|\widetilde{\mathbf{x}}_0)$ are

$$\mathbf{u}_L(\chi_N|\widetilde{\mathbf{x}}_0,i) = \frac{1}{M}B_i\frac{\widetilde{\mathbf{x}}_1-\widetilde{\mathbf{x}}_0}{\Delta t}, \tag{57}$$

where the B_i selects a pair (γ_m,κ_m) based on the corresponding value of $\xi(t/\varepsilon)$, as in (20). Then (55) is

$$\mathbf{u}(\mathbf{x},i) = \frac{1}{M}B_i\frac{\mathbb{E}_{P(\chi_N|\widetilde{\mathbf{x}}_0)}\{\widetilde{\mathbf{x}}_1\}-\mathbf{x}}{\Delta t}. \tag{58}$$

In the formulation, after computing $\mathbf{u}(\mathbf{x}(t),t) = \mathbf{u}(\widetilde{\mathbf{x}}_0,t_0)$, each agent executes only the first increment of that control, at which point the optimal control is recomputed for the next time horizon $[t_0,t_f] = [t,t+t_f]$. In other words, after computing $\mathbb{E}_{P(\chi_N|\widetilde{\mathbf{x}}_0)}\{\widetilde{\mathbf{x}}_1\}$, the AiF applies the control $\mathbf{u}$ as chosen by the B_i in (58) to its

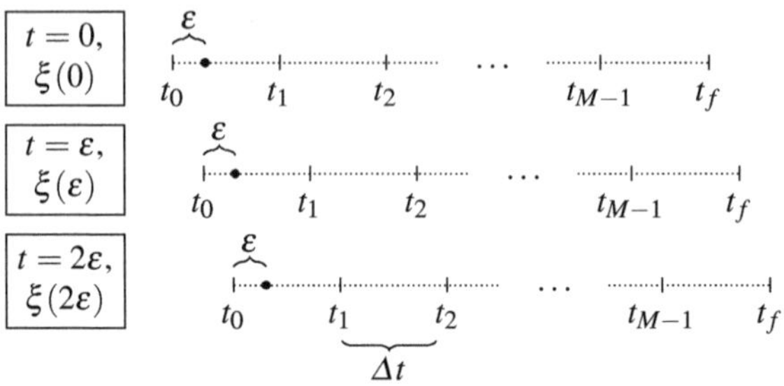

Fig. 2 Receding-horizon timing. At $t = 0$, the AiF computes $\mathbb{E}_{P(\chi_N|\widetilde{\mathbf{x}}_0)}\{\widetilde{\mathbf{x}}_1\}$, and randomly chooses a value for ξ. The control applied is the ξth component of (58). In the next time step ε, the process repeats.

heading angle and speed. Then the problem repeats with a random selection of ξ (see Fig. 2).

The result of this section is a PDF (49) of the system trajectories, including their costs, that is marginalized over each infinitesimal temporal increment of the process under consideration. The optimal control (58) applied by the AiF in state $\mathbf{x}$ is estimated from this trajectory PDF once the joint probability $P(\chi_N|\widetilde{\mathbf{x}}_0)$ has been computed, a nontrivial task to be discussed in the following section.

4 Computing the Control with Kalman Smoothers

In this section we present our approach that estimates the hidden state of a nonlinear stochastic process, which corresponds to the maximally-likely trajectory under state and control costs, using appropriately-chosen noisy measurements.

In general, the marginalization (49) is difficult to evaluate. If one were able to sample K trajectories from the distribution $P(\chi_N|\widetilde{\mathbf{x}}_0)$, approximation of the posterior distribution, i.e., the optimal control (55), would only require a quick calculation:

$$\mathbf{u}(\mathbf{x}) = \frac{1}{K}\sum_{i=1}^{K}\mathbf{u}_L^{(i)}(\widetilde{\mathbf{x}}_1,\ldots,\widetilde{\mathbf{x}}_N|\widetilde{\mathbf{x}}_0),$$

Along these lines, previous works based on the path integral approach to stochastic optimal control use Markov Chain Monte Carlo (MCMC) techniques [12] to sample from $P(\widetilde{\mathbf{x}}_1,\ldots,\widetilde{\mathbf{x}}_N|\widetilde{\mathbf{x}}_0)$. Although MCMC techniques can be used to generate samples of the maximally-likely trajectory, we find them to be slow in practice due to the high dimension of this problem ($\chi_N \in \mathbb{R}^{6NM}$).

Moreover, is is not necessary to sample the entire distribution $P(\chi_N|\widetilde{\mathbf{x}}_0)$ since only the value of $\widehat{\mathbf{x}}_1 \equiv \mathbb{E}_{P(\chi_N|\widetilde{\mathbf{x}}_0)}\{\widetilde{\mathbf{x}}_1\}$ is needed. Note that this estimate is the first infinitesimal increment of the optimal trajectory χ_N^*:

$$\chi_N^* = \underset{\chi_N}{\text{argmax}} \{P(\chi_N|\mu, \widetilde{\mathbf{x}}_0)\}$$
$$\propto \underset{\chi_N}{\text{argmax}} \{P(\mu|\chi_N)P(\chi_N|\widetilde{\mathbf{x}}_0)\}.$$

In this work, we treat the temporal discretization of the optimal trajectory χ_N^* as the hidden state of a stochastic process, where measurements are related to the system goal μ (24). The optimal control can then be computed from the optimal estimate $\widehat{\mathbf{x}}_1$ of the expected value ${}_{P(\chi_N|\widetilde{\mathbf{x}}_0)}\{\widetilde{\mathbf{x}}_1\}$ given the process and measurements over a fixed interval $t_1, \ldots, t_N$. Therefore, we define the following nonlinear smoothing problem.

Nonlinear Smoothing Problem:
Given measurements $\mathbf{y}_k = \mathbf{y}(t_k)$ for $t_k = t_1, \ldots, t_N = t_f$, where $t_{k+1} - t_k = \Delta t$, compute the estimate $\widehat{\mathbf{x}}_{1:N}$ of the hidden state $\widetilde{\mathbf{x}}_{1:N}$ from the nonlinear state-space model:

$$\widetilde{\mathbf{x}}_{k+1} = \widetilde{\mathbf{x}}_k + \Delta t f(\widetilde{\mathbf{x}}_k) + \varepsilon_k \tag{59}$$
$$\mathbf{y}_k = h(\widetilde{\mathbf{x}}_k) + \eta_k, \tag{60}$$

where $f(\cdot)$ and $h(\cdot)$ are as in Section 2, and ε_k and η_k are independent multivariate Gaussian random variables with zero mean and covariances:

$$\mathbb{E}\left(\varepsilon_k \varepsilon_k^T\right) = \Delta t \Sigma \tag{61}$$
$$\mathbb{E}\left(\eta_k \eta_k^T\right) = \frac{\lambda}{\Delta t} A^{-1}. \tag{62}$$

The smoothing is initialized from $\widetilde{\mathbf{x}}_0 = \mathbf{x}$, the current state of the system as viewed by the AiF. Measurements $\mathbf{y}_k$ are always exactly $\mathbf{y}_k = \mu$. If μ is expected to change over time (e.g., a dynamic formation), then $\mathbf{y}_k = \mu_k$.

Note that only the estimate $\widehat{\mathbf{x}}_1$ is needed to compute the control. The measurement noise in the estimation problem (62) is related to the instantaneous state costs (24), and the process noise (61) is related to the instantaneous control costs (25).

To show that the estimation of $\hat{\mathbf{x}}_1$ can be computed based on the nonlinear smoothing, we write the probability of an estimated hidden state $\widehat{\mathbf{x}}_k$ in the filtering algorithm predication/update steps [13], which is proportional to the measurement likelihood $p(\mathbf{y}_k|\widehat{\mathbf{x}}_k)$ and the predicted state $p(\widehat{\mathbf{x}}_k|\widehat{\mathbf{x}}_{k-1})$:

$$p(\widehat{\mathbf{x}}_k|\widehat{\mathbf{x}}_{k-1}) \propto p(\mathbf{y}_k|\widehat{\mathbf{x}}_k) p(\widehat{\mathbf{x}}_k|\widehat{\mathbf{x}}_{k-1}),$$

where

$$p(\mathbf{y}_k|\widehat{\mathbf{x}}_k) \equiv p(\mu_k|\widehat{\mathbf{x}}_k) = \mathcal{N}\left(h(\widehat{\mathbf{x}}_k), \eta_k \eta_k^T\right)$$
$$\propto \exp\left\{-\frac{\Delta t}{2\lambda}(h(\widehat{\mathbf{x}}_k) - \mu)^T A (h(\widehat{\mathbf{x}}_k) - \mu)\right\}$$
$$= \exp\left\{-\frac{\Delta t}{2\lambda} k(\widehat{\mathbf{x}}_k)\right\} \tag{63}$$

and

$$p(\widehat{\mathbf{x}}_k|\widehat{\mathbf{x}}_{k-1}) = \mathscr{N}\left(\widehat{\mathbf{x}}_{k-1} + \Delta t f(\widehat{\mathbf{x}}_{k-1}), \Delta t \Sigma\right)$$
$$\propto \exp\left\{-\frac{1}{2\Delta t}\left(\widehat{\mathbf{x}}_k - \widehat{\mathbf{x}}_{k-1} - \Delta t f(\widehat{\mathbf{x}}_{k-1})\right)^T \right.$$
$$\left. \times \Sigma^{-1}\left(\widehat{\mathbf{x}}_k - \widehat{\mathbf{x}}_{k-1} - \Delta t f(\widehat{\mathbf{x}}_{k-1})\right)\right\}. \quad (64)$$

Comparing the right hand sides of (63-64) with (54), it can be seen that the problm of estimating $\hat{\mathbf{x}}_1$ is equivalent to the estimation of $\hat{\mathbf{x}}_1$ in the smoothing problem.

The most-likely trajectory originating from state $\widehat{\mathbf{x}}_0$, that is, the hidden states $\widehat{\mathbf{x}}_k$, $k = 1,\ldots,N$, can be found by filtering and then smoothing the process given the measurements μ_k using a nonlinear smoother, such as an Extended Kalman RTS Smoother (EKF-RTS) or Unscented Kalman RTS Smoother (UKF-RTS) [34]. A nonlinear Kalman smoothing algorithm assumes that the increments given by (63) and (64) are to some extent Gaussian, but the algorithm is sufficiently fast to be applied in *real-time* by each unicycle in a potentially large group with an even larger state space, motivating its use in this work.

The control to be applied in the current state $\mathbf{x}$ is given by (58), using the $\widehat{\mathbf{x}}_1$ estimated by the smoother. After this increment, the process repeats. When the smoothing is complete and agents have applied their computed control, each agent must then observe the actual states of its neighbors so that the next iteration begins with the correct initial condition. In practice, the controller/smoother must be capable of efficiently filtering and smoothing over the horizon $\left[t_0, t_f\right]$. The computational complexity of such a smoother is analyzed in [34].

The effect of scaling parameter λ becomes clear in in the dual estimation formulation. For $\lambda \gg 1$, the measurement noise is large, and the smoother will place more weight on its predictions. Consequently, the passive components of the system $f(\cdot)$ will dominate, and less control will be applied. Similarly, for $\lambda \ll 1$, the smoother will trust the measurements, and a greater amount of control will be applied. The net effect is that λ decides the fraction of the process noise in the original control problem that is propagated into the estimation problem.

Recall that the spatial boundary condition (43) constrains the speeds of the agents within upper and lower limits, i.e. to remain outside the set $\mathscr{V}$. In the context of the smoothing problem, this requires that the probability of a filter prediction, measurement update, or smoothing update to be zero if the estimate enters the boundary $\mathscr{V}$. To deal with such a problem, the smoothing algorithm should be capable of handling inequality constraints. Several algorithms of this type exist (see [2], for example), but in order to keep computation time at a minimum, we instead employ a more straightforward approach. After each prediction step, if the current estimate $\widehat{\mathbf{x}}_k$ is in violation of the constraints, the estimate is projected in a least-squares sense to lie inside the contraint boundaries using Matlab's `lsqlin`. The same method is also applied if the estimate violates the speed constraints during any of the update or smoothing steps.

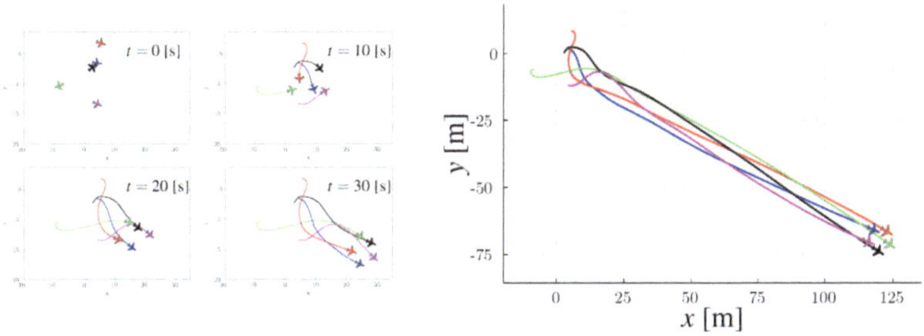

Fig. 3 Five agents, starting from random initial positions and a common speed $v = 2.5$ [m/s], must achieve a regular pentagon formation by an individually-optimal choice of acceleration and turning rate, without any active communication

5 Results

In this section, we apply the methods to a formation control problem in which five agents achieve the formation of a regular pentagon. Each agent is individually estimating the hidden optimal trajectory based on the relative kinematics of all of its neighbors. The instantaneous state cost (24) penalizes the mean squared distance from the unicycle to all of its $M = 4$ neighbors in excess of the side length of the pentagon (5 [m]) or the diagonal of the pentagon, depending on the relative configuration of the pentagon encoded in μ.

The system and control algorithm parameters were chosen as $\lambda = 1$, $\sigma_\theta = \sigma_v = 0.1$, $t_f = 30$ s, $A = 0.1 I_{4\times4}$, $v_{LB} = 1$ [m/s], $v_{UB} = 3$ [m/s], and $\Delta t = \Delta\varepsilon = 0.1$ s. The control was computed from the result of a Discrete-time Unscented Kalman Rauch-Tung-Striebel Smoother [34]. Fig. 3 shows the trajectories of all agents, while the the inter-agent distances can be seen in Fig. 4. With an initial speed of 2.5 [m/s], the agents never hit their limiting speeds v_{LB} or v_{UB}. Once the pentagon has formed, the agents' heading angles are not equal, and the formation rotates. Without

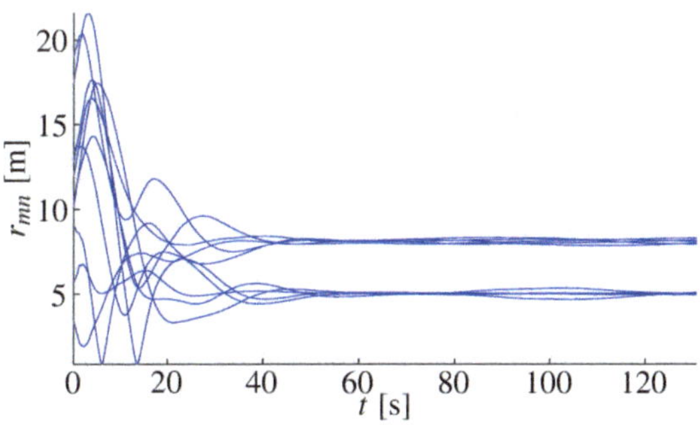

Fig. 4 Inter-agent distance r_{mn} as a function of time. The two radii correspond to the objective pentagon side length (5 [m]) and the pentagon's diagonal ($\frac{5}{2}(1+\sqrt{5}) \approx 8.1$ [m]). The pentagon continues to rotate after forming.

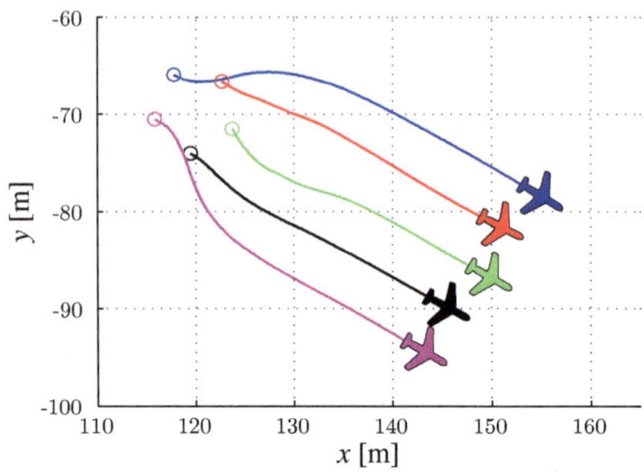

Fig. 5 Transition from a pentagon to a line. The initial condition is the final frame of Fig. 3.

a goal of alignment among the agents, each agent is anticipating the pentagon to rotate (in expected value), and it computes its control so as to maintain its nominal distances in the rotating pentagon.

A dynamic formation was then created by modifying the nominal distances μ during simulation. In Fig. 5, after the pentagon had formed, the formation distances were redefined so that the formation morphed into a line.

6 Discussion

This work considers the problem of unicycle formation control in a distributed optimal feedback control setting. Since this gives rise to a system with a high dimensional state space, we exploit the stochasticity inherent in distributed multi-agent control problems and apply the path integral approach in order to compute the control. The uncertainty in turning rates and accelerations of an agent's neighbors are modeled as stochastic processes, and a fast switching kinematic model links this stochasticity to an agent's control, allowing the optimal control problem to be framed as an estimation problem.

Each agent computes its optimal control in real-time by applying a nonlinear Kalman smoothing algorithm. The measurement noise and process noise of the smoothing problem are created using the structure of the cost function and stochastic kinematics. Aside from mutual observations among agents, the formation is created and maintained without any communication among them.

A number of other goals, e.g., alignment of heading angle, are possible through a simple change in the cost function. More complex aspects of multi-agent formation control, such as collision avoidance and dynamic communication networks, for example, could be handled by a robust smoothing algorithm and will be explored in future research.

Acknowledgements. This work was supported by NSF GRFP 0809125.

References

1. Anderson, B., Fidan, B., Yu, C., Walle, D.: UAV formation control: theory and application. In: Blondel, V., Boyd, S., Kimura, H. (eds.) Recent Advances in Learning and Control, pp. 15–34. Springer (2008)
2. Bell, B.M., Burke, J.V., Pillonetto, G.: An inequality constrained nonlinear Kalman-Bucy smoother by interior point likelihood maximization. Automatica 45(1), 25–33 (2009)
3. Bellingham, J.G., Rajan, K.: Robotics in remote and hostile environments. Science 318(5853), 1098–1102 (2007)
4. Bertsekas, D.P., Tsitsiklis, J.N.: Neuro-dynamic programming. Athena Scientific, Belmont (1996)
5. van den Broek, B., Wiegerinck, W., Kappen, B.: Graphical model inference in optimal control of stochastic multi-agent systems. Journal of Artificial Intelligence Research 32(1), 95–122 (2008)
6. van den Broek, B., Wiegerinck, W., Kappen, B.: Optimal Control in Large Stochastic Multi-agent Systems. In: Tuyls, K., Nowe, A., Guessoum, Z., Kudenko, D. (eds.) Adaptive Agents and MAS III. LNCS (LNAI), vol. 4865, pp. 15–26. Springer, Heidelberg (2008)
7. Bullo, F., Cortes, J., Martinez, S.: Distributed control of robotic networks: A mathematical approach to motion coordination algorithms. Princeton University Press, Princeton (2009)
8. Dimarogonas, D.: On the rendezvous problem for multiple nonholonomic agents. IEEE Transactions on Automatic Control 52(5), 916–922 (2007)
9. Elkaim, G., Kelbley, R.: A Lightweight Formation Control Methodology for a Swarm of Non-Holonomic Vehicles. In: IEEE Aerospace Conference. IEEE, Big Sky (2006)
10. Fleming, W., Soner, H.: Logarithmic Transformations and Risk Sensitivity. In: Controlled Markov Processes and Viscosity Solutions, ch.6. Springer, Berlin (1993)
11. Freidlin, M., Wentzell, A.: Random Perturbations of Dynamical Systems. Springer (1984)
12. Gamerman, D.: Markov Chain Monte Carlo: Stochastic Simulation for Bayesian Inference. Chapman and Hall (1997)
13. Gelb, A.: Applied Optimal Estimation. The M.I.T. Press, Cambridge (1974)
14. Goldstein, H.: Classical Mechanics, 2nd edn. Addison-Wesley (1980)
15. van Kampen, N.G.: Stochastic Processes in Physics and Chemistry, 3rd edn. North-Holland (2007)
16. Kappen, H.: Linear Theory for Control of Nonlinear Stochastic Systems. Physical Review Letters 95(20), 1–4 (2005)
17. Kappen, H.: Path integrals and symmetry breaking for optimal control theory. Journal of Statistical Mechanics: Theory and Experiment 2005, P11,011 (2005)
18. Kappen, H., Wiegerinck, W., van den Broek, B.: A path integral approach to agent planning. In: Autonomous Agents and Multi-Agent Systems, Citeseer (2007)
19. Kappen, H.J.: Path integrals and symmetry breaking for optimal control theory. Journal of Statistical Mechanics, Theory and Experiment 2005, 21 (2005)
20. Kappen, H.J., Gómez, V., Opper, M.: Optimal control as a graphical model inference problem. Machine Learning 87(2), 159–182 (2012)
21. Khas'minskii, R.: On the principle of averaging the Itô's stochastic differential equations. Kybernetika 4(3), 260–279 (1968)
22. Kumar, V., Rus, D., Sukhatme, G.S.: Networked Robotics. In: Sciliano, B., Khatib, O. (eds.) Springer Handbook of Robotics. ch. 41, pp. 943–958 (2008)

23. Kushner, H.J., Dupuis, P.: Numerical Methods for Stochastic Control Problems in Continuous Time, 2nd edn. Springer (2001)
24. Milutinović, D.: Utilizing Stochastic Processes for Computing Distributions of Large-Size Robot Population Optimal Centralized Control. In: Martinoli, A., Mondada, F., Correll, N., Mermoud, G., Egerstedt, M., Hsieh, M.A., Parker, L.E., Støy, K. (eds.) Distributed Autonomous Robotic Systems. STAR, vol. 83, pp. 359–372. Springer, Heidelberg (2013)
25. Milutinović, D., Garg, D.P.: Stochastic model-based control of multi-robot systems. Tech. Rep. 0704, Duke University, Durham, NC (2009)
26. Milutinović, D., Garg, D.P.: A sampling approach to modeling and control of a large-size robot population. In: Proceedings of the 2010 ASME Dynamic Systems and Control Conference. ASME, Boston (2010)
27. Oksendal, B.: Stochastic Differential Equations: An Introduction with Applications, 6th edn. Springer, Berlin (2003)
28. Palmer, A., Milutinović, D.: A Hamiltonian Approach Using Partial Differential Equations for Open-Loop Stochastic Optimal Control. In: Proceedings of the 2011 American Control Conference, San Francisco, CA (2011)
29. Parker, L.E.: Multiple Mobile Robot Systems. In: Sciliano, B., Khatib, O. (eds.) Springer Handbook of Robotics, ch.40, pp. 921–941. Springer (2008)
30. Powell, W.B.: Approximate Dynamic Programming: Solving the Curses of Dimensionality. Wiley Interscience, Hoboken (2007)
31. Ren, W., Beard, R.: Distributed consensus in multi-vehicle cooperative control: Theory and applications. Springer, New York (2007)
32. Ren, W., Beard, R., Atkins, E.: A survey of consensus problems in multi-agent coordination. In: Proceedings of the 2005, American Control Conference, pp. 1859–1864 (2005)
33. Ryan, A., Zennaro, M., Howell, A., Sengupta, R., Hedrick, J.: An overview of emerging results in cooperative UAV control. In: 2004 43rd IEEE Conference on Decision and Control, vol. 1, pp. 602–607 (2004)
34. Särkkä, S.: Continuous-time and continuous-discrete-time unscented Rauch-Tung-Striebel smoothers. Signal Processing 90(1), 225–235 (2010)
35. Singh, A., Batalin, M., Stealey, M., Chen, V., Hansen, M., Harmon, T., Sukhatme, G., Kaiser, W.: Mobile robot sensing for environmental applications. In: Field and Service Robotics, pp. 125–135. Springer (2008)
36. Sutton, R., Barto, A.: Reinforcement Learning: An Introduction. MIT Press, Cambridge (1998)
37. Theodorou, E., Buchli, J., Schaal, S.: A Generalized Path Integral Control Approach to Reinforcement Learning. The Journal of Machine Learning Research 11, 3137–3181 (2010)
38. Theodorou, E., Buchli, J., Schaal, S.: Reinforcement learning of motor skills in high dimensions: A path integral approach. In: 2010 IEEE International Conference on Robotics and Automation (ICRA), vol. 4, pp. 2397–2403. IEEE (2010)
39. Todorov, E.: General duality between optimal control and estimation. In: 47th IEEE Conference on Decision and Control, vol. 5, pp. 4286–4292. IEEE, Cancun (2008)
40. Wang, M.C., Uhlenbeck, G.: On the theory of Brownian Motion II. Reviews of Modern Physics 17(2-3), 323–342 (1945)
41. Wiegerinck, W., van den Broek, B., Kappen, B.: Stochastic optimal control in continuous space-time multi-agent systems. In: Proceedings UAI. Citeseer (2006)
42. Wiegerinck, W., van den Broek, B., Kappen, B.: Optimal on-line scheduling in stochastic multiagent systems in continuous space-time. In: Proceedings of the 6th International Joint Conference on Autonomous Agents and Multiagent Systems, p. 1 (2007)

43. Yin, G.G., Zhu, C.: Hybrid Switching Diffusions. Springer, New York (2010)
44. Yong, J.: Relations among ODEs, PDEs, FSDEs, BDSEs, and FBSDEs. In: Proceedings of the 36th IEEE Conference on Decision and Control, pp. 2779–2784. IEEE, San Diego (1997)

Appendix 1

Here we derive the switching model (15)-(18) from the original model (9)-(14). First, note that the relative angle and relative speed satisfy

$$\gamma_m(t) = (\theta(0) - \theta_m(0)) + \int_0^t \omega(\mathbf{x}(s))ds - \int_0^t \sigma_\theta dw_{m,\theta} \tag{65}$$

$$\kappa_m(t) = (v(0) - v_m(0)) + \int_0^t u(\mathbf{x}(s))ds - \int_0^t \sigma_v dw_{m,v}, \tag{66}$$

from which we may obtain

$$\mathbb{E}(\gamma_m) = \theta(0) - \theta_m(0) + \int_0^t \omega dt \tag{67}$$

$$\mathbb{E}(\gamma_m) - \gamma_m = \int \sigma_\theta dw_{m,\theta} \tag{68}$$

$$\mathbb{E}(\kappa_m) = v(0) - \kappa_m(0) + \int_0^t u dt \tag{69}$$

$$\mathbb{E}(\kappa_m) - \kappa_m = \int \sigma_v dw_{m,v}. \tag{70}$$

Then the heading angles and speeds of the AiF and its neighbor m can both be encoded into γ_m and κ_m by the relations:

$$\theta(t) = \mathbb{E}\left\{\gamma_m(t) + \theta_m(0) + \int_0^t \sigma_\theta dw_{m,\theta}\right\} = \mathbb{E}(\gamma_m(t)) + \theta_m(0) \tag{71}$$

$$\theta_m(t) = -(\gamma_m - \mathbb{E}(\gamma_m(t))) + \theta_m(0) \tag{72}$$

$$v(t) = \mathbb{E}\left\{\kappa_m(t) + v_m(0) + \int_0^t \sigma_\theta dw_{m,v}\right\} = \mathbb{E}(\kappa_m(t)) + v_m(0) \tag{73}$$

$$v_m(t) = -(\kappa_m - \mathbb{E}(\kappa_m(t))) + v_m(0). \tag{74}$$

We assume that only one pair (γ_m, κ_m) evolves at a time. Introducing $\delta_{\xi(t/\varepsilon),m}$ as the Kronecker delta selecting the evolution of the pair m, we would have that

$$d\gamma_m(t) = (\omega dt - \sigma_\theta dw_{\theta,m})\delta_{\xi(t/\varepsilon),m} \tag{75}$$

$$d\kappa_m(t) = (udt - \sigma_v dw_m)\delta_{\xi(t/\varepsilon),m}. \tag{76}$$

However, we wish for the average evolution of the states to be the same as in the original problem formulation. Since each pair m is selected with frequency M^{-1}, we write the evolution of these relative states as

$$d\gamma_m(t) = \left(M\omega dt - \sqrt{M}\sigma_\theta dw_{\theta,m}\right)\delta_{\xi(t/\varepsilon),m} \tag{77}$$

$$d\kappa_m(t) = \left(Mudt - \sqrt{M}\sigma_v dw_m\right)\delta_{\xi(t/\varepsilon),m}. \tag{78}$$

Next, we substitute (72) and (74) into $\theta_m(t)$ and $v_m(t)$, respectively, in the kinematic model for $r_m(t)$ and $\varphi_m(t)$. Finally, we also substitute the averages for $\theta(t)$ and $v(t)$:

$$\theta(t) = \frac{1}{M}\sum_{j=1}^{M}\left(\mathbb{E}(\gamma_j(t)) + \theta_m(0)\right) = \frac{1}{M}\sum_{j=1}^{M}\left(\mathbb{E}(\gamma_j(t))\right) + \overline{\theta_m}(0) \tag{79}$$

$$v(t) = \frac{1}{M}\sum_{j=1}^{M}\left(\mathbb{E}(\kappa_j(t)) + v_m(0)\right) = \frac{1}{M}\sum_{j=1}^{M}\left(\mathbb{E}(\kappa_j(t))\right) + \overline{v_m}(0). \tag{80}$$

Appendix 2

Here we develop a first approximation to (37), reproduced here:

$$\partial_t \Psi_i = \frac{1}{2\lambda}k(x)\Psi_i - f^T\partial_{\mathbf{x}}\Psi_i - \frac{1}{2}\mathrm{Tr}\left(\Gamma_i\Gamma_i^T\partial_{\mathbf{x}}^2\Psi_i\right) - \Psi_i\frac{Q(t)}{\varepsilon}\log\Psi(\mathbf{x},t)(i).$$

This derivation follows closely to that in Chapter 11 of [43].

We seek to find an approximation to $\Psi_i(\mathbf{x},t)$, and begin with an asymptotic expansion to $J_i(\mathbf{x},t)$ of the form

$$J_i(\mathbf{x},t) = A_0(\mathbf{x},t,i) + \varepsilon A_1(\mathbf{x},t,i) + B_0(\mathbf{x},\tau,i) + \varepsilon B_1(\mathbf{x},\tau,i), \qquad i = 1,\ldots,M$$

where $\tau = (t_f - t)/\varepsilon$ is a stretched-time variable, the $A_k(\cdot)$'s are outer expansion terms, and $B_k(\cdot)$'s are terminal layer correction terms. The expansion terms are matched at terminal condition (28) with

$$A_0(\mathbf{x},t_f,i) + B_0(\mathbf{x},0,i) = \phi(\mathbf{x}) = 0 \tag{81}$$

$$A_1(\mathbf{x},t_f,i) + B_1(\mathbf{x},0,i) = 0. \qquad i = 1,\ldots,M \tag{82}$$

From (33), define the transformed expansion terms as

$$a_k(\mathbf{x},t,i) = \exp(-A_k(\mathbf{x},t,i)/\lambda) \tag{83}$$

$$b_k(\mathbf{x},\tau,i) = \exp(-B_k(\mathbf{x},\tau,i)/\lambda), \qquad i = 1,\ldots,M,\ k = 0,1. \tag{84}$$

Substituting the outer expansion terms $a_k(\cdot)$ into (37) and collecting terms by powers of ε, we obtain

$$\varepsilon^0 \ : \ Q(t)a_0(\mathbf{x},t,\cdot)(i) = 0 \tag{85}$$

$$\begin{aligned}\varepsilon^1 \ : \ \partial_t a_0(\mathbf{x},t,i) &= \frac{1}{2\lambda}k(\mathbf{x})a_0(\mathbf{x},t,i) - f^T\partial_{\mathbf{x}}a_0(\mathbf{x},t,i) - \frac{1}{2}\mathrm{Tr}\left(\Gamma_i\Gamma_i^T\partial_{\mathbf{x}}^2 a_0(\mathbf{x},t,i)\right)\\ &- a_0(\mathbf{x},t,i)Q(t)\log a_1(\mathbf{x},t,\cdot)(i).\end{aligned} \tag{86}$$

Writing $a_0(\mathbf{x},t) = [a_0(\mathbf{x},t,1),\ldots,a_0(\mathbf{x},t,M)]^T$, we have from (85) that

$$Q(t)a_0(\mathbf{x},t) = 0.$$

Then from (22)-(23), the rank of $Q(t)$ is $M-1$, implying that the null-space of $Q(t)$ is one dimensional and spanned by a vector of all ones, $\mathbb{1} = [1,\ldots,1]^T$. Then $a_0(\mathbf{x},t)$ must be independent of i, and so

$$a_0(\mathbf{x},t) = \Psi_0(\mathbf{x},t)\mathbb{1}. \tag{87}$$

Note that this condition on $Q(t)$ further implies the existence of a quasi-stationary distribution [43] $\nu(t) = [\nu_1(t),\ldots,\nu_M(t)]$ with the properties that $\sum_{i=1}^M \nu_i = 1$ and $\nu(t)Q(t) = 0$. Substituting (87) into (86), left multiplying by ν_i, and summing over i gives

$$\begin{aligned}\sum_{i=1}^M \nu_i\partial_t\Psi_0(\mathbf{x},t) &= \sum_{i=1}^M \nu_i\frac{1}{2\lambda}k(\mathbf{x})\Psi_0(\mathbf{x},t) - \sum_{i=1}^M \nu_i f^T\partial_{\mathbf{x}}\Psi_0(\mathbf{x},t)\\ &- \sum_{i=1}^M \nu_i\frac{1}{2}\mathrm{Tr}\left(\Gamma_i\Gamma_i^T\partial_{\mathbf{x}}^2\Psi_0(\mathbf{x},t)\right) - \Psi_0(\mathbf{x},t)\sum_{i=1}^M \nu_i Q(t)\log a_1(\mathbf{x},t,\cdot)(i).\end{aligned} \tag{88}$$

The properties of $\nu(t)$ cause the last term to drop out, and the ν_i's in the first three sums add to one.

$$\partial_t\Psi_0(\mathbf{x},t) = \frac{1}{2\lambda}k(\mathbf{x})\Psi_0(\mathbf{x},t) - f^T\partial_{\mathbf{x}}\Psi_0(\mathbf{x},t) - \sum_{i=1}^M \nu_i\frac{1}{2}\mathrm{Tr}\left(\Gamma_i\Gamma_i^T\partial_{\mathbf{x}}^2\Psi_0(\mathbf{x},t)\right). \tag{89}$$

Next, since Γ_i selects the ith pair of $(\gamma_m^{\pm},\kappa_m^{\pm})$ and multiplies them by $\sqrt{M}$, the remaining sum represents a consolidation of the diffusion terms associated with each of the pairs $(\gamma_m^{\pm},\kappa_m^{\pm})$. Then in light of the chosen symmetry of $Q(t)$ (22)-(23), this sum reduces to an average diffusion with covariance

$$\Sigma = \frac{1}{M}\sum_{i=1}^M \Gamma_i\Gamma_i^T,$$

which is (41), and $\Psi_0(\mathbf{x},t)$ satisfies

$$\partial_t \Psi_0(\mathbf{x},t) = \frac{1}{2\lambda} k(\mathbf{x}) \Psi_0(\mathbf{x},t) - f^T \partial_{\mathbf{x}} \Psi_0(\mathbf{x},t) - \frac{1}{2} \mathrm{Tr}\left(\Sigma \partial_{\mathbf{x}}^2 \Psi_0(\mathbf{x},t)\right). \tag{90}$$

which is (37), and with terminal condition $\Psi_0(\mathbf{x},t_f) = \exp(\phi(\mathbf{x})) = 1$, which is (42).

Next we consider the terminal correction terms $b_k(\mathbf{x},\tau,i)$. Rewriting the original PDE in the timescale of τ,

$$-\frac{1}{\varepsilon} \partial_\tau \Psi_i = \frac{1}{2\lambda} k(x) \Psi_i - f^T \partial_{\mathbf{x}} \Psi_i - \frac{1}{2} \mathrm{Tr}\left(\Gamma_i \Gamma_i^T \partial_{\mathbf{x}}^2 \Psi_i\right) - \Psi_i \frac{Q(t_f - \varepsilon\tau)}{\varepsilon} \log \Psi(\mathbf{x},t)(i), \quad i = 1,\ldots,M \tag{91}$$

and expanding $Q(\cdot)$ around t_f,

$$Q(t_f - \varepsilon\tau) \approx Q(t_f) - (\varepsilon\tau)\, Q'(t)\big|_{t=t_f}, \tag{92}$$

we can obtain, using the same method as before,

$$\partial_\tau b_0(\mathbf{x},\tau,i) = b_0(\mathbf{x},\tau,i) Q(t_f) \log b_0(\mathbf{x},\tau,\cdot)(i). \tag{93}$$

From (81) and (42), this implies that $b_0(\mathbf{x},\tau,i) = 1$ for all time and states i. We do not derive asymptotic error bounds here.

M*: A Complete Multirobot Path Planning Algorithm with Optimality Bounds

Glenn Wagner and Howie Choset

Abstract. Multirobot path planning is difficult because the full configuration space of the system grows exponentially with the number of robots. Planning in the joint configuration space of a set of robots is only necessary if they are strongly coupled, which is often not true if the robots are well separated in the workspace. Therefore, we initially plan for each robot separately, and only couple sets of robots after they have been found to interact, thus minimizing the dimensionality of the search space. We present a general strategy called *subdimensional expansion*, which dynamically generates low dimensional search spaces embedded in the full configuration space. We also present an implementation of subdimensional expansion for robot configuration spaces that can be represented as a graph, called M*, and show that M* is complete and finds minimal cost paths.

1 Introduction

Multirobot systems are attractive for surveillance, search and rescue, and warehouse automation applications. Unfortunately, the flexibility and redundancy that make multirobot systems appealing also make assigning robots to tasks and planning collision free paths to perform those tasks difficult. In this work, we describe a novel method, called *subdimensional expansion*, for efficiently generating collision free paths for multiple robots [17]. Subdimensional expansion initially assumes that a path can be found for each robot to the goal in the robots' individual configuration

Glenn Wagner
The Robotics Institute, Carnegie Mellon University, Newell Simon Hall 4221,
5000 Forbes Ave, Pittsburgh, PA 15213
e-mail: gswagner@andrew.cmu.edu

Howie Choset
The Robotics Institute, Carnegie Mellon University, Newell Simon Hall 3205,
5000 Forbes Ave, Pittsburgh, PA 15213
e-mail: choset@cs.cmu.edu

D. Milutinović & J. Rosen (Eds.): Redundancy in Robot Manipulators, LNEE 57, pp. 167–181.
DOI: 10.1007/978-3-642-33971-4_10

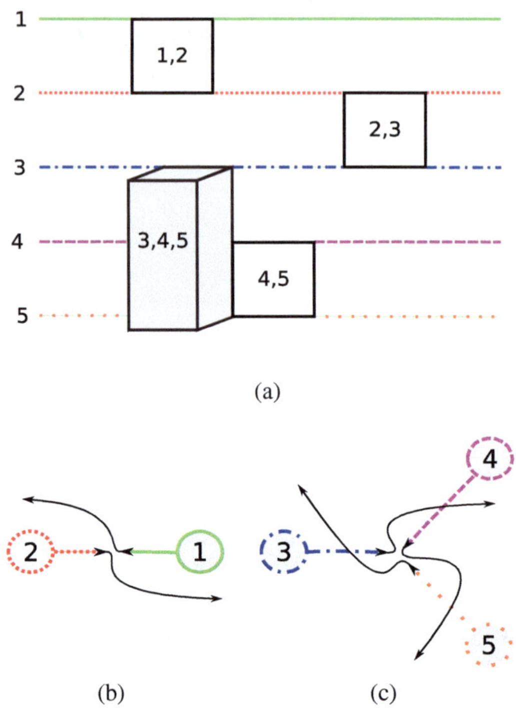

Fig. 1 A conceptual visualization of a variable dimensionality search space for five robots (a). Initially each robot is constrained to its individually optimal path, represented by a single line, but when robots 1 and 2 collide (b), the local dimensionality of the search space must be increased, as represented by a square. When three robots collide while following their individually optimal paths (c), the local dimensionality of the search space must be increased further, represented by the cube, to include all local paths of the three robots.

space, without coordinating with other robots. When the paths of multiple robots intersect, the joint configuration space of those robots is locally constructed and planning occurs in this joint space. Once coordination is no longer necessary, planning reverts to the low dimensional individual spaces until the goal is reached or another collision is found (Fig 1).

2 Prior Work

Multirobot path planning algorithms can be divided into two categories: coupled and decoupled [11]. A coupled algorithm seeks to find a path in the full configuration space of a system [1, 2, 6], which grows exponentially with the number of robots. As a result, coupled planners may be guaranteed to find an optimal path, but are computationally infeasible for systems of many robots.

On the other hand, decoupled algorithms search one or more low dimensional search spaces, which represent a portion of the full configuration space [5, 8, 12, 13, 15]. Searching a lower dimensional representation reduces the computational cost of finding a path, but the representation may not capture some or all of the solutions to the planning problem. As a result, decoupled algorithms generally produce results more quickly, but the quality or existence of the solution is not guaranteed.

Ideally, a multirobot path planner would combine the scalability of decoupled approaches with the completeness and optimality guarantees of a coupled approach. Some prior work seeks to inherit the benefits of both approaches by dynamically fitting the degree of coupling to the specific problem to be solved. Krishna *et al.* developed an approach for decentralized dynamic coupling of robots for velocity planning [10]. In their algorithm, robots first try to resolve a potential collision by independently altering their velocity. If this does not succeed, the robots involved in the collision cooperate to find a safe velocity schedule. If this also fails, they recruit uninvolved robots to alter their velocities to allow for a solution to be found. This approach will never change the spatial path the robots follow, and thus is neither complete nor optimal.

Clark *et al.* introduced dynamic networks, which explicitly search configuration spaces of varying sizes [4]. Joint plans are computed for groups of robots capable of mutual communication. Paths are re-planned whenever a new robot joins the group. This approach will lead to unnecessary coupling, as not all robots that can communicate need to cooperate to find a safe path, and only considers local interactions.

Van den Berg *et al.* [16] developed a planning time algorithm to find a coupling strategy that minimizes the size of the largest set of coupled robots needed to guarantee that a solution will be found. The robots are constrained to move sequentially. Cycles in these constraints can be used to find sets of robots for which coupled planning is necessary. This approach is non-ideal due to the restrictions it places on robot motion, which results in non-optimal paths, and the global nature of the coupling it performs.

There has also been work in the machine learning community to determine when coupling multiple robots is necessary. Kok *et al.* [9] presented an approach which performs Q-learning for robots individually, but stores statistics for the reward of the joint actions that are explored. If these statistics indicate that coordinating actions at a specific space is beneficial, then the algorithm starts learning coordinated actions at that state. This approach has the benefit of being able to handle tasks besides basic path planning, such as capturing targets that required coordinated action by multiple pursers. Melo and Veloso [14] developed a Q-learning algorithm that adds a 'coordinate' action to the set of actions available to each robot, which uses the state of the nearest neighboring robot to help choose the action to perform. Coordination between robots only occurs when a robot learns to take the coordinate action.

3 Problem Statement

The objective of subdimensional expansion is to find an optimal collision free path for a set of n robots, $r^i, i \in I = \{1, \dots, n\}$, in a common workspace $\mathscr{W}$. We denote the start configuration of each robot in its individual configuration space as $q_s^i \in Q^i$. The start configuration of all robots can be described as a point in the full configuration space $q_s \in Q = \prod_{i=1}^n Q^i$. The goal configuration for each robot is denoted $q_f^i \in Q^i$, while the joint goal configuration is denoted $q_f \in Q$. The optimal collision free path from q_s to q_f is denoted $\pi^*(q_s, q_f)$.

The cost $f(\pi)$ of the path π in the full configuration space is assumed to be the sum of the costs $f^i(\pi^i) \geq 0$ of the paths π^i of the individual robots, *i.e.*,

$$f(\pi(q_k, q_l)) = \sum_{i \in I} f^i(\pi^i(q_k^i, q_l^i)) \qquad \pi^i(q_k^i, q_l^i) \subset Q^i, \tag{1}$$

where $\pi(q_i, q_j)$ represents a path from q_i to q_j.

A collision function is defined to represent collisions between robots r^i and r^j,

$$\Psi^{ij}(q) = \begin{cases} \{i, j\}, & A(q^i) \cap A(q^j) \neq \emptyset \\ \emptyset, & \text{otherwise} \end{cases}. \tag{2}$$

where $A(q^i)$ is the subset of $\mathscr{W}$ occupied by r^i when located at $q^i \in Q^i$. We define a global collision function $\Psi : Q \to I$, which is the union of all pairwise collision functions.

$$\Psi(q) = \bigcup_{i \neq j \in I} \Psi^{ij}(q). \tag{3}$$

The collision function is "overloaded" to apply to paths, $\Psi(\pi(.)) = \bigcup_{q \in \pi(.)} \Psi(q)$.

4 Subdimensional Expansion

Multirobot systems which obey (1) and (3) have a natural decoupling between individual robots. (1) guarantees that no joint path can be cheaper than the path found by optimizing for each robot separately, which is thus a good starting point for multirobot path planning. The individual paths are combined to form a joint path for the entire system. When robot-robot collisions are found along the joint path, planning is locally coupled for the involved robots while uninvolved robots proceed independently, which is sufficient to guarantee that a path will be found due to the form of (3). Subdimensional expansion is a method for encoding the dynamic coupling into the geometry of the search space, thereby allowing conventional algorithms to search the necessary portions of the joint configuration space.

$Q^{\#}$ is the variably dimensional search space embedded in Q constructed by subdimensional expansion. Note that $Q^{\#}$ is dynamically constructed as a planner searches $Q^{\#}$. The simultaneous construction and search of $Q^{\#}$ continues until a path is found or determined to be impossible. The construction of $Q^{\#}$ is guided by information about robot-robot collisions found by the planner. Thus, the search space is tailored to the specific problem at hand, allowing the search of a low dimensional space while also guaranteeing the optimal path will be found.

4.1 Approach

Each robot starts with its own individually optimal policy $\phi^i : Q^i \to TQ^i$ which maps the position of a robot to its motion. ϕ^i is chosen such that the path induced

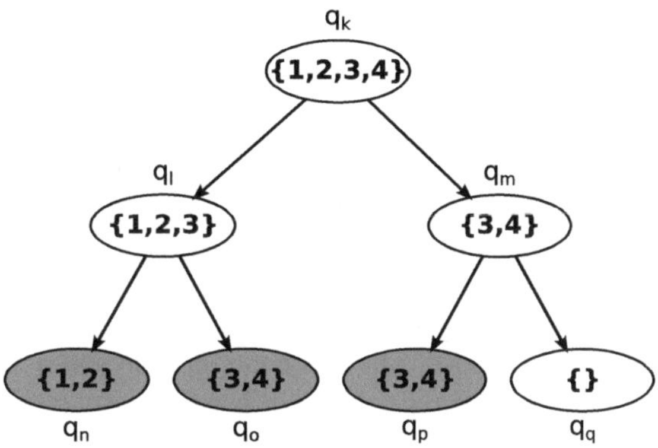

Fig. 2 Representation of a search tree and resultant collision sets. Ovals represent configurations in Q. Arrows represent searched path from the higher to the lower configuration. Gray ovals correspond to states with robot-robot collisions. The set contained inside the oval represents the collision set. Since there is a searched path from q_k to q_n, q_o, and q_p, C_k contains all robots which collide at the aforementioned state. Since the planner has not found a path from q_q to any state with a collision, C_q is empty.

by obeying ϕ^i from any point $q_k^i \in Q^i$ is an optimal path to $q_f^i \in Q^i$, and is denoted $\pi^{\phi^i}(q_k^i, q_f^i)$. The joint path generated by all robots obeying their individually optimal policies is denoted $\pi^{\phi}(q_k, q_f) = \prod_{i \in I} \pi^{\phi^i}(q_k^i, q_f^i)$.

At each instant during the search, the planner initially takes the optimistic view that the individually optimal path from q_k will be collision free, without specific information to the contrary. The planner maintains a *collision set* C_k for each $q_k \in Q$, which is the set of robots for which the optimistic view at q_k has been invalidated. Let $\pi(q_k)$ be the set of paths the planner has searched that pass through q_k. Then C_k is defined as

$$C_k = \bigcup_{\pi \in \pi(q_k)} \Psi(\pi) \tag{4}$$

The collision set C_k thus consists of all robots r^i for which the planner has found a path from q_k to a collision containing r^i (Figure 2).

Initially, $Q^\#$ is $\pi^{\phi}(q_s, q_f)$, with each robot restricted to following its individually optimal path. When the planner finds a collision, it expands $Q^\#$ by locally allowing the robots involved in the collision to deviate from their individually optimal policies. Naturally, the set of robots not in the collision set, $\bar{C} = I \setminus C$ remain restricted to their individually optimal paths, in line with the optimistic belief that this portion of the path is collision free. These constraints are encoded in the geometry of the search space $Q^\#$ by proper choice of the tangent space $T_{q_k}Q^\#$ of $Q^\#$ at q_k.

$$T_{q_k}Q^\# = \mathbf{t}^{\bar{C}_k}(q_k) \times \prod_{j \in C_k} T_{q_k^j}Q^j \tag{5}$$

where $\mathbf{t}^{\bar{C}_k}(q_k) \in T_{q_k^{\bar{C}_k}} Q^{\bar{C}_k}$ is tangent to the joint individually optimal path for the robots in $\bar{C}_k$. $Q^{\#}$ is then grown differentially along $T_{q_k} Q^{\#}$ from q_k, expanding into a higher dimensional space when the collision set is large, or along the one-dimensional individual optimal path when the collision set is empty.

4.2 *M**

M* is an implementation of subdimensional expansion for cases where the configuration space of each robot r^i can be represented by a directed graph $G^i = \{V^i, E^i\}$. V^i is the set of vertices in G^i that represent positions in Q^i, while E^i is the set of directed edges e^i_{kl} which represent valid transitions connecting $v^i_k \in V^i$ to $v^i_l \in V^i$. We make no assumption about the representation used, so G^i may be an approximate cellular decomposition, a generalized Voronoi diagram, or other graph representation of the configuration space. The full configuration space of the system is represented by the graph $G = \{V, E\} = \prod_{i \in I} G^i$. The Cartesian product of two graphs, $G^i \times G^j$, has the vertex set $V^i \times V^j$, and the edge e_{kl} is in the edge set if $e^i_{kl} \in E^i$ and $e^j_{kl} \in E^j$. The vertex in G which represents the start configuration of the system is denoted v_s, while the goal configuration is denoted v_f.

Representing the configuration space as a graph converts the path planning problem into a graph search problem. This allows us to base M* on A*, a complete and optimal graph search algorithm [7]. Recall that A* maintains an *open list* of vertices v_k to explore. These are sorted based on the sum of the cost of the cheapest path $\pi(v_s, v_k)$ and a *heuristic cost*, which is a lower bound on the cost of any path $\pi(v_k, v_f)$. At each iteration, the most promising vertex, v_k, from the open list is expanded. For each neighbor v_l of v_k, A* checks whether reaching v_l via v_k is the cheapest path found thus far to v_l. If so, v_l is added to the open list. This continues until v_f is expanded, indicating that an optimal path to the goal has been found.

M* is similar to A*, however the expansion step is a little different: M* only considers the *limited neighbors* of v_k, a subset of the neighbors of v_k in G, determined by C_k, providing the benefit of only exploring the "necessary" subspace of the configuration space. The set of limited neighbors $\hat{V}_k$ is the set of vertices v_l which can be reached from v_k while moving each robot $r^i \in \bar{C}_k$ according to its individually optimal policy $\phi^i(v^i_k)$, where v^i_k is the position of r^i when the system is at v_k. Conversely, the robots $r^j \in C_k$ are allowed to move to *any* neighbor of v^j_k in Q^j

$$\hat{V}_k = \left\{ v_l | \forall i \in I,\ v^i_l \ s.t. \begin{cases} e^i_{kl} \in E^i, & i \in C_k \\ v^i_l = \phi^i(v^i_k), & i \notin C_k \end{cases} \right\} \tag{6}$$

If $\Psi(v_k) \neq \emptyset$, we set $\hat{V}_k = \emptyset$, to prevent M* from considering paths which pass through collisions.

Information about collisions must be passed back along all searched paths that reach them. To this end, the planner maintains a *backpropagation set* for each vertex v_k, which is the set of all vertices v_l that were expanded while v_k was in $\hat{V}_l$. The

Algorithm 1. backprop(v_k, C_l, open):
v_k- vertex in the backpropagation set of v_l
C_l- the collision set of v_l
open- the open list for M*

```
if C_l ⊄ C_k then
   C_k ← C_k ∪ C_l
   if ¬(v_k ∈ open) then
      open.append(v_k) {If the collision set changed, we will need to re-expand v_k}
   for v_m ∈ v_k.back_set do
      {Iterate over the backpropagation set}
      backprop(v_m, C_k, open)
```

backpropagation set is thus the set of neighbors of v_k through which the planner has found a path to v_k. The planner propagates information about a collision at v_k by adding $C_k = \Psi(v_k)$ to C_l for each v_l in the backpropagation set of v_k. The planner then adds C_l to the collision set of each vertex in the backpropagation set of v_l, and repeats this process until a collision set is encountered which contains C_k. Since $\hat{V}_l$ is dependent on C_l, changing C_l adds new paths through v_l to the search space. As a result, v_l must be added back to the open list so that these new paths can be searched (See Algorithm 1).

Finally, since $f(\pi^\phi(v_k, v_f))$ is a lower bound on the cost of all paths $\pi(v_k, v_f)$, we use it as the heuristic function for M*. Denote the heuristic function

$$h(v_k) = f(\pi^\phi(v_k, v_f)) \leq f(\pi^*(v_k, v_f)). \tag{7}$$

M* is described in algorithm 2.

We developed two variants of M*, inflated M* and recursive M* (rM*). Inflated M* multiplies the heuristic by inflation factor $\varepsilon > 1$ to find a suboptimal path quickly. rM* is a hierarchical planner that breaks the path planning problem into multiple sub-problems, by separating the planning for non-interacting groups of mutually interfering robots. A sub-planner is recursively generated for each such group to find a path for the colliding robots to the goal. Sub-planners continue to be generated for smaller groups of colliding robots until the collision involves every robot handled by the sub-planner.

4.3 Graph-Centric Description

The description of M* in 4.2 provides a local description of the search process, which is useful for implementation. However, the local description makes proving the global properties of M* difficult. We now present an alternative description of M* which better captures the global properties, but is not appropriate for implementation.

When examining algorithm 2 we see that M* differs from A* in the existence of the **backprop** function, and the neighbors added to the open list during the

Algorithm 2. Pseudocode for M*

---

for all $v_k \in V$ **do**
  v_k.cost ← MAXCOST
  $C_k \leftarrow \emptyset$
v_s.cost ← 0
v_s.back_ptr = $\emptyset$
open = $\{v_s\}$
while True **do**
  open.sort() {Sort in ascending order by v.cost + h(v)}
  v_k = open.pop(0)
  if $v_k = v_s$ **then**
    {We have found a solution}
    return back_track(v_k) {Reconstruct the optimal path by following v_k.back_ptr}
  if $\Psi(v_k) \neq \emptyset$ **then**
    CONTINUE {Skip vertices in collision}
  for $v_l \in \hat{V}_k$ **do**
    v_l.back_set.append(v_k) {Add v_k to the back propagation list}
    $C_l \leftarrow C_l \bigcup \Psi(v_l)$
    {Update collision sets, and add vertices whose collision set changed back to open}
    backprop(v_k, C_l, open)
    if v_k.cost+$f(e_{kl}) <$ v_l.cost **then**
      {We have found a cheaper path to v_l}
      v_l.cost ← v_k.cost+$f(e_{kl})$
      v_l.back_ptr ← v_k {Keep track of the best way to get here}
return No path exists

---

expansion of a vertex. The backprop function only has a non-trivial result when a new path to one or more collisions is found. Therefore, M* behaves exactly like A* running on a graph $G^{\#}$ where the neighbors of v_k in $G^{\#}$ are the vertices in $\hat{V}_k$, until a new robot-robot collision is found. By thinking of M* as alternating between running A* on $G^{\#}$ and updating $G^{\#}$ based on partial results, we can exploit the optimality and completeness of A* to prove similar properties of M*.

$G^{\#}$ consists of three subgraphs: G', $\hat{G}$, and G^{ϕ}. G' is the portion of $G^{\#}$ which has been searched by M*, $\hat{G}$ represents the limited neighbors of the vertices in G' and G^{ϕ} connects the vertices in $\hat{G}$ to v_f by obeying ϕ.

$G' = \{V', E'\}$ represents the portion of G which has been searched by M*. V' is the set of vertices which have been added to the open list. E' consists of the directed edges e_{kl} connecting each vertex v_k which has been expanded by M* to the vertices $v_l \in \hat{V}_k$. Since G' represents all paths which have been explored by the planner, we can use G' to define the collision set

$$C_k = \begin{cases} \Psi(v_k) \bigcup_{v_l \text{ s.t. } \exists \pi(v_k, v_l) \subset G'} \Psi(v_l) & v_k \in G' \\ \emptyset & v_k \notin G' \end{cases} \tag{8}$$

If $v_k \notin G'$, then M* has never visited v_k. Thus in accordance with the optimistic view that v_k and $\pi^\phi(v_k, v_f)$ are collision free, C_k is set to the empty set until v_k is added to the open list.

$\hat{G}_k$ represents the portion of the graph which will be explored when v_k is expanded, and is the graph formed from v_k, its limited neighbors $\hat{V}_k$, and the edges connecting v_k to the vertices in $\hat{V}_k$. Let $\hat{G} = \bigcup_{v_k \in G'} \hat{G}_k$.

Since $C_k = \emptyset$ for all v_k which are not in G', we know that the search from $v_k \in \hat{G}$ will be constrained to $\pi^\phi(v_k, v_f)$ as long as this path lies entirely outside of G'. Let the graph G_k^ϕ represent the portion of $\pi^\phi(v_k, v_f)$ from v_k to the first vertex along the path in G', or v_f if $\pi^\phi(v_k, v_f)$ never reenters G'.

$G^\#$ can now be defined as

$$G^\# = G' \bigcup_{v_k \in G'} \left(\hat{G}_k \bigcup_{v_l \in \hat{G}_k \setminus G'} G_l^\phi \right) \tag{9}$$

As a result of the definitions of G', $\hat{G}$ and G^ϕ, vertices and edges shift from G^ϕ to $\hat{G}$, and from $\hat{G}$ to G' as M* searches $G^\#$. See Figure 3 for an illustration of how the subgraphs change over time. However, $G^\#$ as a whole only changes when the collision set of a vertex in $G^\#$ changes.

4.4 Completeness and Cost-Optimality

A path planning algorithm is *complete* if it is guaranteed to either find a path or to determine that no path exists in finite time [3]. M* will be shown to be both complete and will find a minimal cost path. As demonstrated in 4.3, M* can be treated as alternating between running A* search on $G^\#$ and modifying $G^\#$ based on the partial search results. Since A* is complete and cost optimal [7], M* is complete and cost optimal if $G^\#$ will contain $\pi^*(v_s, v_f)$ after a finite number of modifications or, if $\pi^*(v_s, v_f)$ does not exist, $G^\#$ will be modified at most a finite number of times.

We start by assuming that no solution exists, and show that M* will terminate in finite time without returning a path. $G^\#$ is only modified when the collision set of at least one vertex in $G^\#$ is modified. Each modification adds one or more robot to the collision set, thus each collision set can be modified at most $n-1$ times, as the first modification must add at least two robots. Therefore, $G^\#$ can be modified at most $(n-1) * |G|$ times. A* will expand each vertex in a graph at most once [7]. Thus, M* will always terminate in finite time.

M* will never return a path containing a robot-robot collision. A vertex v_k at which a robot-robot collision takes place can only have outneighbors if $v_k \notin G'$. However, before a path through v_k can be returned, v_k must be added to G', which will trigger a modification of $G^\#$ that removes the outneighbors of v_k. Therefore, M* will never return a path containing a robot-robot collision. Thus M* will correctly determine that no valid path exists in finite time.

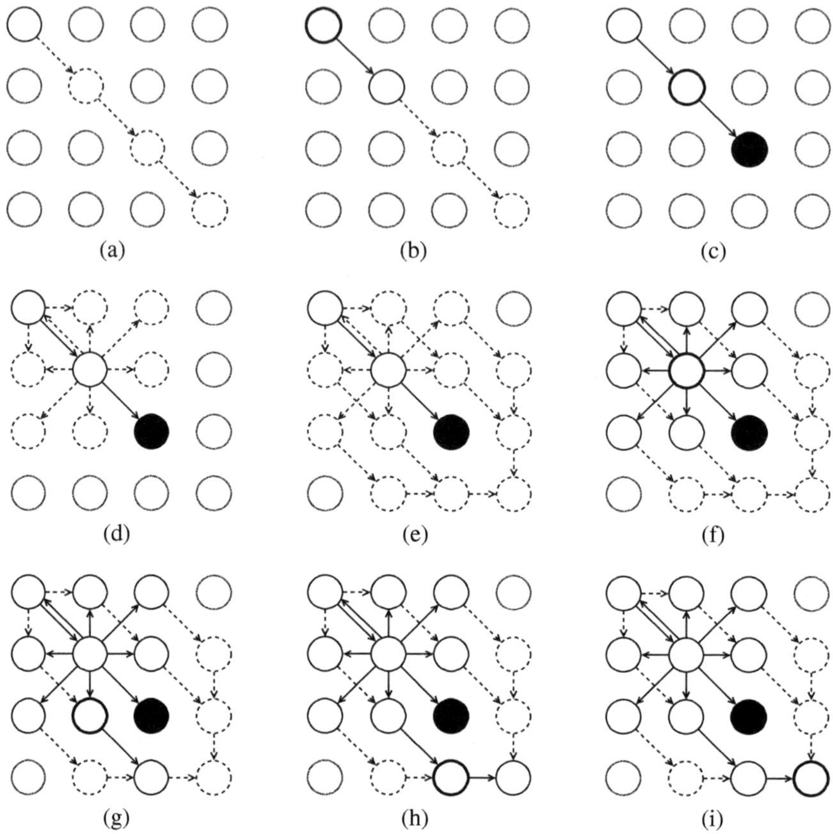

Fig. 3 The above figure shows how G' and $G^{\#}$ evolve in the configuration space of two one-dimensional robots. Vertices are represented as circles, with arrows representing directed edges. G' is denoted by solid lines, while $G^{\#} \backslash G'$ is shown as dashed lines. $G \backslash G^{\#}$ is represented by dotted lines, with edges suppressed for clarity. A vertex is given a bold outline when it is expanded, while filled circles represent vertices with known robot-robot collisions. v_s is in the upper left, while v_f is in the bottom right. In (a), (b), and (c), the most promising vertex in the open list is expanded, until a collision is found. $\hat{G}$ is updated to reflect the new collision sets in (d). G^{ϕ} is then updated in (e). In (f) a vertex is re-expanded, having been added back to the open list when its collision set was changed. (g), (h), and (i) see the most promising vertices in the open list expanded, until v_F is expanded, indicating that a path has been found.

Next, assume that a path from v_s to v_f exists. We will show that M* will find $\pi^*(v_s,v_f)$ as long as one of two cases is always true. We will then prove that one of these two cases must always hold. Assume that $G^{\#}$ always contains either

Case 1: an optimal, collision free path, $\pi^*(v_s,v_f)$, or
Case 2: a path $\pi(v_s,v_c)$ s.t. $f(\pi(v_s,v_c))+h(v_c)\leq f(\pi^*(v_s,v_f))$, and $\exists\, v_d\in\pi(v_s,v_c)$ s.t. $\Psi(v_c)\not\subset C_d$

If case 1 holds, running A* on $G^{\#}$ will find $\pi^*(v_s,v_f)$, unless there exists a cheaper path $\tilde{\pi}(v_s,v_f)\subset G^{\#}$. By the definition of $\pi^*(.)$, there must be a vertex $v_k\in\tilde{\pi}(v_s,v_f)$ *s.t.* $\Psi(v_k)\neq\emptyset$, and by (7) $f(\tilde{\pi}(v_s,v_k))+h(v_k)<f(\pi^*(v_s,v_f))$. $v_k\notin G'$, as otherwise it would have no outneighbors, which implies $C_k=\emptyset$. As a result, v_k fulfills the roles of both v_c and v_d in the definition of case 2. Therefore M* will find $\pi^*(v_s,v_f)$ if case 1 holds, unless case 2 also holds.

If case 2 holds, then v_c will be added to G' before A* finds any path to v_f that costs more than $f(\pi^*(v_s,v_f))$ [7]. Adding v_c to G' will modify C_d, which will in turn change $G^{\#}$ to reflect the new $\hat{V}_d$ and restart A* search. Therefore, M* will never return a suboptimal path as long as case 2 holds.

For case 2 to hold, there must be at least one vertex v_d such that C_d is a strict subset of I. $G^{\#}$ can be modified at most $(n-1)*|G|$ times before all collision sets are equal to I. Therefore, case 2 can only hold for a finite number of modifications to $G^{\#}$. By hypothesis, either case 1 or case 2 holds, which implies that within finite time only case 1 will be true. M* will thus find $\pi^*(v_s,v_f)$ in finite time.

We will now show that case 1 or case 2 must always hold. We proceed by showing that we can always find a path $\pi'(v_k,v_f)$, $f(\pi(v_k,v_f))\leq f(\pi^*(v_k,v_f))$ under the restriction that the robots in $\bar{C}_k$ obey their individually optimal policies.

First note that, by the form of (3), if $\pi^*(v_k,v_f)$ exists, then for any subset of robots, $\Omega\subset I$, there exists an optimal, collision free path $\pi^{*\Omega}(v_k^{\Omega},v_f^{\Omega})$, which may not be the same as the paths taken by the robots in Ω in $\pi^*(v_k,v_f)$. Therefore a path $\pi'(v_k,v_f)=\pi^{*C_k}(v_k^{C_k},v_f^{C_k})\times\pi^{\phi^{\bar{C}_k}}(v_k^{\bar{C}_k},v_f^{\bar{C}_k})$ can be constructed which costs no more than $f(\pi^*(v_k,v_f))$.

$$f(\pi'(v_k,v_f))=f^{C_k}(\pi^{*C_k}(v_k^{C_k},v_f^{C_k}))+\sum_{j\in\bar{C}_k}f^j(\pi^{\phi^j}(v_k^j,v_f^j)) \tag{10}$$

$$=\min_{\pi^{C_k}(v_k^{C_k},v_f^{C_k})\text{ s.t. }\Psi(\pi^{C_k}(v_k^{C_k},v_f^{C_k}))=\emptyset}f^{C_k}(\pi^{C_k}(v_k^{C_k},v_f^{C_k}))+\min\sum_{j\in\bar{C}_k}f^j(\pi^j(v_k^j,v_f^j)) \tag{11}$$

$$=\min_{\pi(v_k,v_f)\text{ s.t. }\Psi(\pi^{C_k}(v_k^{C_k},v_f^{C_k}))=\emptyset}f(\pi(v_k,v_f)) \tag{12}$$

$$\leq\min_{\pi(v_k,v_f)\text{ s.t }\Psi(\pi^{C_l}(v_k^{C_l},v_f^{C_l}))=\emptyset,C_k\subset C_l}f(\pi(v_k,v_f)) \tag{13}$$

$$\leq\min_{\pi(v_k,v_f)\text{ s.t.}\Psi(\pi(v_k,v_f))=\emptyset}f(\pi(v_k,v_f)) \tag{14}$$

$$\leq f(\pi^*(v_k,v_f)) \tag{15}$$

The successor v_l of v_k along $\pi'(v_k, v_f)$ is in $\hat{V}_k$ by (6). Furthermore, $C_l \subset C_k$ by (8), so by (13) and (14)

$$f(\pi'(v_k, v_l)) + f(\pi'(v_l, v_f)) \leq f(\pi'(v_k, v_f)) \leq f(\pi^*(v_k, v_f)) \tag{16}$$

Using the above two facts, a path $\pi''(v_s, v_f) \in G^{\#}$ can be constructed which satisfies case 1 or case 2. Starting from v_s, the successor of the m'th vertex $v_m \in \pi''(v_s, v_f)$ is the successor of v_m in $\pi'(v_m, v_f)$. Applying (16) backwards from the last vertex from v_f to v_s guarantees that $f(\pi''(v_s, v_f)) \leq f(\pi'_I(v_s, v_f)) \leq f(\pi^*(v_s, v_f))$. If $\pi''(v_s, v_f) = \pi^*(v_s, v_f)$ then case 1 is satisfied. Otherwise, there is a vertex $v_k \in \pi''(v_s, v_f)$ such that $\Psi(v_k) \neq \emptyset$. By construction, $\Psi(v_k) \not\subset C_l$, where v_l is the predecessor of v_k. By (7), $f(\pi''(v_s, v_k)) + h(v_k) \leq f(\pi''(v_s, v_f)) \leq f(\pi^*(v_s, v_f))$, so case 2 is satisfied. It has now been shown that case 1 or case 2 must always hold. Therefore M* will find $\pi^*(v_s, v_f)$, if it exists, in finite time. Since M* is guaranteed to find the optimal collision free path, or to determine that no valid path exists in finite time, M* is complete and optimal with respect to $f(\pi(.))$.

5 Results

We tested the path planning performance of M* and its variants with randomly assigned goals. Our simulations were run on a Core i7 processor at 2.8 GHz with 12 Gb of RAM. All simulations are implemented in unoptimized Python. We chose a square, four-connected grid with a density of 104 cells per robot as our workspace, allowing the number of robots to vary without changing the level of congestion. Each cell in the workspace has an independent 35% chance of being an obstacle. Start and goal positions for each robot are chosen randomly, but such that a path always exists from the start position of a robot to its goal position (Figure 4). Each robot incurred a cost of one for each time step for which it was not at its goal. We tested 100 random environments for a given number of robots, and each trial was given at most five minutes to find a solution.

The time required to find solutions using A* shows the expected exponential growth with the number of robots. M* and rM* show performance substantially superior to A* ,which was unsuccessful for problems involving more than 6 robots. rM* has roughly three times the success rate of M* for the non-inflated case at 10 robots. Using an inflated heuristic, rM* has a greater performance increase, with run times of approximately one and a half orders of magnitude less than basic M* for systems of 20 robots, and scaling to twice as many robots with reasonable success rates (Figure 5). Most importantly, the time to solution plots for inflated rM* are sublinear on a logarithmic axis. This indicates that for the environments we investigated, the average computational cost of rM* grows sub-exponentially with the number of robots.

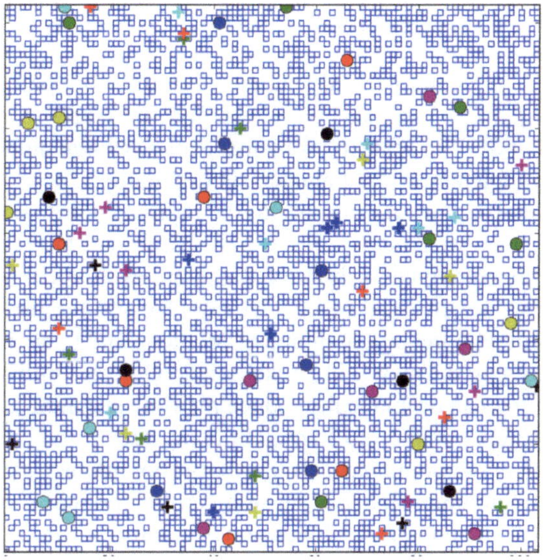

Fig. 4 A typical configuration for a 40 robot test run. Circles represent start positions of the robots, squares represent obstacles, and crosses represent goal positions. We tested 100 such randomly generated environments for each number of robots.

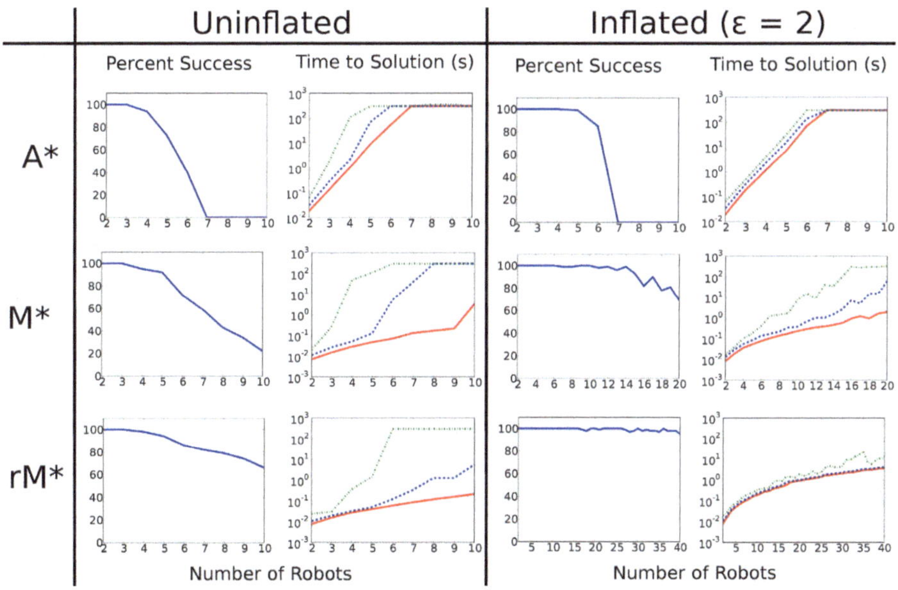

Fig. 5 We plot the percent of trials in which each algorithm was able to find a solution within 5 minutes, and the 10'th, 50'th, and 90'th percentile of times required to find a solution for A*, M*, and recursive M* with both non-inflated and inflated heuristics. The time to solution plots flatten out when a sufficient number of trials are halted by the time limit. A* and inflated A* are only simulated to 8 robots, because they always timed out for 7 or more robots. To allow A*, M*, and rM* to be plotted over similar domains, we assumed that A* and inflated A* would always time out for systems of 9 and 10 robots. Inflated M* and inflated rM* were able to solve 20 and 40 robot problems respectively, which is reflected in the domains of their plots.

6 Conclusions

We present a general approach to multirobot path planning, called subdimensional expansion, and an implementation for graph search, called M*. We demonstrate that M* can scale to problems involving large numbers of robots, while maintaining completeness and bounded suboptimality. These results illustrate the advantage of tailoring the search space to the individual problem being solved.

One weakness of M* is that search will fail if a sufficient number of robots are concentrated at a single choke point, as this will force M* to search an excessively high dimensional space. One possible solution is to couple the path planning problem with the task assignment problem, and to avoid task assignments which require passing through said choke points. Our preliminary results indicate that such coupling can dramatically reduce the time required to find solutions, as well as reduce the cost of the resultant path compared to the optimal path for a task assignment found without considering robot-robot interactions.

Subdimensional expansion can be applied to path planning algorithms besides A*. In general, subdimensional expansion can be applied to nearly any path planning algorithm that produces a search tree which can be used to define collision sets. In particular, subdimensional expansion can be applied to RRTs and PRMs [18].

References

1. Ayanian, N., Kumar, V.: Decentralized feedback controllers for multi-agent teams in environments with obstacles. In: Proceedings of IEEE International Conference on Robotics and Automation, pp. 1936–1941 (May 2008)
2. Carpin, S., Pagello, E.: On parallel RRTs for multi-robot systems. In: Proceedings of 8th Conference Italian Association for Artificial Intelligence, pp. 834–841. Citeseer (2002)
3. Choset, H.M.: Principles of robot motion: theory, algorithms, and implementation. The MIT Press (2005)
4. Clark, C.M., Rock, S.M., Latombe, J.C.: Motion planning for multiple robot systems using dynamic networks. In: Proceedings of IEEE International Conference on Robotics and Automation, pp. 4222–4227 (2003)
5. Erdmann, M., Lozano-Perez, T.: On multiple moving objects. Algorithmica 2(1), 477–521 (1987)
6. Ghrist, R.W., Koditschek, D.E.: Safe cooperative robot dynamics on graphs. SIAM Journal on Control and Optimization 40(5) (2002)
7. Hart, P.E., Nilsson, N.J., Raphael, B.: A formal basis for the heuristic determination of minimum cost paths. IEEE Transactions on Systems Science and Cybernetics 4(2) (July 1968)
8. Kant, K., Zucker, S.W.: Toward efficient trajectory planning: The path-velocity decomposition. The International Journal of Robotics Research 5(3), 72 (1986)
9. Kok, J.R., Hoen, P.J., Bakker, B., Vlassis, N.: Utile coordination: Learning interdependencies among cooperative agents. In: Proceedings of the IEEE Symposium on Computational Intelligence and Games (2005)
10. Krishna, K.M., Hexmoor, H., Chellappa, S.: Reactive navigation of multiple moving agents by collaborative resolution of conflicts. Journal of Robotic Systems, 249–269 (2005)

11. LaValle, S.M.: Planning algorithms. Cambridge Univ. Pr. (2006)
12. Leroy, S., Laumond, J.-P., Siméon, T.: Multiple path coordination for mobile robots: A geometric algorithm. In: Proceedings of the Sixteenth International Joint Conference on Artificial Intelligence, pp. 1118–1123. Morgan Kaufmann Publishers Inc., San Francisco (1999)
13. Malcom, R.: Multi-robot path-planning with subgraphs. In: Australasian Conference on Robotics and Automation (2006)
14. Melo, F.S., Veloso, M.: Learning of coordination: Exploiting sparse interactions in multiagent systems. In: Proc. of 8th Int. Conf. on Autonomous Agents and Multiagent Systems (May 2009)
15. Saha, M., Isto, P.: Multi-robot motion planning by incremental coordination. In: Proceedings of IEEE/RSJ International Conference on Intelligent Robots and Systems, pp. 5960–5963 (October 2006)
16. Jur van den Berg, J., Snoeyink, M., Lin, D.: Manocha. Centralized path planning for multiple robots: Optimal decoupling into sequential plans. In: Proceedings of Robotics: Science and Systems (2009)
17. Wagner, G., Choset, H.: M*: A complete multirobot path planning algorithm with performance bounds. In: Proceedings of IEEE International Conference on Intelligent Robots and Systems 2011 (September 2011)
18. Wagner, G., Kang, M., Choset, H.: Probabilistic path planning for multiple robots with subdimensional expansion. In: Proceedings of IEEE/RSJ International Conference on Robotics and Automation (May 2012)

Individual Control of Redundant Skeletal Muscles Using an Exoskeleton Robot

Jun Ueda and Ming Ding

Abstract. Individual muscle control involves the use of an exoskeleton to induce specific muscle activation patterns during the wearer's voluntary movement. A pneumatically powered wearable exoskeleton has been developed for the upper extremities. It is comprised of multiple pneumatic artificial muscles, force transducers, and custom-designed software that utilize a musculoskeletal model to predict and modify the wearer's muscle activities (e.g. activation level, torque). The interaction with such a wearable robot induces muscle activities that may not be achieved in the standard exercise environment. The software uses a computational algorithm to plan an adequate motor-task by modeling not only a complex coupling between multiple muscles and joints, but also the optimization principle in the neuromuscular system that coordinates redundant muscles.

1 Introduction

Neurological disorders, such as stroke, spinal cord injury (SCI), Parkinson's disease, and multiple sclerosis (MS), affect a number of individuals every year [16]. According to World Health Organization, approximately 15 million individuals suffer stroke each year [16]. Such neurological disorders are widely seen or diagnosed among senior citizens as well as veterans. A huge amount of lost productivity is expected in the world due to such neurological disorders. Several countries across the

Jun Ueda
Mechanical Engineering, Georgia Institute of Technology, 771 Ferst Drive, Atlanta, GA, 30322-0405, United States
e-mail: jun.ueda@me.gatech.edu

Ming Ding
RIKEN-TRI Collaboration Center for Human-Interactive Robot Research,
Advanced Science Institute, RIKEN, 2271-130, Anagahora, Shimoshidami Moriyama-ku, Nagoya, Aichi 463-0003, Japan
e-mail: mingding@naogya.riken.jp

D. Milutinović & J. Rosen (Eds.): Redundancy in Robot Manipulators, LNEE 57, pp. 183–199.
DOI: 10.1007/978-3-642-33971-4_11

world are currently experiencing an aging population and are anticipating losses in productivity due to age related neurological disorders. Consequently, there is a need for healthcare technologies that help senior citizens find neuromotor impairments at an early stage to prevent further functional losses. In addition, medical advances are saving more lives in the battlefield, which in turn has created a greater push for technologies to rehabilitate the wounded soldiers that are returning home.

The authors' project is intended to be a preliminary step toward diagnosis and rehabilitation of neuromotor disorders. The project involved the development of a robotic device that is designed and programed to induce a specific muscle activation pattern in a wearer. This robotic device modulates muscle activities based on a mathematical solution of the human-robot physical interaction. Healthy individuals modulate muscle activation patterns according to intended movement and environment. In contrast, neurological patients with movement disorders have problems in movement control due primarily to their inability to modulate their muscle activation pattern in an appropriate manner [5, 11, 21, 9]. The modulation of muscle activation patterns in patients is expected to be different from those in healthy subjects; the characteristics of the modulation are expected to be dependent on the degree of impairment in movement disorders.

The ultimate goal of this research is to quantitatively characterize the inability of patients by applying unique load compensation from a wearable robot. Detection and diagnosis of potentially serious movement disorders by using a robotic device would enable us to choose an appropriate treatment at an early stage, which would lead to the reduction of medical expenses. To accomplish these objectives, the mechanisms of redundant muscle coordination during human-robot physical interaction need to be investigated and understood. The authors' approach is to mathematically formulate this control problem based on the Optimization Principle in the neuromuscular system [19, 6, 18]. Computed forces applied by a wearable robot would alter the coordination of individual muscles in the wearer and induce a specific muscle adjustment.

This approach is different from approaches in the literature that apply force to an end-point to guide the end-point along a trajectory [20]. Note that this robot-assisted muscle isolation technology is a comprehensive concept; neuromuscular function test, therapeutic training, power-assisting, and muscle fitness training, can be boiled down to a single question:

How can we determine an adequate exercise that induces a desired change in a target muscle force in a *pinpointed* manner?

More specifically, the goal is to determine an adequate magnitude and direction of force that a subject must exert for a desired effect. With the current state of the art, procedures rely heavily on therapists' knowledge [12, 2]. The proposed approach addresses this issue and provides a computational method by integrating the robot and human models for planning an adequate motor-task. Induced muscle activity would help a clinician differentiate muscle activation patterns in patients from stereotypical patterns in healthy adults.

This chapter intends to give an overview of the individual muscle control methodology as implemented with an exoskeleton. First, the concept and a possible application to muscle function testing will be presented. Second, the exoskeleton robot system and its system configuration will be explained. Third, the mathematical formulation and closed-form solution will be given. Fourth, some experimental results for healthy volunteers will be presented. Finally, limitations and future work will be discussed.

2 Concept

As shown in Fig. 1, an exoskeleton applies joint torques to a subject via multiple interfaces attached to the upper limb while a subject performs a motor task. In this case, the subject applies a force to a handle via the robots' end effector. From an observation of a muscle activity that is different from what is expected in a healthy muscle, one would expect impairment in the observed muscle. The above discussion implies the need for inducing a specific muscle activation pattern in persons. The use of a wearable exoskeleton-type robot is well-suited to obtain a wider variety of muscle activity data than performing conventional motor tasks; a patient is asked to perform a task (e.g., pushing a handle) by opposing the exoskeleton robot as the robot assists or resists the movement by applying a various combination of forces/torques against the joints. Because of this physical interaction between the robot and person, a desired muscle activation pattern is expected to be induced in subject's muscles. Our previous study confirmed that the interaction with a robot can induce muscle activities that may not be achieved in the standard exercise environment, thus increasing the breadth of tests that can be used for diagnostic testing [24].

3 Muscle Control Exoskeleton

Figure 2 (a) shows a proof-of-concept wearable exoskeleton robot. Figure 1 (b) shows its schematic diagram. Although the robotic device is referred to as an "exoskeleton," this particular device, unlike other exoskeleton mechanisms, does not have any rigid links. Instead, the device has 8 flexible pneumatic actuators for the sake of safety. This robot is capable of applying torques to 3 joints with a total of 4 degrees of freedom: flexion/extension of the elbow joint, supination/pronation of the forearm, flexion/extension of the wrist joint, and radial and ulnar deviation of the wrist. Adjustment holes on the actuator attachments provide the adaptability to different body sizes. Each actuator is equipped with a force transducer to apply force feedback control. To analyze the physical interaction between the robot and wearer at the level of muscles, a musculoskeletal human model of the upper-right limb with 51 muscles [23] has been developed and integrated with a dynamic model of the robot (see Figures 1 (c) and (d)).

The overall system configuration is shown in Fig. 3(a). The system composed of 4 modules: human posture measurement module, muscle force estimation module,

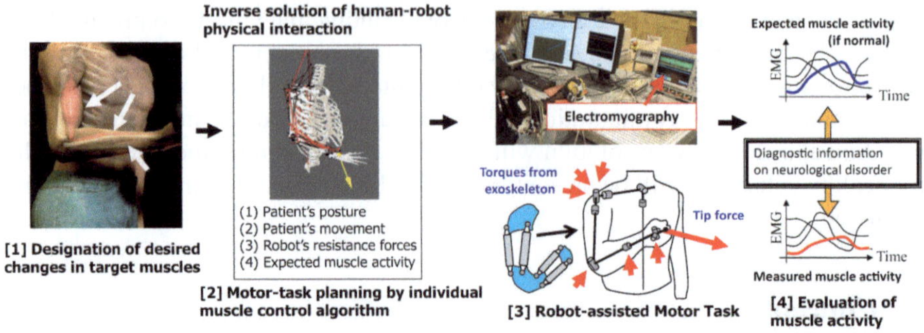

Fig. 1 Concept of individual muscle control for neuromuscular function test by using an exoskeleton: [Step1] Determine target muscles of interest and designate desired changes. [Step2] The physiology-based optimality criterion checks the feasibility of given changes in target muscle forces. The force that the subject must to exert to the handle is calculated. Control commands to the exoskeleton robot are calculated. [Step3] Exoskeleton robot applies joint torques against the subject. Subjects perform a force-matching task to match the hand-force with the instructed force by a monitor by opposing the exoskeleton robot. Electromyographic measurement equipment records muscle activities. [Step4] The physical interaction induces a desired muscle activation pattern in subject's muscles. Comparisons between the recorded muscle activation pattern and stereotypical (normal) pattern predicted by the musculoskeletal human model will be made.

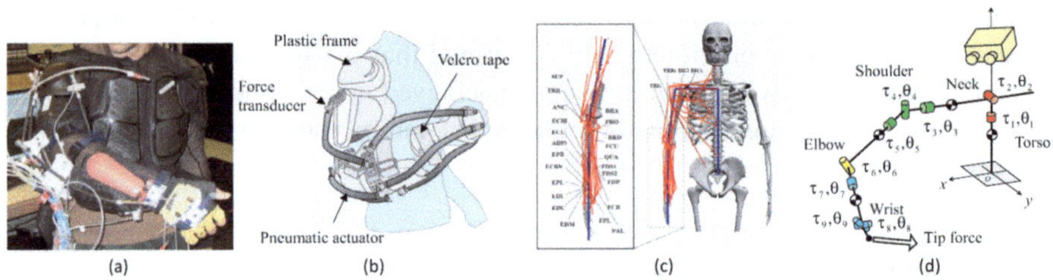

Fig. 2 Exoskeleton and Human Kinematic Models: (a) robot developed at Georgia Tech; (b) the robot assists/resists 4 degrees of freedom in the human arm; 1 DOF for the flexion/extension of the elbow joint, 1 DOF for the supination/pronation of the forearm, and 2 DOF for the flexion/extension and adduction/abduction of the wrist joint, by using 10 pneumatic artificial actuators; (c) 51 Muscles modeled in the musculoskeletal model (not all of them are labeled); (d) Joint condiguration.

muscle force control module, and power-assisting device control module. A graphical user interface provides an easy operation to designate desired forces for target muscles, and to view the resultant distribution of the muscle forces.

Posture Measurement Module. A target posture is measured by a motion capture system (Mac3D system, Motion Analysis Corporation). Measurement software (EvaRT) reproduces the posture of the subject from 3 dimensional positions of the markers attached on the subject and measures each of the joint angles.

Muscle Force Estimation Module. Joint torques are calculated by substituting the obtained joint data from the musculoskeletal human model. The skeletal model provides the moment arms and lengths of muscles. Then a physiologically based criterion of muscle forces [19] estimates the human muscle forces for a given posture. This estimation is solved as a quadratic programming problem using MATLAB.

Muscle Force Control Module. An operator designates desired muscle forces for target muscles using a graphical user interface as shown in Fig. 3(b). The ratio of change of target muscles is designated based on the nominal muscle forces specified by a slide bar interface.

Device Control Module. The human joint torques, when assisted, are calculated from the resultant muscle forces. By subtracting the resultant human joint torques from the nominal joint torques, the torques that need to be generated by the power-assisting device are calculated. This device control module calculates the pressure for each pneumatic actuator to realize the resultant torques. The pressure reference commands are then sent to electropneumatic regulators and execute the assist.

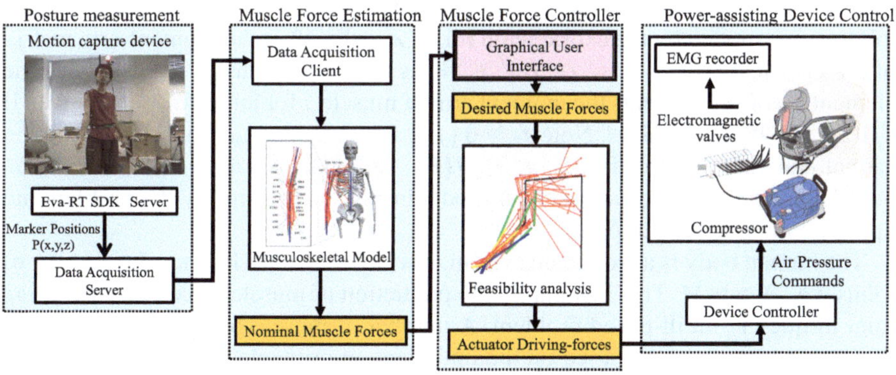

(a) Individual muscle control algorithm

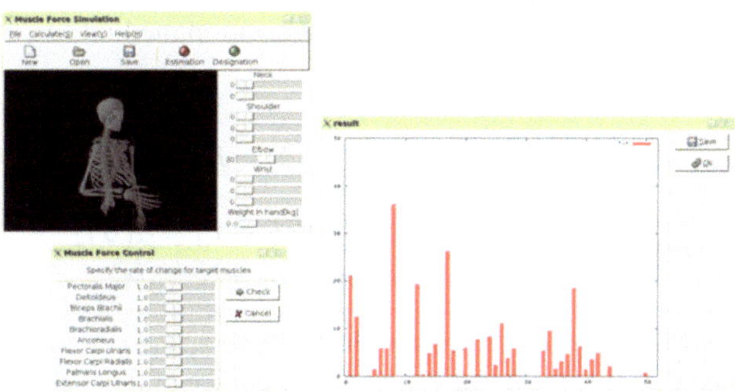

(b) Graphical user interface

Fig. 3 System configuration [14]

4 Optimization Principle in the Neuromuscular System

Individual muscle-force control during static tasks is considered. It is assumed that a subject does not change his/her posture during a task and all muscle contractions are isometric. The dynamics of the body and exoskeleton robot are neglected. This is primarily due to the difficulties in verifying non-isometric contractions by the measurement of electromyographic signals. Consider a human musculoskeletal model that has M joints and N muscles. The static equation of this musculoskeletal system (e.g., see Fig. 1 (c) and (d)) is given by

$$\begin{aligned}\boldsymbol{\tau}_h &= \boldsymbol{g}(\boldsymbol{\theta}) + \boldsymbol{J}(\boldsymbol{\theta})^T \boldsymbol{F} - \boldsymbol{\tau}_a = \boldsymbol{A}(\boldsymbol{\theta})\boldsymbol{f} \\ &= \begin{bmatrix} a_{11} & \cdots & a_{1N} \\ \vdots & \ddots & \vdots \\ a_{M1} & \cdots & a_{MN} \end{bmatrix} \begin{bmatrix} f_1 \\ \vdots \\ f_N \end{bmatrix}.\end{aligned} \tag{1}$$

where $\boldsymbol{\tau}_h \in \Re^M$ is a vector of human joint torques, $\boldsymbol{\theta} = [\theta_1, \cdots, \theta_M]^T \in \Re^M$ is a vector of joint angles, $\boldsymbol{F} = [F_x, F_y, F_z]^T$ is translational force at the tip, $\boldsymbol{J}(\boldsymbol{\theta})$ is the Jacobian between the joints and end-point, $\boldsymbol{g}(\boldsymbol{\theta})$ is the gravity force, $\boldsymbol{\tau}_a \in \Re^M$ is the joint torque applied by the exoskeleton robot, $\boldsymbol{A} \in \Re^{M\times N}$ is the moment-arm matrix of the muscles, and $\boldsymbol{f} = [f_1, \cdots, f_N]^T \in \Re^N$ is the human muscle force vector. The element a_{ij} of $\boldsymbol{A}$ represents the moment arm of muscle j for joint i. $a_{ij} = 0$ is given if f_j does not affect on joint i. Note $f_j \geq 0$ $(j = 1, \cdots, N)$ because muscle contraction cannot generate negative forces. $\boldsymbol{g}(\boldsymbol{\theta})$, $\boldsymbol{J}(\boldsymbol{\theta})$, and $\boldsymbol{A}(\boldsymbol{\theta})$ for a given posture $\boldsymbol{\theta}$ can be calculated by the musculoskeletal model in Fig. 1. To simplify the problem, no upper-bound is assumed for $\boldsymbol{\tau}_a$.

The human body is a redundant system, having more muscles than the number of joints, i.e., $N >> M$. This fact makes the prediction of muscle forces $\boldsymbol{f}$ by knowing joint torques $\boldsymbol{\tau}_h$ an ill-posed problem. A number of muscle-force prediction methods have been presented based on the optimization principle in the neuromuscular system [19, 3, 4, 18]. In the literature that deals with isometric or relatively slow motions, the cost functions have a general form comprised of the sum of muscular stress or force raised to a power. The static optimization method can be formulated as follows.

$$\text{Minimize} \quad u(\boldsymbol{f}) = \sum_{j=1}^{N} c_j f_j^r \tag{2}$$

$$\text{subject to} \begin{cases} \boldsymbol{\tau}_h = \boldsymbol{g}(\boldsymbol{\theta}) + \boldsymbol{J}(\boldsymbol{\theta})^T \boldsymbol{F} - \boldsymbol{\tau}_a = \boldsymbol{A}(\boldsymbol{\theta})\boldsymbol{f} \\ 0 \leq f_j \leq f\max_j (j = 1, \cdots, N) \end{cases} \tag{3}$$

where $u(\boldsymbol{f})$ is a cost function, c_j's are weighting factors, and r is an integer number. It should be noted that arguments still exist on the choice of the weighting factors c_j and the integer r of the power [19, 3, 4, 18]. There are still arguments and criticism of the neurological background of this muscle force prediction; however, the effectiveness of this approach for predicting stereotyped motor performances has been reported in the literature.

5 Problem Formulation

The proposed individual muscle control requires us to solve the inverse problem of the aforementioned muscle force prediction in (2) to control N muscles by adjusting M joint torque inputs. Note that this inverse solution is not straightforward since any induced muscle activation pattern must satisfy the physiology-based optimality criterion. The first condition in (3) represents the balance of joint torques. The second condition means that each muscle can only "pull", exerting positive contracting force limited by its maximum voluntary force. Mathematically, this can be treated as finding an adequate equality constraint for the joint torques such that desired muscle forces are obtained as a result of the cost function minimization. As described earlier, the exoskeleton robot device merely modifies human joint torques, which is equivalent to the modification of the first equality condition in (3).

Let $\boldsymbol{f}_0$ be the nominal muscle forces obtained in Step 1 in Fig. 1 when performing a nominal task. The human force vector $\boldsymbol{f}_0$ may be permutated according to three groups of muscles: target muscles, non-target muscles, and inactive muscles. Let $\tilde{N} \leq N$ be the number of the active muscles, and $N - \tilde{N}$ be the number of the inactive muscles. The active muscles have elements with nonzero values in $\boldsymbol{f}_0$, and the inactive muscles have zeros. The active muscle group consists of target muscles $\boldsymbol{f}_t \in \Re^{N_t}$ and non-target muscles $\boldsymbol{f}_n \in \Re^{N_n}$ where $N_t + N_n = \tilde{N}$. Without the loss of generality, the order of the N muscles may be permutated according to these three groups for the simplicity of description.

$$\boldsymbol{f} \triangleq \left[\begin{array}{c} \boldsymbol{f}_t \\ \boldsymbol{f}_n \\ \hline \boldsymbol{0} \end{array}\right] \begin{array}{l} \cdots \text{target muscles} \\ \cdots \text{non-target muscles} \\ \cdots \text{inactive muscles} \end{array} \tag{4}$$

The above permutation for $\boldsymbol{f}$ is also applied to the moment-arm matrix $\boldsymbol{A}$ accordingly:

$$\boldsymbol{A}^T = \left[\begin{array}{c} \boldsymbol{A}_t \\ \boldsymbol{A}_n \\ \hline \boldsymbol{A}_v \end{array}\right] \begin{array}{l} \cdots \text{target muscles} \\ \cdots \text{non-target muscles} \\ \cdots \text{inactive muscles} \end{array} \tag{5}$$

Hereafter these permutated vectors and matrices will be used.

The desired muscle forces $\boldsymbol{f}_{td}$ are given as follows by explicitly specifying the change ratio for each of the target muscles:

$$\boldsymbol{f}_{td} = \text{diag}([\gamma_1, \gamma_2, \cdots, \gamma_{N_t}])\boldsymbol{f}_{t0} \tag{6}$$

where $\gamma_j (> 0)$ is the change ratio of the j-th target muscle. Hereafter the subscript d denotes the desired muscle forces, and 0 denotes the nominal muscle forces. Let the total external torque $\boldsymbol{\tau}_{ex} \in \Re^M$ be

$$\boldsymbol{\tau}_{ex} = \boldsymbol{J}^T \boldsymbol{F} - \boldsymbol{\tau}_a. \tag{7}$$

Equation (7) shows how the joint toques can be adjusted by applying $\boldsymbol{\tau}_a$ from the robot and by performing a motor task exerting $\boldsymbol{F}$ at the tip. Note that there is a certain freedom in choosing $\boldsymbol{F}$ and $\boldsymbol{\tau}_a$ for a given $\boldsymbol{\tau}_{ex}$.

The problem to realize $\boldsymbol{f}_{td}$ can be formulated as follows.

Perfect Individual Muscle Control Problem

Find $\boldsymbol{F}$ and $\boldsymbol{\tau}_a$ in (3) such that the solution of (2) includes a desired muscle activation pattern for the target muscles, $\boldsymbol{f}_{td}$, i.e.,

$$[\boldsymbol{f}_{td}^T, \boldsymbol{f}_{nd}^T, \boldsymbol{0}^T]^T = \underset{\boldsymbol{f}(\boldsymbol{\tau}_{ex})}{\text{argmin}}\, u(\boldsymbol{f})$$

with minimum changes in non-target muscle forces, i.e., $|\boldsymbol{f}_{nd} - \boldsymbol{f}_{n0}| \to min$.

6 KKT Conditions

For simplicity, the condition on the upper-bounds of muscle forces is neglected and only the condition $0 \leq f_j$ is considered. The feasibility of the individual muscle force control is analyzed as a constrained optimization problem. Since $\boldsymbol{f}$ is a solution of (2), it must satisfy Kuhn-Tucker theorem [17, 1]:

$$\nabla u(\boldsymbol{f}) + \sum_{i=1}^{M} \mu_i \nabla h_i(\boldsymbol{f}) + \sum_{j=1}^{N} \lambda_j \nabla g_j(\boldsymbol{f}) = 0, \tag{8}$$

$$h_i(\boldsymbol{f}) = 0 \quad (i = 1, \cdots, M), \tag{9}$$

$$\lambda_j g_j(\boldsymbol{f}) = 0, \lambda_j \geq 0, g_j(\boldsymbol{f}) \leq 0 \; (j = 1, 2, \cdots, N), \tag{10}$$

where $h_i(\boldsymbol{f}) = \tau_i - \boldsymbol{a}_i^T \boldsymbol{f}$ and $g_j(\boldsymbol{f}) = -f_j$. $\boldsymbol{a}_j \in \Re^N$ is a column vector of $\boldsymbol{A}$. Recall

$$q_j \triangleq \frac{\partial u(\boldsymbol{f})}{\partial f_j} = r c_j {f_j}^{r-1}, (j = 1, \cdots, N) \tag{11}$$

$$\frac{\partial h_i(\boldsymbol{f})}{\partial f_j} = a_{ij}, \tag{12}$$

$$\frac{\partial g_{j_1}}{\partial f_{j_2}} = \begin{cases} -1, & j_1 = j_2 \\ 0, & j_1 \neq j_2 \end{cases}. \tag{13}$$

Therefore, (8) is written as

$$\boldsymbol{q} = w(\boldsymbol{f}) = \boldsymbol{A}^T \mu + \boldsymbol{\lambda} \tag{14}$$

where $\boldsymbol{q} = [q_1, \cdots, q_N]^T$, $\boldsymbol{\mu} = [\mu_1, \cdots, \mu_M]^T$, and $\boldsymbol{\lambda} = [\lambda_1, \cdots, \lambda_N]^T$. $w(*)$ is a function that converts the muscle force vector $\boldsymbol{f}$ to a new vector $\boldsymbol{q}$, i.e., $\boldsymbol{q} = w(\boldsymbol{f})$, where the j-th elements of $\boldsymbol{q}$ is given by

$$q_j \triangleq \frac{\partial u(\boldsymbol{f})}{\partial f_j} = r c_j f_j^{r-1}, (j = 1, \cdots, N). \tag{15}$$

From (10), $\lambda_j = 0$ if $f_j > 0$. Using (4) and (5), (14) can be rewritten as

$$\begin{bmatrix} \boldsymbol{q}_t \\ \boldsymbol{q}_n \\ \hline \boldsymbol{0} \end{bmatrix} = \begin{bmatrix} \boldsymbol{A}_t \\ \boldsymbol{A}_n \\ \hline \boldsymbol{A}_v \end{bmatrix} \boldsymbol{\mu} + \begin{bmatrix} \boldsymbol{0} \\ \boldsymbol{0} \\ \hline \boldsymbol{\lambda}_v \end{bmatrix}. \tag{16}$$

For the normal muscle force $\boldsymbol{f}_0 (= w^{-1}(\boldsymbol{q}_0))$, (16) is given as

$$\boldsymbol{q}_0 = \boldsymbol{A}^T \boldsymbol{\mu}_0 + \boldsymbol{\lambda}_0 \to \begin{bmatrix} \boldsymbol{q}_{0t} \\ \boldsymbol{q}_{0n} \\ \hline \boldsymbol{0} \end{bmatrix} = \begin{bmatrix} \boldsymbol{A}_t \\ \boldsymbol{A}_n \\ \hline \boldsymbol{A}_v \end{bmatrix} \boldsymbol{\mu}_0 + \begin{bmatrix} \boldsymbol{0} \\ \boldsymbol{0} \\ \hline \boldsymbol{\lambda}_{0v} \end{bmatrix}. \tag{17}$$

Similarly, for the desired muscle force $\boldsymbol{f}_d (= w^{-1}(\boldsymbol{q}_d))$,

$$\boldsymbol{q}_d = \boldsymbol{A}^T \boldsymbol{\mu}_d + \boldsymbol{\lambda}_d \to \begin{bmatrix} \boldsymbol{q}_{dt} \\ \boldsymbol{q}_{dn} \\ \hline \boldsymbol{0} \end{bmatrix} = \begin{bmatrix} \boldsymbol{A}_t \\ \boldsymbol{A}_n \\ \hline \boldsymbol{A}_v \end{bmatrix} \boldsymbol{\mu}_d + \begin{bmatrix} \boldsymbol{0} \\ \boldsymbol{0} \\ \hline \boldsymbol{\lambda}_{dv} \end{bmatrix}. \tag{18}$$

7 Solution of Perfect Individual Muscle Control by Using an Ideal Exoskeleton

The authors have analyzed so-called *ideal* individual muscle control [23, 24] with assumptions that (1) $\boldsymbol{f}_{td}$ is physiologically realizable and (2) the exoskeleton robot has means to adjust all of the joints involved in a task, i.e., $\boldsymbol{\tau}_h$ is *fully* modifiable by $\boldsymbol{\tau}_{ex}$, and a combination of $\boldsymbol{F}$ and $\boldsymbol{\tau}_a$ always exists that realizes a given $\boldsymbol{\tau}_{ex}$. A lack of means to apply joint toques may result in errors in individual muscle forces. This issue has been partly addressed in [22] and will be further examined in future publications.

Since the number of control inputs (i.e., the number of the elements of $\boldsymbol{\tau}_{ex}$) is, in general, less that the number of muscles, the priority-based approach is applied. The first priority is to perfectly realize the desired forces $\boldsymbol{f}_{td}$ of the target muscles. The second priority is to minimize the changes of the non-target muscles since the non-target muscle forces will be influenced by the first-priority muscle control due to the physical coupling among the muscles.

Theorem: Perfect individual muscle control [24]

The external torque $\boldsymbol{\tau}_{ex}$ is obtained as

$$\boldsymbol{\tau}_{ex} = \left[\boldsymbol{A}_t^T\ \boldsymbol{A}_n^T\right] w^{-1}\left(\begin{bmatrix}\boldsymbol{A}_t\\ \boldsymbol{A}_n\end{bmatrix}\boldsymbol{\alpha}\right) \tag{19}$$

where $\boldsymbol{f} = w^{-1}(\boldsymbol{q})$ is the inverse function of $w(*)$. Also, $\boldsymbol{\alpha}$ is given as $\boldsymbol{\alpha} = \boldsymbol{A}_t^+\left[w(\boldsymbol{f}_{td}) - w(\boldsymbol{f}_{t0})\right] + (\boldsymbol{I} - \boldsymbol{A}_t^+\boldsymbol{A}_t)\boldsymbol{\beta}$, where $\boldsymbol{I}$ is the identity matrix, and $\boldsymbol{\beta}$ is a free parameter that represents the remaining redundancy for controlling the non-target muscles as the second priority. To minimize the influence on the non-target muscles in terms of the root-mean-square (RMS) change, $\boldsymbol{\beta}$ may be given as $\boldsymbol{\beta} = \left[-\boldsymbol{A}_n(\boldsymbol{I} - \boldsymbol{A}_t^+\boldsymbol{A}_t)\right]^+ \boldsymbol{A}_n\boldsymbol{A}_t^+\left[w(\boldsymbol{f}_{td}) - w(\boldsymbol{f}_{t0})\right]$.

The computed net joint torque in (19) is realized by an appropriate choice of $\boldsymbol{F}$ and $\boldsymbol{\tau}_a$ (see (7)). It should be noted that (7) and (19) imply the necessity of the use of an exoskeleton-type robot; the sole application of the subject's voluntary force (i.e., the choice of the three parameters in $\boldsymbol{F} = [F_x, F_y, F_z]^T$ for conventional motor tasks) may not be sufficient in terms of the number of control degrees of freedom when $M > 3$. The existence of $\boldsymbol{\alpha}$ for a given $\boldsymbol{f}_{td}$ can be checked by the following three feasibility conditions.

Feasibility conditions for perfect individual muscle control

1. $\boldsymbol{f}_{td}$ for the target muscles is perfectly realized if

$$\text{rank}(\ \boldsymbol{A}_t\ ) = \text{rank}(\ \left[\boldsymbol{A}_t,\ w(\boldsymbol{f}_{td}) - w(\boldsymbol{f}_{t0})\right]\ ). \tag{20}$$

2. The inactive muscles keep inactive if

$$-\boldsymbol{A}_v \begin{bmatrix}\boldsymbol{A}_t\\ \boldsymbol{A}_n\end{bmatrix}^+ \begin{bmatrix}\boldsymbol{q}_{t0}\\ \boldsymbol{q}_{n0}\end{bmatrix} - \boldsymbol{A}_v\boldsymbol{\alpha} > 0. \tag{21}$$

3. The resultant muscle forces of the non-target muscles remain positive if

$$\boldsymbol{A}_n\boldsymbol{\alpha} + w(\boldsymbol{f}_{n0}) > 0. \tag{22}$$

If all of the conditions are not satisfied, the control of the designated target muscles for given $\boldsymbol{f}_{td}$ is not physiologically realizable, i.e., the violation of the Optimality Principle. If not feasible, the target muscle forces or the choice of the target muscles must be modified.

Proof: Let $\boldsymbol{\mu}_d = \boldsymbol{\mu}_0 + \boldsymbol{\alpha}, \boldsymbol{\lambda}_{dv} = \boldsymbol{\lambda}_{0v} - \boldsymbol{A}_v\boldsymbol{\alpha}$ where $\boldsymbol{\alpha}$ is a control input vector in the Kuhn-Tucker form. $\boldsymbol{q}_d$ can be represented as

$$\boldsymbol{q}_d = \begin{bmatrix} \boldsymbol{q}_{dt} \\ \boldsymbol{q}_{dn} \\ \hline \boldsymbol{0} \end{bmatrix} = \begin{bmatrix} \boldsymbol{q}_{0t} \\ \boldsymbol{q}_{0n} \\ \hline \boldsymbol{0} \end{bmatrix} + \begin{bmatrix} \boldsymbol{A}_t \\ \boldsymbol{A}_n \\ \hline \boldsymbol{A}_v \end{bmatrix} \boldsymbol{\alpha}$$
$$= \left(\begin{bmatrix} \boldsymbol{A}_t \\ \boldsymbol{A}_n \\ \hline \boldsymbol{A}_v \end{bmatrix} \boldsymbol{\mu}_0 + \begin{bmatrix} \boldsymbol{0} \\ \boldsymbol{0} \\ \hline \boldsymbol{\lambda}_{0v} \end{bmatrix} \right) + \left(\begin{bmatrix} \boldsymbol{A}_t \\ \boldsymbol{A}_n \\ \hline \boldsymbol{A}_v \end{bmatrix} \boldsymbol{\alpha} + \begin{bmatrix} \boldsymbol{0} \\ \boldsymbol{0} \\ \hline -\boldsymbol{A}_v \boldsymbol{\alpha} \end{bmatrix} \right)$$
$$= \begin{bmatrix} \boldsymbol{A}_t \\ \boldsymbol{A}_n \\ \hline \boldsymbol{A}_v \end{bmatrix} (\boldsymbol{\mu}_0 + \boldsymbol{\alpha}) + \begin{bmatrix} \boldsymbol{0} \\ \boldsymbol{0} \\ \hline \boldsymbol{\lambda}_{0v} - \boldsymbol{A}_v \boldsymbol{\alpha} \end{bmatrix} \tag{23}$$

The first row of (23) provides the condition to perfectly realize the desired forces for the target muscles, i.e.,

$$\Delta \boldsymbol{q}_t = \boldsymbol{q}_{td} - \boldsymbol{q}_{t0} = w(\boldsymbol{f}_{td}) - w(\boldsymbol{f}_{t0}) = \boldsymbol{A}_t \boldsymbol{\alpha}. \tag{24}$$

A solution for $\boldsymbol{\alpha}$ exists if the following condition holds:

$$\text{rank}(\ \boldsymbol{A}_t\ ) = \text{rank}(\ [\boldsymbol{A}_t\ w(\boldsymbol{f}_{td}) - w(\boldsymbol{f}_{t0})]\ ) \tag{25}$$

which gives the condition 1).

In order not to induce any forces in inactive muscles, $\boldsymbol{\lambda}_{dv} > \boldsymbol{0}$ must be satisfied. Therefore,

$$\begin{aligned} \boldsymbol{\lambda}_{dv} &= \boldsymbol{\lambda}_{0v} - \boldsymbol{A}_v \boldsymbol{\alpha} \\ &= -\boldsymbol{A}_v \begin{bmatrix} \boldsymbol{A}_t \\ \boldsymbol{A}_n \end{bmatrix}^+ \begin{bmatrix} \boldsymbol{q}_{0t} \\ \boldsymbol{q}_{0n} \end{bmatrix} - \boldsymbol{A}_v \boldsymbol{\alpha} > \boldsymbol{0} \end{aligned} \tag{26}$$

which gives the condition 2).

To maintain non-target muscles positive, i.e., $\boldsymbol{f}_{nd}$, the condition is obtained from the second row of (23) as follows:

$$w(\boldsymbol{f}_{nd}) = \boldsymbol{A}_n \boldsymbol{\alpha} + w(\boldsymbol{f}_{n0}) > \boldsymbol{0} \tag{27}$$

which gives the condition 3).

The solution of (24) is given by

$$\boldsymbol{\alpha} = \boldsymbol{A}_t^+ [w(\boldsymbol{f}_{td}) - w(\boldsymbol{f}_{t0})] + (\boldsymbol{I} - \boldsymbol{A}_t^+ \boldsymbol{A}_t) \boldsymbol{\beta} \tag{28}$$

where $\boldsymbol{\beta}$ is a free parameter that indicates the redundancy of the solution; $\boldsymbol{\beta}$ determines the distribution of resultant non-target muscle forces while maintaining the complete realization of desired forces for target muscles. Note that the choice of $\boldsymbol{\beta}$ is generally arbitrary. One reasonable choice may be to minimize the change of non-target muscles. The change of non-target muscles is represented by

$$\begin{aligned}\Delta\boldsymbol{q}_n &= w(\boldsymbol{f}_{nd}) - w(\boldsymbol{f}_{n0}) = \boldsymbol{A}_n\boldsymbol{\alpha} \\ &= \boldsymbol{A}_n\{\boldsymbol{A}_t^+[w(\boldsymbol{f}_{td}) - w(\boldsymbol{f}_{t0})] + (I - \boldsymbol{A}_t^+\boldsymbol{A}_t)\boldsymbol{\beta}\} \\ &= \boldsymbol{A}_n\boldsymbol{A}_t^+[w(\boldsymbol{f}_{td}) - w(\boldsymbol{f}_{t0})] - [-\boldsymbol{A}_n(I - \boldsymbol{A}_t^+\boldsymbol{A}_t)]\boldsymbol{\beta}\end{aligned} \tag{29}$$

The following $\boldsymbol{\beta}$ may minimize $||\Delta\boldsymbol{q}_n||$ to avoid unnecessary influences on the non-target muscles:

$$\boldsymbol{\beta} = [-\boldsymbol{A}_n(\boldsymbol{I} - \boldsymbol{A}_t^+\boldsymbol{A}_t)]^+\boldsymbol{A}_n\boldsymbol{A}_t^+[w(\boldsymbol{f}_{td}) - w(\boldsymbol{f}_{t0})] \tag{30}$$

Note that the resultant $\boldsymbol{\alpha}$ must satisfy (22). The following changes are created for the target and non-target muscles at the level of individual muscles

$$w^{-1}\left(\begin{bmatrix}\boldsymbol{q}_{dt}\\ \boldsymbol{q}_{dn}\end{bmatrix} - \begin{bmatrix}\boldsymbol{q}_{0t}\\ \boldsymbol{q}_{0n}\end{bmatrix}\right) = w^{-1}\left(\begin{bmatrix}\boldsymbol{A}_t\\ \boldsymbol{A}_n\end{bmatrix}\boldsymbol{\alpha}\right). \tag{31}$$

Therefore, the application of the torques represented by (19) realizes target muscle forces.

8 Violation of the Feasibility Conditions

According to the optimization principle in the neuromuscular system, any distribution of muscle forces must not violate the physiology-based optimization criterion; achievable changes in muscles are not completely arbitrary, but physiologically realizable muscle activation patterns need to be specified. However, this designation of physiologically realizable muscle forces is not easy due to the presence of a number of muscles and intricate coupling between joint-torques and muscle-forces [13, 8]. An operator may give a muscle activation pattern that is not physiologically realizable. An inappropriate choice of $\boldsymbol{f}_{td}$ violates the condition (20) and may result in producing errors in target muscles even if the exoskeleton robot is capable of controlling all of the joint torques:

$$\text{rank}(\ \boldsymbol{A}_t\ ) \neq \text{rank}(\ [\boldsymbol{A}_t,\ w(\boldsymbol{f}_{td}) - w(\boldsymbol{f}_{t0})]\ ). \tag{32}$$

Instead of finding a solution that perfectly realizes a desired muscle activation pattern, it is desirable that a computational algorithm finds a quasi-optimized input that can minimize the errors in individual muscle forces. This approach can be represented by $||\boldsymbol{f}_{td} - \boldsymbol{f}_t|| \to \min$. It is still possible to determine $\boldsymbol{\alpha}$ by using (28) such that the norm of the error in the target muscles is minimized, i.e., $||\boldsymbol{q}_{td} - \boldsymbol{q}_t|| \to \min$, with first priority and the norm of the error in the non-target muscles is minimized with second priority, i.e., $||\boldsymbol{q}_n - \boldsymbol{q}_{n0}|| \to \min$. In a practical usage, a certain magnitude of error may be accepted since force generation in biological systems produces certain variability. The relaxation of (21) will be presented in our future publication. The condition (22) would not be relaxed since positive muscle force generation is due to the biomechanism of skeletal muscles.

9 Experiment

The brachialis muscle (BRA), brachioradialis muscle (BRD), and flexor carpi ulnaris muscle (FCU) are chosen as target muscles. Biceps muscle is chosen as a representative non-target muscle. A total of 11 tasks with various desired ratios of change shown in Table 1 were tested for the posture shown in Fig. 4 by maintaining the elbow angle approximately 90 degrees with holding a weight of 3 kg. Since the brachialis muscle and brachioradialis muscle are known to be anatomically coupled, these muscles are treated as one group and the same ratios of change were given. Tasks I and J assist only Flexor Carpi Ulnaris muscle, Tasks C and G assist only brachialis and brachioradialis muscles, and other trials are the mixture of assisting and resisting for the three target muscles.

The individual muscle control algorithm computed robot torques that realize these desired changes in the three target muscles and minimize the change in the non-target muscles. Table 2 shows the robot torques of the elbow and wrist joints. Due to the antagonistic arrangement of pneumatic actuators in this exoskeleton, only 4 actuators are controlled at the same time.

A total of eight healthy male volunteers were involved in the experiment. Subjects were asked to repeat each trial five times with a rest between trials. Recorded electromyographic signals of the four muscles for five seconds were filtered and rectified. Figure 5 shows the results. White bars show desired changes of the target muscles and black bars show measured changes of electromyographic signals.

It has been reported that the magnitude of electromyographic signals is approximately, but not strictly, proportional to isometric contraction force. Therefore, only

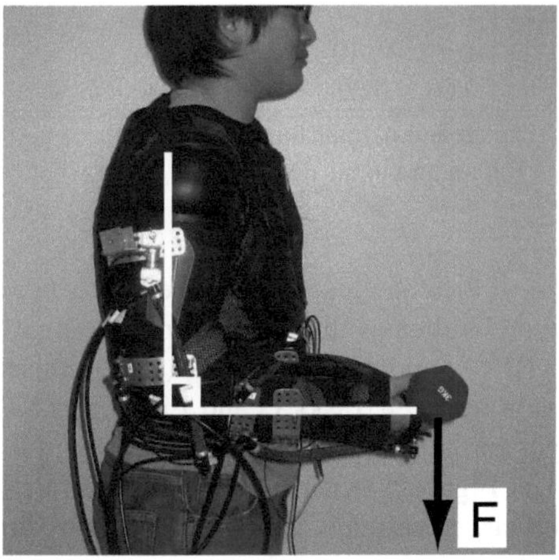

Fig. 4 Snapshot of the experiment: The exoskeleton robot applies computed torques against 4 joints from the elbow and wrists. A subject is asked to hold a weight of 3 by maintaining the elbow angle approximately 90 degrees. Surface electromyographic signals of brachialis (BRA), brachioradialis (BRD), flexor carpi ulnaris (FCU) and biceps muscles are measured. [7]

Table 1 Desired change rates of target muscle force

Task	BRA & BRD	FCU	BIC
A	×50%	×50%	×101%
B	×50%	×75%	×101%
C	×50%	×100%	×101%
D	×50%	×125%	×101%
E	×75%	×50%	×101%
F	×75%	×75%	×101%
G	×75%	×100%	×101%
H	×75%	×125%	×101%
I	×100%	×50%	×100%
J	×100%	×75%	×100%
K	×100%	×100%	×100%

Table 2 Target joint torque [Nm]

Task	τ_{elbow}	τ_{wrist1}*	τ_{wrist2}**
A	-1.27	-0.15	-0.22
B	-1.15	-0.05	-0.08
C	-1.02	0.05	0.07
D	-0.90	0.16	0.21
E	-0.76	-0.18	-0.26
F	-0.64	-0.08	-0.12
G	-0.51	0.03	0.03
H	-0.39	0.13	0.18
I	-0.25	-0.21	-0.29
J	-0.12	-0.10	-0.14
K	0.00	0.00	0.00

* Dorsal and palmar flexion joint of the wrist
** Radial and ulnar flexion joint of the wrist

the observed tendencies of the changes are discussed here. For example, a comparison between Task A and C indicates that the flexor carpi ulnaris (FCU) muscle and the group of brachialis (BRA) and brachioradialis (BRD) muscles can be controlled separately. A comparison between Tasks G and I indicates a similar result. In this experiment, the changes in the non-target muscle, biceps muscle (BIC), were also strictly zero as shown in Figure 5 (k). However, these errors were smaller overall than the changes in the target muscles. These observations support the efficacy of the individual muscle control algorithm. In contrast, there is still an issue in the accuracy of control. For example, no significant difference was observed between Tasks A and B or between Tasks E and F. This is partly because of the difficulty of precise force application against the body by using a wearable device.

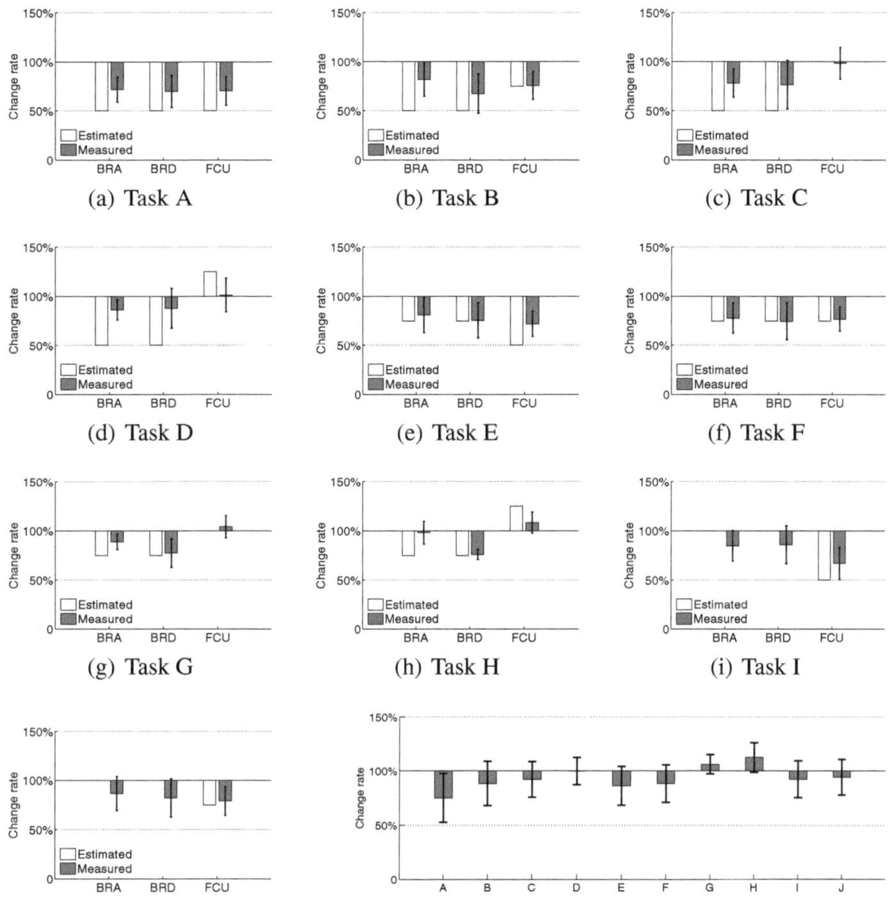

(a) Task A (b) Task B (c) Task C
(d) Task D (e) Task E (f) Task F
(g) Task G (h) Task H (i) Task I
(j) Task J (k) Changes of non-target muscle (BIC)

Fig. 5 Experimental results [7]

10 Conclusion

An overview of the individual muscle control methodology by using an exoskeleton has been presented. One of the future goals is to examine the age-associated and disease-associated muscle adjustments induced by the physical interaction between a person and a wearable exoskeleton-type robot device with integrated individual muscle controlling capability. There are still technical and scientific challenges that need to be overcome for practical and clinical use. The mathematical structure of the individual muscle control needs to be better understood. Specifically, the obtained feasibility conditions are restrictive and need to be relaxed. More muscle activity data need to be collected and compared for various motor tasks to identify the applicability as well as the performance limit. The collected data must then be interpreted from clinical perspectives. With further development, we imagine that a portable device will be developed for home use or in non-medical facilities.

Acknowledgements. This work was supported in part by the National Science Foundation (NSF) under Grant: IIS-1142438 and Japan Science and Technology Agency (JST) - National Science Foundation (NSF) Strategic International Cooperative Program.

References

1. Bertsekas, D.P.: Nonlinear programming. Athena Scientific, Belmont (1999)
2. Biel, A.: Trail guide to the body: How to locate muscles, bones and more. Andrew Biel (1997)
3. van Bolhuis, B., Gielen, C.: A comparison of models explaining muscle activation patterns for isometric contractions. Biological Cybernetics 81(3), 249–261 (1999)
4. Buchanan, T., Shreeve, D.: An Evaluation of Optimization Techniques for the Prediction of Muscle Activation Patterns During Isometric Tasks. Journal of Biomechanical Engineering 118, 565 (1996)
5. Davies, P.M.: Steps to Follow: The Comprehensive Treatment of Patients with Hemiplegia. Springer (2000)
6. Delp, S., Loan, J., Hoy, M., Zajac, F., Topp, E., Rosen, J., Center, V., Alto, P.: An interactive graphics-based model of the lower extremity to studyorthopaedic surgical procedures. IEEE Transactions on Biomedical Engineering 37(8), 757–767 (1990)
7. Ding, M., Ueda, J., Ogasawara, T.: Pinpointed muscle force control using a power-assisting device. Journal of the Robotics Society of Japan 27(9), 1037–1045 (2009)
8. Jacob, S.W., Francone, C.A., Lossow, W.J.: Structure and Function in Man. WB Saunders Company (1982)
9. Jesunathadas, M., Marmon, A., Gibb, J., Enoka, R.: Recruitment and derecruitment characteristics of motor units in a hand muscle of young and old adults. Journal of Applied Physiology 108(6), 1659–1667 (2010)
10. Karlsson, D., Peterson, B.: Towards a model for force predictions in the human shoulder. Journal of Biomechanics 25(2), 189–199 (1992)
11. Krishnamoorthy, V., Goodman, S., Zatsiorsky, V., Latash, M.: Muscle synergies during shifts of the center of pressure by standing persons: identification of muscle modes. Biological Cybernetics 89(2), 152–161 (2003)
12. Larsen, B.: Muscles: Testing and function with posture and pain. Medicine & Science in Sports & Exercise 37(8), 1447 (2005)
13. Martini, F., Bartholomew, E.: Essentials of Anatomy & Physiology. Prentice Hall (2000)
14. Ming, D., Ueda, J., Ogasawara, T.: Pinpointed muscle force control using a power-assisting device: System configuration and experiment. In: 2nd IEEE RAS & EMBS International Conference on Biomedical Robotics and Biomechatronics, BioRob 2008, pp. 181–186 (2008), doi:10.1109/BIOROB.2008.4762829
15. MotCo Project: `http://motco.info/data/pcsa.html`
16. National Institute of Neurological Disorders and Stroke: National Institute of Neurological Disorders and Stroke (2012), `http://www.ninds.nih.gov/`
17. Pardalos, P., Rosen, J.: Constrained global optimization: algorithms and applications (1987)
18. Prilutsky, B.: Coordination of two-and one-joint muscles: functional consequences and implications for motor control. Motor Control 4(1), 48–52 (2000)

19. Crowninshield, R., et al.: A physiologically based criterion of muscle force prediction in locomotion. J. Biomechanics 14, 793–801 (1981)
20. Reinkensmeyer, D., Aoyagi, D., Emken, J., Galvez, J., Ichinose, W., Kerdanyan, G., Maneekobkunwong, S., Minakata, K., Nessler, J., Weber, R., et al.: Tools for understanding and optimizing robotic gait training. Journal of Rehabilitation Research and Development 43(5), 657 (2006)
21. Reisman, D., Scholz, J.: Aspects of joint coordination are preserved during pointing in persons with post-stroke hemiparesis. Brain 126(11), 2510 (2003)
22. Ueda, J., Hyderabadwala, M., Krishnamoorthy, V., Shinohara, M.: Motor task planning for neuromuscular function tests using an individual muscle control technique. In: IEEE 11th International Conference on Rehabilitation Robotics (2009)
23. Ueda, J., Matsugashita, M., Oya, R., Ogasawara, T.: Control of muscle force during exercise using a musculoskeletal-exoskeletal integrated human model. In: Experimental Robotics, pp. 143–152 (2008)
24. Ueda, J., Ming, D., Krishnamoorthy, V., Shinohara, M., Ogasawara, T.: Individual muscle control using an exoskeleton robot for muscle function testing. IEEE Transactions on Neural Systems and Rehabilitation Engineering 18(4), 339–350 (2010)

Appendix

Crowninshield's Static Optimization

Crowninshield's method [19] is a special case presented in (2) that predicts human muscle forces by minimizing a physiologically based criterion $u(\boldsymbol{f})$:

$$u(\boldsymbol{f}) = \sum_{j=1}^{N} \left(\frac{f_j}{\mathrm{PCSA}_j} \right)^r \to \min \tag{33}$$

$$\text{subject to} \begin{cases} \boldsymbol{\tau}_h = \boldsymbol{A}\boldsymbol{f} \\ f_{\min_j} \leq f_j \leq f_{\max j}(j = 1, \cdots, N) \end{cases},$$

where PCSA_j is the physiological cross sectional area (PCSA), and $f_{\max j} = \varepsilon \cdot \mathrm{PCSA}_j$ is the maximum muscle force of the j-th muscle. In this paper, $\varepsilon = 0.7 \times 10^6 [\mathrm{N/m^2}]$ is given according to [10]. PCSA_j's are given according to [15]. $f_{\min j} = 0, \forall j$ and $r = 2$ are used. See [19] for the choice of r.

Synthesizing Redundancy Resolution Criteria of the Human Arm Posture in Reaching Movements

Zhi Li, Hyunchul Kim, Dejan Milutinović, and Jacob Rosen

Abstract. The aim of this work is to characterize the regularity and variability of human arm movements. The arm posture is quantified by a swivel angle that is experimentally measured when a healthy subject is reaching for targets in a spherical workspace. It is shown that without specific instruction, a subject moving his/her arm tends to use a consistent arm posture with small variations when reaching the same target position, regardless of whether the subject is moving toward or away from the target. This observed posture consistency indicates that human motor control chooses a unique arm posture associated to a 3D hand position. From the perspective of posture consistency, this work further examines the posture predictions based on two hypotheses on human arm control strategies: one that maximizes the movement efficiency towards the head, particularly toward the mouth; and the other that minimizes the power consumption in joint space. The arm posture predictions based on these two control strategies are compared and the prediction errors for each control strategy are analyzed.

1 Introduction

The study of human arm movements is complex due to the kinematic redundancy in the human arm. The human arm processes seven degrees of freedom (DOFs)

Zhi Li · Hyunchul Kim · Dejan Milutinović · Jacob Rosen
University of California, Santa Cruz, 1156 High Street, Santa Cruz, CA 95064, USA.
Bionics lab at UCSC
e-mail: {zhil,dejan}@soe.ucsc.edu,
hyunchul78@gmail.com,
rosen@ucsc.edu
http://bionics.soe.ucsc.edu/

D. Milutinović & J. Rosen (Eds.): Redundancy in Robot Manipulators, LNEE 57, pp. 201–240.
DOI: 10.1007/978-3-642-33971-4_12

while tasks in three-dimensional (3D) space only require six DOFs to be specified. A healthy human motor system provides natural resolutions to the inverse kinematics of the arm, with its control of the extra DOF outperforming that of wearable robotic arms (e.g., upper limb exoskeletons). Existing research has been focused on characterizing human arm movements in order to reveal a control strategy of the healthy human motor system, and to facilitate the design and control of upper limb exoskeletons. Enhanced synergy between the human arm and upper limb exoskeletons is expected to benefit applications such as the rehabilitation of stroke patients via upper limb exoskeletons [39, 25].

1.1 Characteristics of Human Arm Movements

Research on the characterization of human arm movements falls into two categories: from the perspective of regularity, the movements of healthy human arms demonstrate significant similarity when completing daily-life tasks, within and across human individuals; from the perspective of variability, it has be observed that the arm movements of each human individual are not exactly the same even when repeating the same task. These characteristics, namely regularity and variability universally exist in human movement, including human arm movement. Both of them contribute to the higher performance of the human arm, compared to that of existing robotic arms.

Regularity in Human Arm Movements: For decades, continuous research efforts on the regularity of human movements intended to reveal the control strategy of the healthy human motor system. According to Donders' law, the central nervous system (CNS) chooses a unique eye orientation for each gaze direction. When applied to human arm movements, Donders' law predicts that every position of the hand in 3D space naturally corresponds to a unique posture of the arm, which can be parameterized by joint angles at the shoulder and elbow. The unique pointing direction of the human arm corresponding to a given hand position (denoted by $\mathbf{r}$) can be expressed by a rotation axis $\mathbf{n}$ and a rotation angle α [17].

$$\mathbf{r} = \tan\frac{\alpha}{2}\mathbf{n} \qquad (1)$$

However, there have been experimental results that contradicted Donders' law. The law is obeyed more strictly for pointing movements with straight arms than for pointing movements with less restriction. According to Soechting et al [53], the arm posture corresponding to a given hand location is not independent of its previous posture. Furthermore, it has been confirmed that Donders's law is violated in some 3D space tasks [29]. The upper arm torsion varies widely when the pointing target is specified, yet the variation of torsion can be reduced by specifying the elbow angle. Note that while Donders's law is partially valid for reaching/pointing movements, it is generally invalidated by grasping experiments [57, 54]. Due to the limited motion range at the wrist, the posture of the upper arm is strongly affected by hand orientation and therefore the violation of Donders's law is significant. Without restricting the precise orientation of the hand, Donders' law might still be applicable to the arm motions [9].

The posture consistency in reaching movements of the human arm revealed by Donders's law results in a category of posture-based movement planning strategies. These movement planning strategies assume that there exists an optimal final posture for each target position at the end of the trajectory. This assumption contradicts the prediction of trajectory-based movement planning strategies, which may lead to various arm postures at the end of the trajectory. It was proposed that posture-based strategies plan the movements at a kinematic level, while trajectory-based strategies plan the movements at a dynamic level [43]. Other approaches of combining movement planning at kinematic and dynamic levels with the posture at the end of the trajectory are described in [18, 22]

Variability in Human Arm Movements: Variability is another universal characteristic of the motor control of human arm movements [63, 66]. Early experimental studies (e.g. recording of hammering movements by Bernstein [5]) in human motor control find that human movements do not repeat in exactly the same way for the same task, even with intention. It has been found that this variability can be used as a signature to distinguish skilled from unskilled task performance. A lower level of the variability may indicate the existence of control, while its absence may indicate diseases [34]. The redundancy in the human motor system may contribute to the variability of human movements, though it is not necessarily the source of the variability [13, 31, 59].

1.2 Redundancy Resolution Based on Performance Optimization

By controlling redundant degrees of freedom, the resolution of inverse kinematics or inverse dynamics can satisfy additional task-based constraints and/or achieve an optimized performance. Existing research has considered performance optimization from the perspectives of manipulability, energy consumption, smoothness of movement, task accuracy and control complexity. Task-based redundancy resolutions are more straightforward since the control of the extra DOFs can be generally achieved by integrating the task-dependent constraints into an augmented Jacobian matrix [7, 50, 51]. The redundancy resolutions based on performance optimization tend to be more flexible, given that there are many performance indices in consideration.

Manipulability Performance: At a singular configuration, a manipulator can only execute motion and/or resist wrenches in limited directions. Keeping the manipulator away from its singularities is convenient for task operation in general, and this can be achieved either by mechanical design and/or motion planning.

Manipulability was originally defined either as the distance from the singularity [2], or as the efficiency of velocity/force transmission [6]. The Jacobian matrix, denoted as $\mathbf{J}$ (or the matrix $\mathbf{J}\mathbf{J}^{\mathrm{T}}$ if $\mathbf{J}$ is a lower-rectangular matrix), has been used to quantify manipulability. Singular value decomposition (SVD) can be applied to the Jacobian matrix of manipulators, in order to construct the manipulability ellipsoid [45]. Possible manipulability indices for performance optimization are mostly based on the measures of the manipulability ellipsoid. The radii of the manipulability ellipsoid are frequently considered, either for the maximum/minimum singular values, or for their ratio (e.g., condition number [3], isotropy [1]). The determinant of the Jacobian matrix or dynamic Jacobian is also considered, resulting in performance indices such as manipulability and dynamic manipulability [67, 68].

The directions of the principle axes of the manipulability ellipsoid have rarely been considered as manipulability performance indices. It is worth noting that the direction of the principle axes indicate the movement efficiency of a manipulator configuration. For a given uniform effort (measured by joint velocity) in all the applicable directions in the joint space, the most efficient movement in task space is in the direction of the major principle axis of the manipulability ellipsoid, while the least efficient movement in task space is in the direction of the minor principle axis. With regards to global manipulability, indices such as condition number, isotropy can be integrated for the measurements of the workspace [14, 27].

Energy Performance: Minimization of energy, either in joint space or task space, implies that the final arm posture depends on both the initial arm posture and the trajectory. As a consequence, the arm postures for a given 3D hand position are not unique. It has been shown that energy minimization can not account for the average behavior of the arm movement [36], of eye movements [16] and of some full-body movements (e.g., standing up from a chair [37]). However, the consideration of energy performance can not be ruled out given the effects of dynamics. Instead, it should be integrated into other performance considerations such as the smoothness of motion, which reduces energy consumption by penalizing joint torque [62, 35], muscle forces [37], or time-derivatives of end-effector acceleration (i.e. jerk) [19, 11, 60, 52].

Smoothness of Movement: The idea of optimization for the smoothness of movement was first introduced as the minimization of jerk [19, 11], to account for the straight path and bell-shaped velocity of task-space trajectories in reaching movements, as well as for trajectories of "via-point" tasks, in which the hand is instructed to pass a sequence of positions. For arbitrary arm movements, minimizing the jerk along the trajectory accurately predict the speed profile of the trajectory [60] compared to the 2/3 power law [56, 64, 46]. The minimization of jerk has been also extended to account for movements in grasping tasks [52]. An alternative of jerk minimization in task space is to minimize the jerk in the joint space [65].

Performance optimization for the smoothness of motion can also be achieved at a dynamic level by minimizing the time-derivative of joint torque [62, 35]. This minimization also accounts for the slight asymmetry observed in some via-point tasks [62, 35], which cannot be addressed by kinematic motion strategies that ignore the nonlinear arm dynamics.

Task Accuracy: Motor noise is considered to contribute to the variance of end-effector position across repetitions of the same task. It is know that motor noise is dependent on control, with its magnitude proportional to muscle activations [58, 55, 47]. As a consequence, the choice of control signals will affect the variability of a movement.

Within an open-loop control framework, the control strategy of minimizing the variance intends to optimize a sequence of muscle activations, for reduced variances in the end-effector positions and improved task accuracy [15]. The minimum variance model produces an accurate prediction of eye movements at the level of muscle activations, yet its prediction accuracy is not clear for human arm movements. Movements with longer durations can not be addressed by minimum-variance control, since the movement variability is strongly affected by sensory feedback, which is not considered in open-loop control [44].

However, considering the universal existence of the motor noise in biological systems, it makes sense to assume that there exists a general control strategy so that the relation between a trajectory and its velocity profile can be addressed.

Control Complexity: Control strategies yield different performance in the presence of noise, even if the averaged behavior is the same [59]. Optimal feedback controllers can resolve the redundancy in real-time according to the minimum intervention principle: make no effort to correct deviations away from the average behavior unless the task performance is affected.

As demonstrated in [59], the minimal intervention principle pushes the state vector orthogonally to the redundant direction, in which performance is maintained and corresponding states are equivalent to each other. In the redundant direction, which has been quantified as an "uncontrolled manifold", the probability distribution of observed states scatters in a wider range, compared to the non-redundant direction. A wide range of behaviors [59, 5, 48, 49, 26] have provided evidence of the minimal intervention principle.

Integrating Multiple Criteria for a Better Estimation of the Arm Posture: Existing hypotheses, either working collaboratively or individually, have not been able to fully predict the natural movements of human arms. However, the integration of multiple hypotheses for better prediction can help in understanding the control strategy of natural human arm movements. In this case, the challenge of formulating a cost function is that performance indices have different units, and therefore it is not trivial to combine them in a single criterion. Having this in mind, a appropriately chosen intermediate variable may help the integration of different indices into a single criterion. As shown in [23], the swivel angle is chosen as the intermediate variable, to merge two performance indices of different units (manipulability and energy). However, the chosen intermediate variable may have different levels of sensitivity to changes in different performance indices.

The optimization of a comprehensive cost function that integrates various types of performance indices cannot be simply extended from the optimization of a single performance index, particularly with the presence of noise and disturbance in the implementation of the movement plan [42, 11]. The optimization of a single performance index along a deterministic trajectory can be constrained by task-dependent constraints, such as end-effector position, velocity and acceleration specified for the beginning or ending state. Such constraints are not valid for stochastic problems, in which the final state is affected by noise.

2 Kinematic and Dynamic Modeling of the Human Arm

2.1 *Kinematic Modeling of the Human Arm*

The kinematics and dynamics of the human arm during activities of daily living (ADL) have been studied to determine specifications for exoskeleton design (Figure 1) [40][38]. Articulation of the exoskeleton is achieved by seven single-axis revolute joints which support 99% of the range of motion required to perform daily activities [40]. Three revolute joints are responsible for shoulder abduction-adduction, flexion-extension and internal-external rotation. A single rotational joint is employed at the elbow, creating elbow flexion-extension. Finally, the lower arm and hand are connected by a three-axis spherical joint resulting in wrist pronation-supination, flexion-extension, and radial-ulnar deviation. As a human-machine interface (HMI), four six-axis force/torque sensors (ATI Industrial Automation, model-Mini40) are attached to the upper arm, the lower arm, the hand and the tip of the exoskeleton [32]. The force/torque sensor at the tip of the exoskeleton allows measurement of interactions between the exoskeleton and the environment.

2.1.1 Forward Kinematics

This section derives the forward kinematics for the upper limb exoskeleton.

Base Rotation for Singularity Avoidance: The bases of the two robotic arms of the upper limb exoskeleton are rotated according to Table 1, in order to move the singularity out of the range of the daily movements of the human arm.

Table 1 Base rotation of upper limb exoskeleton

	about X-axis (θ_X)	about Y-axis (θ_Y)	about Z-axis (θ_Z)
Left arm	132.5°	45°	90°
Right arm	132.5°	−45°	90°

Fig. 1 The upper limb exoskeleton with seven DOFs, supporting 99% of the range of motion required to preform daily activities

The transformation matrix for base rotation can be represented as Equation (2). Note that $\sin\theta_i$ is denoted as s_i, $\cos\theta_i$ is denoted as c_i, $\sin\alpha_i$ is denoted as $s\alpha_i$, $\cos\alpha_i$ is denoted as $c\alpha_i$.

$$T_{base} = Rotx(\theta_X)Rotz(\theta_Y)Rotz(\theta_Z)$$
$$= \begin{bmatrix} 1 & 0 & 0 & 0 \\ 0 & c\theta_X & -s\theta_X & 1 \\ 0 & s\theta_X & c\theta_X & 1 \\ 0 & 0 & 0 & 1 \end{bmatrix} \begin{bmatrix} c\theta_Y & 0 & s\theta_Y & 1 \\ 0 & 1 & 0 & 0 \\ -s\theta_Y & 0 & c\theta_Y & 1 \\ 0 & 0 & 0 & 1 \end{bmatrix} \begin{bmatrix} c\theta_Z & -s\theta_Z & 0 & 1 \\ s\theta_Z & c\theta_Z & 0 & 1 \\ 0 & 0 & 1 & 0 \\ 0 & 0 & 0 & 1 \end{bmatrix} \quad (2)$$

For the left arm,

$$T_{base} = \begin{bmatrix} 0.0000 & -0.7071 & 0.7071 & 0 \\ -0.6756 & -0.5213 & -0.5213 & 0 \\ 0.7373 & -0.4777 & -0.4777 & 0 \\ 0 & 0 & 0 & 1.0000 \end{bmatrix}, \quad (3)$$

for the right arm,

$$T_{base} = \begin{bmatrix} 0.0000 & -0.7071 & -0.7071 & 0 \\ -0.6756 & 0.5213 & -0.5213 & 0 \\ 0.7373 & 0.4777 & -0.4777 & 0 \\ 0 & 0 & 0 & 1.0000 \end{bmatrix}. \quad (4)$$

Denavit-Hartenberg (DH) Parameters: The Denavit-Hartenberg (DH) parameters of the upper limb exoskeleton (shown in Table 2) are derived in the standard method defined by [8].

Table 2 Denavit-Hartenberg (DH) Parameters for upper limb exoskeleton

Robot	$i-1$	i	α_i	a_i	d_i	θ_i
Left Arm	0	1	$\pi/2$	0	0	$\theta_1+\pi-32.94°$
	1	2	$\pi/2$	0	0	$\theta_2+\pi/2-28.54°$
	2	3	$\pi/2$	0	0	$\theta_3+\pi-53.6°$
	3	4	$\pi/2$	0	L_1	θ_4
	4	5	$-\pi/2$	0	0	$\theta_5-\pi/2$
	5	6	$-\pi/2$	0	L_2	$\theta_6+\pi/2$
	6	7	$\pi/2$	0	0	$\theta_7+\pi$
Right Arm	0	1	$\pi/2$	0	0	$\theta_1-32.94°$
	1	2	$\pi/2$	0	0	$\theta_2-\pi/2-28.54°$
	2	3	$-\pi/2$	0	0	$\theta_3-\pi-53.6°$
	3	4	$-\pi/2$	0	$-L_1$	θ_4
	4	5	$\pi/2$	0	0	$\theta_5+\pi/2$
	5	6	$-\pi/2$	0	$-L_2$	$\theta_6+\pi/2$
	6	7	$\pi/2$	0	0	$\theta_7+\pi$

Note that L_1 and L_2 are the length of the upper and lower arms, respectively.

By direct kinematics, we can derive the transformation matrix ${}^0_7T$, which includes the position and the orientation of the wrist of the exoskeleton with respect to the base frame:

$$ {}^{base}_{7}T = T_{base} \cdot {}^0_1T \cdot {}^1_2T \cdot {}^2_3T \cdot {}^3_4T \cdot {}^4_5T \cdot {}^5_6T \cdot {}^6_7T = \begin{bmatrix} r_{11} & r_{12} & r_{13} & P_{wx} \\ r_{21} & r_{22} & r_{23} & P_{wy} \\ r_{31} & r_{32} & r_{33} & P_{wz} \\ 0 & 0 & 0 & 1 \end{bmatrix} \quad (5) $$

For reaching movements, the three DOFs at the wrist are not considered. Therefore, the forward kinematics that involves four DOFs of the human arm (three DOFs at the shoulder and one DOF at the elbow) becomes:

$$ {}^{base}_{7}T = T_{base} \cdot {}^0_1T \cdot {}^1_2T \cdot {}^2_3T \cdot {}^3_4T \cdot {}^4_5T \quad (6) $$

2.1.2 Inverse Kinematics

With the specification of the transformation matrix ${}^0_7T$, the inverse kinematics of the exoskeleton can be derived for the left and the right arms, respectively. The redundant DOF of the human arm can be constrained by specifying the elbow position ($P_e = [Pe_x, Pe_y, Pe_z]^T$).

Based on shoulder position P_s, elbow position P_e, and wrist position P_w, θ_4 can be derived as:

$$W = ||P_w - P_s|| \tag{7}$$

$$c_4 = \frac{L_1^2 + L_2^2 - W^2}{2L_1L_2} \tag{8}$$

$$s_4 = \sqrt{1 - c_4^2} \tag{9}$$

$$\theta_4 = \pi - Atan2(s_4, c_4) \tag{10}$$

The transformation matrix ${}^3_4T$ and its inverse ${}^3_4T^{-1}$ can be found based on θ_4.

The transformation matrix without the base rotation, denoted ${}^{base}_7T$, can be found by:

$${}^0_7T = T_0^{-1} \cdot {}^{base}_7 T = \begin{bmatrix} r'_{11} & r'_{12} & r'_{13} & {}^0_7P_{wx} \\ r'_{21} & r'_{22} & r'_{23} & {}^0_7P_{wy} \\ r'_{31} & r'_{32} & r'_{33} & {}^0_7P_{wz} \\ 0 & 0 & 0 & 1 \end{bmatrix} \tag{11}$$

Thus, the wrist position with respect to the rotated base is ${}^0_7P_w = [{}^0_7P_{wx}, {}^0_7P_{wy}, {}^0_7P_{wz}]^T$.

Similarly, the elbow position with respect to the rotated base, denoted by ${}^0_7P_e = [{}^0_7P_{ex}, {}^0_7P_{ey}, {}^0_7P_{ez}]^T$, is:

$$\begin{bmatrix} {}^0_7P_{ex} \\ {}^0_7P_{ey} \\ {}^0_7P_{ez} \\ 1 \end{bmatrix} = T_0^{-1} \cdot \begin{bmatrix} {}^{base}_7P_{ex} \\ {}^{base}_7P_{ey} \\ {}^{base}_7P_{ez} \\ 1 \end{bmatrix} \tag{12}$$

Note that ${}^0_7P_e = {}^0_4P_e$ and

$${}^0_4T = {}^0_1T \cdot {}^1_2T \cdot {}^2_3T \cdot {}^3_4T = \begin{bmatrix} & {}^0_4R & & {}^0_4P_{ex} \\ & & & {}^0_4P_{ey} \\ & & & {}^0_4P_{ez} \\ 0 & 0 & 0 & 1 \end{bmatrix} = \begin{bmatrix} & {}^0_4R & & L_1c_1s_2 \\ & & & L_1c_2 \\ & & & L_1s_1s_2 \\ 0 & 0 & 0 & 1 \end{bmatrix} \tag{13}$$

For the both arms,

$$c_2 = \frac{{}^0_4P_{ey}}{L_1} \tag{14}$$

For the left arm,

$$s_2 = \sqrt{(1 - c_2^2)} \tag{15}$$

for the right arm,

$$s_2 = -\sqrt{(1 - c_2^2)} \tag{16}$$

Thus, θ_2 can be resolved as:

$$\theta_2 = Atan2(s_2, c_2) - (\pi/2 - 28.54^\circ) \tag{17}$$

To resolve θ_1, for the both arms,

$$c_1 = \frac{{}^0_4P_{ex}}{L_1 s_2} \tag{18}$$

$$s_1 = \frac{{}^0_4P_{ez}}{L_1 s_2} \tag{19}$$

Thus, for the left arm,

$$\theta_1 = Atan2(s_1, c_1) - (\pi - 32.94^\circ) \tag{20}$$

for the right arm,

$$\theta_1 = Atan2(s_1, c_1) + 32.94^\circ \tag{21}$$

The transformation matrices ${}^0_1T$ and ${}^1_2T$ and their inverses ${}^0_1T^{-1}$ and ${}^1_2T^{-1}$ can be found accordingly.

Thus, the wrist position with respect to Frame 2, denoted ${}^2_7P_w = [{}^2_7P_{wx}, {}^2_7P_{wy}, {}^2_7P_{wz}]^T$, can be found:

$${}^2_7T = {}^1_2T^{-1} \cdot {}^0_1T^{-1} \cdot {}^0_7T = \begin{bmatrix} & {}^2_7R & & {}^2_7P_{wx} \\ & & & {}^2_7P_{wy} \\ & & & {}^2_7P_{wz} \\ 0 & 0 & 0 & 1 \end{bmatrix} \tag{22}$$

For the left arm,

$${}^2_7P_w = \begin{bmatrix} -L_2c_3s_4 \\ -L_1 - L_2c_4 \\ -L_2s_3s_4 \end{bmatrix} \tag{23}$$

for the right arm,

$${}^2_7P_w = \begin{bmatrix} -L_2c_3s_4 \\ -L_1 - L_2c_4 \\ L_2s_3s_4 \end{bmatrix} \tag{24}$$

To resolve θ_3, for the both arms,

$$c_3 = \frac{{}^2_7P_{wx}}{-L_2 s_4} \tag{25}$$

For the left arm,

$$s_3 = \frac{{}^2_7P_{wz}}{L_2 s_4} \tag{26}$$

$$\theta_3 = Atan2(s_3, c_3) - (\pi - 53.6^\circ) - 2\pi \tag{27}$$

for the left arm,

$$s_3 = \frac{{}^2_7P_{wz}}{-L_2 s_4} \theta_3 = Atan2(s_3, c_3) + (\pi + 53.6^\circ) \tag{28}$$

The transformation matrix ${}^2_3T$ and its inverse ${}^2_3T^{-1}$ can be found accordingly.

θ_5, θ_6 and θ_7 can be derived from the transformation matrices from Frame 4 to Frame 7 ${}^4_7T$.

$${}^4_7T = {}^3_4T^{-1} \cdot {}^2_3T^{-1} \cdot {}^1_2T^{-1} \cdot {}^0_1T^{-1} \cdot {}^0_7T = \begin{bmatrix} {}^4_7r_{11} & {}^4_7r_{12} & {}^4_7r_{13} & {}^4_7P_{wx} \\ {}^4_7r_{21} & {}^4_7r_{22} & {}^4_7r_{23} & {}^4_7P_{wy} \\ {}^4_7r_{31} & {}^4_7r_{32} & {}^4_7r_{33} & {}^4_7P_{wz} \\ 0 & 0 & 0 & 1 \end{bmatrix} \tag{29}$$

For the left arm,

$$\begin{aligned} {}^4_7T &= {}^3_4T^{-1} \cdot {}^2_3T^{-1} \cdot {}^1_2T^{-1} \cdot {}^0_1T^{-1} \cdot {}^0_7T \\ &= \begin{bmatrix} c_5c_6c_7 - s_5s_7 & -c_7s_5 - c_5c_6s_7 & c_5s_6 & 0 \\ -c_7s_6 & s_6s_7 & c_6 & L_2 \\ -c_5s_7 - c_6c_7s_5 & c_5c_7 - c_6s_5s_7 & -s_5s_6 & 0 \\ 0 & 0 & 0 & 1 \end{bmatrix} \end{aligned} \tag{30}$$

for the right arm,

$$\begin{aligned} {}^4_7T &= {}^3_4T^{-1} \cdot {}^2_3T^{-1} \cdot {}^1_2T^{-1} \cdot {}^0_1T^{-1} \cdot {}^0_7T \\ &= \begin{bmatrix} c_5c_6c_7 - s_5s_7 & -c_7s_5 - c_5c_6s_7 & c_5s_6 & 0 \\ c_7s_6 & -s_6s_7 & -c_6 & L_2 \\ c_5s_7 + c_6c_7s_5 & c_5c_7 - c_6s_5s_7 & s_5s_6 & 0 \\ 0 & 0 & 0 & 1 \end{bmatrix} \end{aligned} \tag{31}$$

Thus, for the left arm,

$$c_6 = {}^4_7r_{23} \tag{32}$$

$$s_6 = \sqrt{1-c_6^2} \tag{33}$$

$$c_5 = \frac{{}^4_7r_{13}}{s_6} \tag{34}$$

$$s_5 = -\frac{{}^4_7r_{33}}{s_6} \tag{35}$$

$$c_7 = -\frac{{}^4_7r_{21}}{s_6} \tag{36}$$

$$s_7 = \frac{{}^4_7r_{22}}{s_6} \tag{37}$$

for the right arm,

$$c_6 = -{}^4_7r_{23} \tag{38}$$

$$s_6 = \sqrt{1-c_6^2} \tag{39}$$

$$c_5 = -\frac{{}^4_7r_{13}}{s_6} \tag{40}$$

$$s_5 = -\frac{{}^4_7r_{33}}{s_6} \tag{41}$$

$$c_7 = -\frac{{}^4_7r_{21}}{s_6} \tag{42}$$

$$s_7 = -\frac{{}^4_7r_{22}}{s_6} \tag{43}$$

For the left arm,

$$\theta_5 = Atan2(s_5, c_5) + \pi/2 \tag{44}$$

$$\theta_6 = Atan2(s_6, c_6) - \pi/2 \tag{45}$$

$$\theta_7 = Atan2(s_7, c_7) - \pi + 2\pi \tag{46}$$

for the right arm,

$$\theta_5 = Atan2(s_5, c_5) - \pi/2 \tag{47}$$

$$\theta_6 = Atan2(s_6, c_6) - \pi/2 \tag{48}$$

$$\theta_7 = Atan2(s_7, c_7) - \pi + 2\pi \tag{49}$$

For reaching movements, the four DOFs in consideration (three DOFs at the shoulder and one DOF at the elbow) can be resolved based on the wrist position P_w and the elbow position P_e: θ_4 is resolved according to Equation (10); θ_1 and θ_2 are resolved according to Equations (13) to (21). With regards to θ_3,

For the left arm,

$$ {}^{2}_{5}P_{w} = \begin{bmatrix} -L_2 c_3 s_4 \\ -L_1 - L_2 c_4 \\ -L_2 s_3 s_4 \end{bmatrix} \quad (50) $$

for the right arm,

$$ {}^{2}_{5}P_{w} = \begin{bmatrix} -L_2 c_3 s_4 \\ -L_1 - L_2 c_4 \\ L_2 s_3 s_4 \end{bmatrix} \quad (51) $$

Therefore, θ_3 can be resolved as Equations (25) to (28).

2.2 Jacobian Matrix

The Jacobian matrix denotes the mapping from joint space to task space at the velocity level.

$$ \dot{P}_w = J\dot{\theta} \quad (52) $$

For the seven-DOF arm model involving wrist orientation,

$$ \dot{P}_w = J_{3\times 7}\dot{\theta} \quad (53) $$

where $\theta = [\theta_1, \theta_2, \theta_3, \theta_4, \theta_5, \theta_6, \theta_7]^T$, and

$$ J_{3\times 7} = \begin{bmatrix} J_1 \ J_2 \ J_3 \ J_4 \ \mathbf{0} \ \mathbf{0} \ \mathbf{0} \end{bmatrix} \quad (54) $$

The arm model for reaching movements only involves four DOFs and therefore

$$ \dot{P}_w = J_{3\times 4}\dot{\theta} \quad (55) $$

where $\theta = [\theta_1, \theta_2, \theta_3, \theta_4]^T$ and

$$ J_{4\times 7} = \begin{bmatrix} J_1 \ J_2 \ J_3 \ J_4 \end{bmatrix} \quad (56) $$

For the right arm, given that

$$ P_w = \begin{bmatrix} L_2(s_4(s_1 s_3 - c_1 c_2 c_3) + c_1 c_4 s_2) + L_1 c_1 s_2 \\ L_1 c_2 + L_2(c_2 c_4 + c_3 s_2 s_4) \\ L_1 s_1 s_2 - L_2(s_4(c_1 s_3 + c_2 c_3 s_1) - c_4 s_1 s_2) \end{bmatrix} \quad (57) $$

we have

$$ J_1 = \begin{bmatrix} L_2(s_4(c_1 s_3 + c_2 c_3 s_1) - c_4 s_1 s_2) - L_1 s_1 s_2 \\ L_1 c_2 + L_2(c_2 c_4 + c_3 s_2 s_4) \\ L_2(s_4(s_1 s_3 - c_1 c_2 c_3) + c_1 c_4 s_2) + L_1 c_1 s_2 \end{bmatrix} \quad (58) $$

$$J_2 = \begin{bmatrix} L_2(s_4(s_1s_3 + c_1c_3s_2) + c_1c_2c_4) + L_1c_1c_2 \\ -L_1s_2 - L_2(c_4s_2 - c_2c_3s_4) \\ L_1c_2s_1 - L_2(s_4(c_1s_3 - c_3s_1s_2) - c_2c_4s_1) \end{bmatrix} \tag{59}$$

$$J_3 = \begin{bmatrix} L_2(s_4(c_3s_1 + c_1c_2s_3) + c_1c_4s_2) + L_1c_1s_2 \\ L_1c_2 + L_2(c_2c_4 - s_2s_3s_4) \\ L_1s_1s_2 - L_2(s_4(c_1c_3 - c_2s_1s_3) - c_4s_1s_2) \end{bmatrix} \tag{60}$$

$$J_4 = \begin{bmatrix} L_2(c_4(s_1s_3 - c_1c_2c_3) - c_1s_2s_4) + L_1c_1s_2 \\ L_1c_2 - L_2(c_2s_4 - c_3c_4s_2) \\ L_1s_1s_2 - L_2(c_4(c_1s_3 + c_2c_3s_1) + s_1s_2s_4) \end{bmatrix} \tag{61}$$

For the left arm, given that

$$P_w = \begin{bmatrix} L_2(s_4(s_1s_3 - c_1c_2c_3) + c_1c_4s_2) + L_1c_1s_2 \\ L_1c_2 + L_2(c_2c_4 + c_3s_2s_4) \\ L_1s_1s_2 - L_2(s_4(c_1s_3 + c_2c_3s_1) - c_4s_1s_2) \end{bmatrix} \tag{62}$$

we have

$$J_1 = \begin{bmatrix} -L_2(s_4(c_1s_3 - c_2c_3s_1) + c_4s_1s_2) - L_1s_1s_2 \\ L_1c_2 + L_2(c_2c_4 + c_3s_2s_4) \\ L_1c_1s_2 - L_2(s_4(s_1s_3 + c_1c_2c_3) - c_1c_4s_2) \end{bmatrix} \tag{63}$$

$$J_2 = \begin{bmatrix} L_1c_1c_2 - L_2(s_4(s_1s_3 - c_1c_3s_2) - c_1c_2c_4) \\ -L_1s_2 - L_2(c_4s_2 - c_2c_3s_4) \\ L_2(s_4(c_1s_3 + c_3s_1s_2) + c_2c_4s_1) + L_1c_2s_1 \end{bmatrix} \tag{64}$$

$$J_3 = \begin{bmatrix} L_1c_1s_2 - L_2(s_4(c_3s_1 - c_1c_2s_3) - c_1c_4s_2) \\ L_1c_2 + L_2(c_2c_4 - s_2s_3s_4) \\ L_2(s_4(c_1c_3 + c_2s_1s_3) + c_4s_1s_2) + L_1s_1s_2 \end{bmatrix} \tag{65}$$

$$J_4 = \begin{bmatrix} L_1c_1s_2 - L_2(c_4(s_1s_3 + c_1c_2c_3) + c_1s_2s_4) \\ L_1c_2 - L_2(c_2s_4 - c_3c_4s_2) \\ L_2(c_4(c_1s_3 - c_2c_3s_1) - s_1s_2s_4) + L_1s_1s_2 \end{bmatrix} \tag{66}$$

2.3 Redundancy Representation by Swivel Angle

In addition to the elbow position, the extra degree of freedom can be constrained by specifying the swivel angle. When the arm is not straight, the positions of the shoulder (P_s), the elbow (P_e) and the wrist (P_w) form a triangle. With the spherical joins at both the shoulder and wrist, the elbow position P_e can only rotate around the vector $(P_w - P_s)$ (see Figure 2). A local coordinate system at the center of the elbow circle (P_c) gives a reference for measuring the swivel angle (ϕ) of the elbow. A normal vector that points in the direction of $(P_w - P_s)$ is defined as:

$$\mathbf{n} = \frac{P_w - P_s}{||P_w - P_s||} \tag{67}$$

A normalized vector projected onto the plane orthogonal to $\mathbf{n}$ is given by:

$$\mathbf{u} = \frac{\mathbf{a} - (\mathbf{a} \cdot \mathbf{n})\mathbf{n}}{||\mathbf{a} - (\mathbf{a} \cdot \mathbf{n})\mathbf{n}||} \tag{68}$$

where $\mathbf{a}$ is the vector to be projected. Badler and Torlani [4] suggest that $\mathbf{a}$ should be $[0, 0, -1]^T$. This selection has real physical meaning. When ϕ is equal to zero, the elbow is at its lowest possible point. The last vector of the coordinate system ($\mathbf{v}$), is found by taking the cross product of $\mathbf{n}$ and $\mathbf{u}$. Vectors $\mathbf{n}$, $\mathbf{u}$ and $\mathbf{v}$ form an orthogonal coordinate system, where $\mathbf{u}$ and $\mathbf{v}$ are in the plane of the elbow circle (Figure 2(b)). The radius (R) and center (P_c) of the circle are easily found through geometry.

$$\cos(\alpha) = \frac{L_1^2 - L_2^2 - ||P_w - P_s||^2}{-2L_2^2||P_w - P_s||} \tag{69}$$

$$\sin(\alpha) = \sqrt{1 - \cos(\alpha)} \tag{70}$$

$$R = L_1 \sin(\alpha) \tag{71}$$

$$P_c = P_s + L_1 \cos(\alpha) \cdot \mathbf{n} \tag{72}$$

, where L_1 and L_2 are the lengths of the upper and lower arm segments (Figure 2(a)). The position of the elbow can now be expressed as a parametrization of ϕ [61].

$$P_e = R\left[\cos(\phi)\mathbf{u} + \sin(\phi)\mathbf{v}\right] + P_c \tag{73}$$

2.4 Dynamic Modeling of the Human Arm

The dynamic models of the left and right human arms are built up by integrating the kinematic model with the estimates of mass, the center of mass and the moment of inertia. Dynamic models of the human arm are rendered via the Autolev package [12], which generates the motion equation by Kane's method [20]. Each arm model processes seven DOFs (three DOFs for the shoulder, three DOFs for the wrist and one DOF for the elbow motion), with the frame setup in accordance with the EXO-UL7. Since the analysis of reaching movements in free space does not specify the wrist posture, the orientation of the hand in the dynamic model is pre-specified by locking the three DOFs at the wrist joint.

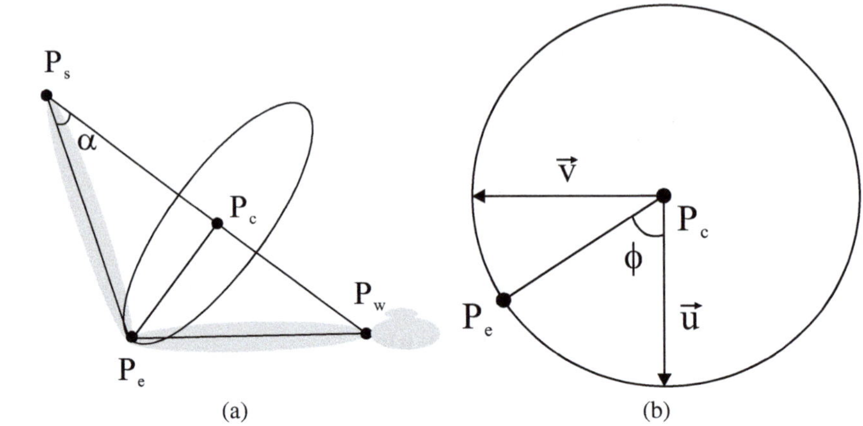

Fig. 2 Swivel angle

Given the initial condition of the human arm, the dynamic model can respond to external forces (such as gravity) and provide an analytical calculation of the joint space variables (i.e., joint angles, velocities and accelerations), as well as the kinetic energy and potential energy. The analytical calculation of joint torque is integrated in the dynamic model and can be extracted to compute work in the joint space. The dynamic model can also respond to external joint torques and generate the resulting joint space values accordingly.

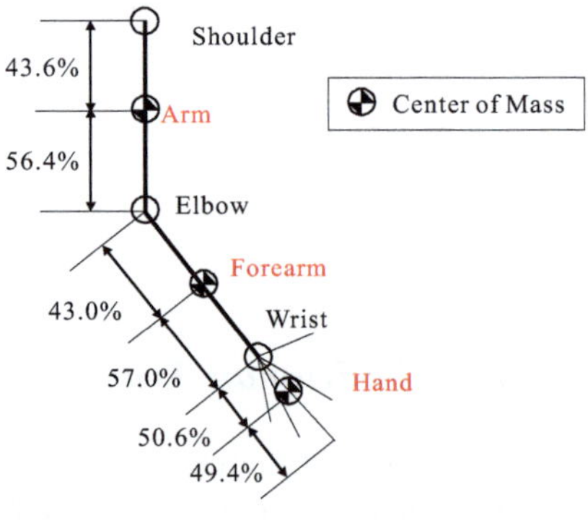

Fig. 3 The distribution of the center of mass (COM) for the human arm segments [41]

The estimation of the mass, the center of mass and the moment of inertia of each arm segment is required to customize the dynamic model for each subject. Figure 3 shows the distribution of the center of mass (COM) for the human arm segments. On average, the human arm contributes 4.8% of the total body weight. The mass of arm segments and their inertia matrices are calculated based on the weight of subjects according to the regression equations in [30].

3 Swivel Angle Estimation for Reaching Movements

3.1 Criterion I: Swivel Angle Estimation Using a Biologically-Based Kinematic Constraint

Given the role of the head as a cluster of sensing organs and the importance of arm manipulation to deliver food to the mouth, we hypothesize that the swivel angle is determined by the human motor control system to efficiently retract the hand to the head region. This hypothesis is supported by intracortical stimulation experiments that evoked coordinated forelimb movements in conscious primates [10][33]. It has been reported that each stimulation site produced a stereotyped posture in which the arm moved to the same final position regardless of its posture in the initial stimulation. In the most complex example, the monkey formed a frozen pose with the hand in a grasping position in front of the open mouth. This implies that during the arm movement toward an actual target, the virtual target point at the head can be set for the potential retraction of the palm to the virtual target as shown in Figure 4.

Manipulability Ellipsoid: According to the above notion of efficient arm movements toward a virtual target at the head, the redundancy of the human arm can be closely associated with the manipulability ellipsoid. Let P_m denote the virtual target position at the center of the head in Figure 5(a). When we consider the combinations of joint velocities satisfying the condition in which $\Sigma_{i=1}^{n} \dot{\theta}_i^2 = 1$, the hand velocity as a function of the joint velocity is described by an ellipsoid that defines the arm's scaled Jacobian. The longest principle axis of the manipulability ellipsoid (i.e., the major principle axis) defines the direction of the highest sensitivity where the end effector velocity varies in response to the joint space velocity (see Figure 5(b)) [28]. Assuming that the virtual hand movement follows the shortest path connecting P_w to P_m, the swivel angle is chosen such that the projection of the major principle axis of the manipulability ellipsoid onto $(P_m - P_w)$ is maximized.

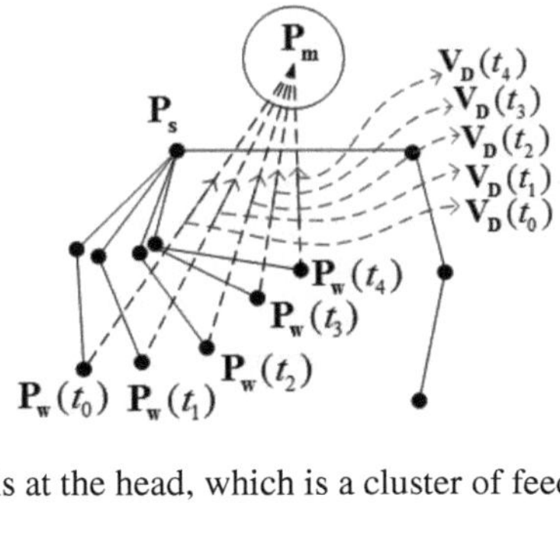

Fig. 4 Virtual destination is at the head, which is a cluster of feedback sensors

Lemma 1. *Given the inequality $\|P_w - P_s\| > \|P_w - P_e\|$, the major principle axis of the manipulability ellipsoid is coplanar with plane S, defined by P_w, P_e and P_s, and its magnitude σ_1 is expressed as*

$$\sigma_1 = \sqrt{\lambda_1} = \sqrt{((L_{ws}^2 + L_{we}^2) + (L_{ws}^2 + L_{we}^2)c_1)/2} \tag{74}$$
$$c_1 = \sqrt{1 - c_2},\ c_2 = 4L_{we}^2 L_{ws}^2 \sin^2 \varphi / (L_{ws}^2 + L_{we}^2)^2$$

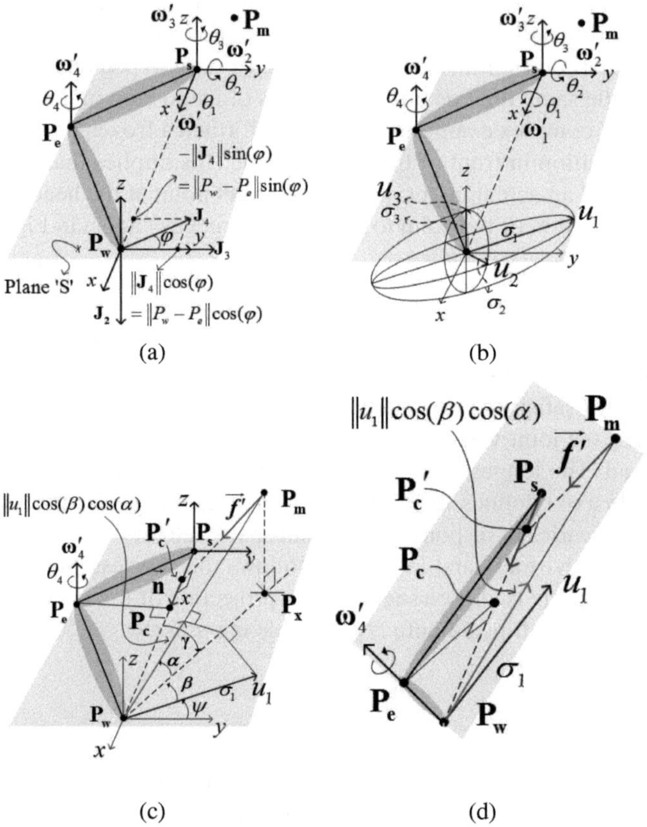

Fig. 5 The new coordinate system composed of P_w, P_e, P_s and P_m. (a) Each element J_i in the Jacobian matrix is defined with respect to the newly defined frame on the shoulder where the x axis is defined as $(P_w - P_s)/\|P_w - P_s\|$ and the y axis sits on the plane S composed of P_w, P_e and P_s. The new frame on the shoulder is defined for the convenience of the calculation. (b) Manipulability ellipsoid at the wrist position. u_1,u_2 and u_3 indicate the three principle axes of the ellipsoid with magnitudes σ_1,σ_2 and σ_3. (c) The direction of the largest manipulability (i.e., vector u_1) projected on the $(P_m - P_w)/\|P_m - P_w\|$ is marked as an arrow along $(P_m - P_w)$ and its magnitude can be represented as $\|u_1\|\cos(\alpha)\cos(\beta)$. (d) It shows the specific elbow position for the given wrist position that maximizes the manipulability projected on the direction from the hand toward the virtual target. In this configuration, P_m,P_s,P_e and P_w are on the same plane.

Proof. As the human arm moves, the wrist position P_w and elbow position P_e change and therefore a time-varying Plane 'S' can be defined by three points P_s, P_e and P_w. As shown in Figure 5(a), a frame attached to the time-varying Plane S, with its origin located at P_s. The x-axis, denoted by ω_1, is defined along the vector $|P_w - P_s|$, while the z-axis, denoted by ω_3, is orthogonal to the plane S'. Accordingly, the y-axis, denoted by ω_2 is within the time-varying Plane S. With P_e as the origin of the frame of elbow flexion, θ_4 represents the elbow flexion within the Plane S, with ω_4 as the axis to rotate about.

The relationship between the end-effector velocity $\dot{\mathbf{P}} = [\dot{P}_{wx}\ \dot{P}_{wy}\ \dot{P}_{wz}]^T$ and joint velocity $\dot{\theta} = [\dot{\theta}_1\ \dot{\theta}_2\ \dot{\theta}_3\ \dot{\theta}_4]$ is defined as:

$$\begin{aligned}\dot{\mathbf{P}} &= \mathbf{J}\dot{\theta} = [\mathbf{J}_1\ \mathbf{J}_2\ \mathbf{J}_3\ \mathbf{J}_4]\dot{\theta} \\ &= \mathbf{J}_1\dot{\theta}_1 + \mathbf{J}_2\dot{\theta}_2 + \mathbf{J}_3\dot{\theta}_3 + \mathbf{J}_4\dot{\theta}_4\end{aligned} \tag{75}$$

$$\mathbf{J}_i = \begin{cases} \omega_i \times (P_w - P_s), & i = 1,2,3 \\ \omega_i \times (P_w - P_e), & i = 4 \end{cases} \tag{76}$$

Note that $\mathbf{J}_1 = \omega_i \times (P_w - P_s) = \mathbf{x} \times (P_s - P_w) = 0$. Therefore, we have

$$\dot{\mathbf{P}} = +\mathbf{J}_2\dot{\theta}_2 + \mathbf{J}_3\dot{\theta}_3 + \mathbf{J}_4\dot{\theta}_4 \tag{77}$$

With respect to the time-varying Plane S, the attached frame has its x-axis, y-axis, and z-axis in a fixed direction, i.e., $\mathbf{x} = \omega_1 = [1,0,0]^T$, $\mathbf{y} = \omega_2 = [0,1,0]^T$ and $\mathbf{z} = \omega_3 = [0,0,1]^T$. ω_4 is in parallel with ω_3 and therefore we have $\omega_4 = [0,0,1]^T$. Using the cross product, the direction of $\mathbf{J}_i$ for $i = 2,3,4$ can be determined: $\mathbf{J}_2$ is in the direction $[0,0,-1]^T$; $\mathbf{J}_3$ is in the direction $[0,1,0]^T$; the direction of $\mathbf{J}_4$ is perpendicular to $P_w - P_e$ and ω_4 and therefore in the Plane S, as shown in Figure 5(a). Here, we define the angle between $\mathbf{J}_3$ and $\mathbf{J}_4$ as φ. Since all the ω_is are unit vectors, the magnitude of $\mathbf{J}_i$ can be determined:

$$||\mathbf{J}_i|| = \begin{cases} ||P_w - P_s||, & i = 1,2,3 \\ ||P_w - P_e||, & i = 4 \end{cases} \tag{78}$$

Based on the above definitions, we can explicitly express each vector of the Jacobian matrix with respect to the frame attached to Plane S:

$$\mathbf{J}_1 = [0,0,0]^T \tag{79}$$
$$\mathbf{J}_2 = ||P_w - P_s|| \cdot [0,0,-1]^T \tag{80}$$
$$\mathbf{J}_3 = ||P_w - P_s|| \cdot [0,1,0]^T \tag{81}$$
$$\mathbf{J}_4 = ||P_e - P_s|| \cdot [-\sin\varphi, \cos\varphi, 0]^T \tag{82}$$

Thus, the Jacobian matrix is:

$$\mathbf{J} = \begin{pmatrix} 0 & 0 & 0 & -L_{we}\sin\varphi \\ 0 & 0 & L_{ws} & L_{we}\cos\varphi \\ 0 & -L_{ws} & 0 & 0 \end{pmatrix} \tag{83}$$

where $L_{ws} = ||P_w - P_s||$ and $L_{we} = ||P_e - P_s||$.

By singular value decomposition, we can find $\mathbf{J} = \mathbf{U}\Sigma\mathbf{V}^T$, where $\mathbf{U} = [u_1, u_2, u_3]^T$, $\Sigma = diag(\sigma_1, \sigma_2, \sigma_3)$ and $\mathbf{V} = [v_1, v_2, v_3]^T$. The vectors u_i determine the direction of principle axes of the manipulability ellipsoid, and σ_i determine the radii, as shown in Figure 5(b). By resolving $\det(\mathbf{J}\mathbf{J}^T - \lambda\mathbf{I}) = 0$, we can obtain $\sigma_i = \sqrt{\lambda_i}$. Based on Sarrus's rule, the eigen-values λ_i can be found as:

$$\lambda_{1,2} = \frac{(L_{ws}^2 + L_{we}^2) \pm (L_{ws}^2 + L_{we}^2)c_1}{2}, (\lambda_1 > \lambda_2) \tag{84}$$

$$\lambda_3 = L_{ws}^2 \tag{85}$$

with

$$c_1 = \sqrt{1 - c_2} \tag{86}$$

$$c_2 = \frac{4L_{ws}^2 L_{we}^2 \sin\varphi^2}{(L_{ws}^2 + L_{we}^2)^2} \tag{87}$$

Note that $0 < c_1 < 1$ and $0 < c_2 < 1$ and therefore $\lambda_{1,2}$ are not complex numbers. Knowing that $\lambda_1 > \lambda_2$, the following proof will show $\lambda_1 > \lambda_3$ and therefore the eigen-vector u_1 corresponding to λ_1 is the longest eigen-vector.

case1: $L_{ws} \geq L_{we}$

$$\begin{aligned} \lambda_1 - \lambda_3 &= \frac{(L_{ws}^2 + L_{we}^2) + (L_{ws}^2 + L_{we}^2)c_1}{2} - L_{ws}^2 \\ &\geq \frac{(L_{we}^2 - L_{ws}^2) + (L_{ws}^2 + L_{we}^2)c_{min1}}{2} \\ &= \frac{(L_{we}^2 - L_{ws}^2) + (L_{ws}^2 + L_{we}^2)\sqrt{1 - c_{max2}}}{2} \\ &= \frac{(L_{we}^2 - L_{ws}^2) + \sqrt{(L_{ws}^2 + L_{we}^2)^2 - 4L_{ws}^2 L_{we}^2}}{2} \\ &= \frac{(L_{we}^2 - L_{ws}^2) + \sqrt{(L_{ws}^2 - L_{we}^2)^2}}{2} = 0 \end{aligned} \tag{88}$$

where c_{min1} is the minimum of c_1; c_{max2} is the maximum of c_2 and

$$c_{max2} = \frac{\max(4L_{we}^2 L_{ws}^2 \sin(\varphi)^2)}{(L_{ws}^2 + L_{we}^2)^2} = \frac{4L_{we}^2 L_{ws}^2}{(L_{ws}^2 + L_{we}^2)^2} \tag{89}$$

case2: $L_{ws} < L_{we}$

$$\begin{aligned}
\lambda_1 - \lambda_3 &= \frac{(L_{ws}^2 + L_{we}^2) + (L_{ws}^2 + L_{we}^2)c_1}{2} - L_{ws}^2 \\
&\geq \frac{(1 + c_{min1})(L_{ws}^2 + L_{we}^2)}{2} - L_{ws}^2 \\
&\geq \frac{(L_{ws}^2 + L_{we}^2)}{2} - L_{ws}^2 \\
&= \frac{(L_{we}^2 - L_{ws}^2)}{2} \geq 0
\end{aligned} \tag{90}$$

$\lambda_1 > \lambda_3$ for all the possible L_{ws}, corresponding to wrist positions for the reaching movements of the human arm. Thus, we can conclude that the longest eigen-vector of the manipulability ellipsoid is u_1, corresponding to the largest eigen-value of

$$\sigma_1 = \sqrt{\lambda_1} = \sqrt{((L_{ws}^2 + L_{we}^2) + (L_{ws}^2 + L_{we}^2)c_1)/2} \tag{91}$$

The direction of the longest eigen-vector can be found by:

$$(\mathbf{J} \cdot \mathbf{J}^T)\mathbf{X} = \lambda \mathbf{X} \tag{92}$$

where $\mathbf{X} = [x_1, x_2, x_3]^T$.

$$(\mathbf{J} \cdot \mathbf{J}^T)\mathbf{X} = \begin{pmatrix} L_{we}^2 \sin\varphi^2 & -L_{we}^2 \cos\varphi \sin\varphi & 0 \\ -L_{we}^2 \cos\varphi \sin\varphi & L_{ws}^2 + L_{we}^2 \cos\varphi^2 & 0 \\ 0 & 0 & L_{ws}^2 \end{pmatrix} \begin{pmatrix} x_1 \\ x_2 \\ x_3 \end{pmatrix} = \lambda_1 \begin{pmatrix} x_1 \\ x_2 \\ x_3 \end{pmatrix} \tag{93}$$

Therefore

$$\begin{aligned}
(L_{we}^2 \sin\varphi^2 - \lambda_1)x_1 &= L_{we}^2 \cos\varphi \sin\varphi x_2 \\
L_{we}^2 \cos\varphi \sin\varphi x_1 &= (L_{ws}^2 + L_{we}^2 \cos\varphi^2 - \lambda_1)x_2 \\
L_{ws}^2 x_3 &= \lambda_1 x_3
\end{aligned} \tag{94}$$

For the solution, we have

$$\begin{aligned}
x_1 &= x_1 \\
x_2 &= -\frac{\lambda_1 - L_{we}^2 \sin\varphi^2}{L_{we}^2 \cos\varphi \sin\varphi} x_1 \\
x_3 &= 0
\end{aligned} \tag{95}$$

Due to the joint limit, $0° < \varphi < 90°$. When $\varphi = 0°$, the arm is fully extended and therefore at its singular position; $\varphi = 90°$ cannot be achieved since upper and the lower arms can not fully overlap each other.

Note that

$$
\begin{aligned}
\lambda_1 - L_{we}^2 \sin\varphi^2 &= \frac{(L_{ws}^2 + L_{we}^2) + (L_{ws}^2 + L_{we}^2)c_1 - 2L_{we}^2 \sin\varphi^2}{2} \\
&= \frac{(L_{ws}^2 + L_{we}^2) + (L_{ws}^2 + L_{we}^2)\sqrt{1 - \frac{4L_{ws}^2 L_{we}^2 \sin\varphi^2}{(L_{ws}^2 + L_{we}^2)^2}} - 2L_{we}^2 \sin\varphi^2}{2} \\
&\geq \frac{(L_{ws}^2 + L_{we}^2) + (L_{ws}^2 + L_{we}^2)\sqrt{1 - \frac{4L_{ws}^2 L_{we}^2}{(L_{ws}^2 + L_{we}^2)^2}} - 2L_{we}^2 \sin\varphi^2}{2} \\
&= \frac{(L_{ws}^2 + L_{we}^2) + ||L_{ws}^2 - L_{we}^2|| - 2L_{we}^2 \sin\varphi^2}{2} \\
&= \frac{(L_{ws}^2 - L_{we}^2) + ||L_{ws}^2 - L_{we}^2|| + 2L_{we}^2 \cos\varphi^2}{2} \\
&\geq \frac{2L_{we}^2 \cos\varphi^2}{2} = L_{we}^2 \cos\varphi^2 \geq 0
\end{aligned} \tag{96}
$$

With $0 < \cos\varphi \leq 1$ and $0 \leq \sin\varphi < 1$,

$$
-\frac{\lambda_1 - L_{we}^2 \sin\varphi^2}{L_{we}^2 \cos\varphi \sin\varphi} x_1 \leq -\frac{L_{we}^2 \cos\varphi^2}{L_{we}^2 \cos\varphi \sin\varphi} = -\frac{1}{\tan\varphi} < 0 \tag{97}
$$

The direction of u_1 is as shown in Figure 5(c).

Optimization of Swivel Angle Using Movement Efficiency: The control of the extra DOF in the human arm, specified by the swivel angle ϕ, can be optimized for best movement efficiency. The proposed biologically-based hypothesis considers the human head (particularly the position of human mouth) as the virtual target for human arm movements. Thus, an optimum swivel angle can be determined such that the projection of the major principle axis u_1 on the vector $P_m - P_w$ is maximized for a given wrist position, since in the direction of the major principle axis, the Jacobian matrix provides the most efficient mapping from joint space velocity to task space velocity.

$$
\begin{aligned}
\phi &= \underset{\alpha,\beta \in [0\ \pi/2]}{\arg\max}\ [u_1^T (P_m - P_w)] \\
&= \underset{\alpha,\beta \in [0\ \pi/2]}{\arg\max}\ [||u_1|| ||P_m - P_w|| \cos(\alpha)\cos(\beta)]
\end{aligned} \tag{98}
$$

By Equation (98), the brute force method is used to explore all the possible swivel angles for a wrist position, to determine the optimum swivel angle. Figure 5(c) demonstrates the geometry of finding the projection of u_1 on $P_m - P_w$. In Figure 5(c), α denotes the angles between $(P_m - P_w)$ and plane S; β denotes the angle between u_1 and the projection of $(P_m - P_w)$ onto S. The component of u_1 projected onto $(P_m - P_w)$ is represented by $||u_1|| \cos(\alpha)\cos(\beta)$, marked by the green arrow. It is expected that Equation (98) is maximized when $\alpha = 0°$, regardless of the

β determined by the given wrist position; when $\alpha = 0°$, plane S is coplanar with the plane composed by P_m, P_s and P_w, as shown in Figure 5(d). The optimum swivel angle following the proposed hypothesis can be determined given the positions of P_m, P_w and P_s. Define $\mathbf{f} = P_w - P_m$ and $\mathbf{f}'$ to be the projection of $\mathbf{f}$ on the direction of $P_w - P_c'$. Since $\mathbf{f}'$ is parallel to vector $P_e(\phi) - P_c$ when $\alpha = 0°$, the optimum swivel angle is estimated as:

$$\phi_{kin} = \arctan 2(\mathbf{n} \cdot (\mathbf{f}' \times \mathbf{u}), \mathbf{f}' \cdot \mathbf{u}) \tag{99}$$

This algorithm provides a real-time estimate of the swivel angle and therefore a real-time solution to the inverse kinematics of the human arm. The performance of the ϕ_{est} estimation has been evaluated in [24] and compared to a dynamic model in [23]. This chapter intends to examine this algorithm (referred to as **Criterion I** in the following section) in a more general experimental setup, in comparison with another swivel angle estimation algorithm which addresses the effect of the dynamics of the human arm.

3.2 Criterion II: Swivel Angle Estimation by a Dynamic Constraint

According to [24], the biologically-based swivel angle estimation algorithm using purely kinematic constraints can provide a good estimation. However, the effect of dynamics on human arm movements cannot be underestimated. [23] integrates a dynamic criterion in order to provide an improved estimation of swivel angles and to reveal the effect of the dynamics on human arm movements. The dynamic criterion, proposed in [21] and referred to as **Criterion II** in the following sections, resolves the inverse kinematics by minimizing the magnitude of total work done by joint torques for each time step. It has generated satisfactory predictions of the joint space trajectory for the fundamental motions of the human arm, such as shoulder adduction/abduction, shoulder flexion/extension, shoulder internal/external and elbow flexion/extension. Note that there exist other dynamic criteria which can also be used to improve the estimation performance.

Optimization of Swivel Angles by Minimizing Work in Joint Space: For reaching movements in a 3D workspace, the wrist position of the human arm can be uniquely defined by three variables in the task space, while in the joint space there are four joint angles (three for the shoulder motion and one for the elbow motion) available for configuration. Accordingly, the relationship between movements and muscle forces in a musculoskeletal model is based on the four dynamic equations [21]:

$$T = M\ddot{Q} + C\left(Q, \dot{Q}\right) + G(Q) \tag{100}$$

In Equation (100), $\ddot{Q} = [\ddot{q}_1, \ddot{q}_2, \ddot{q}_3, \ddot{q}_4]$ and $\dot{Q} = [\dot{q}_1, \dot{q}_2, \dot{q}_3, \dot{q}_4]$, where q_i represents the joint angle for the i-th DOF. M, $C\left(Q, \dot{Q}\right)$ and $G(Q)$ represent the matrix of the moment of inertia, the centrifugal/coriolis forces and the gravity force respectively. The external force is represented by E and this is regarded as zero in this paper since the given task does not involve interacting with an external load. The active and passive joint torque rendered by musculotendinous forces are represented by T. The calculation of work in the joint space for each time step depends on (1) the joint torques and (2) the difference in joint angles. Therefore, the work in the joint space during the movement interval $[t_k, t_{k+1}]$ can be computed for two different conditions. The dynamic constraint adopted in this chapter is from the original work done by [30]. Here, we briefly include the essential parts of the algorithm for the integrity of the chapter.

if $T_{i,t_k} \cdot T_{i,t_{k+1}} > 0$,

$$W_i = \frac{(T_{i,t_k} + T_{i,t_{k+1}}) \cdot \Delta q_i}{2} \tag{101}$$

where T_{i,t_k} and $T_{i,t_{k+1}}$ are the joint torques of the i-th joint at the time t_k and t_{k+1}. $\Delta q_i = (q_{i,t_{k+1}} - q_{i,t_k})$ is the difference of the i-th joint angle during the time interval $[t_k, t_{k+1}]$.

When $T_{i,t_k} \cdot T_{i,t_{k+1}} < 0$,

$$W_i = \frac{(|\Delta q_i| - h_i) \cdot T_{i,t_{k+1}}}{2} - \frac{h_i \cdot T_{i,k}}{2} \tag{102}$$

where $h_i = (|T_{i,t_k}| \cdot |\Delta q_i|)/|T_{i,t_{k+1}} - T_{i,t_k}|$ and denotes the difference of the i-th joint angle from q_{i,t_k} to the value corresponding to zero crossing of joint torque.

To minimize the work done in joint space at each time step (E.g. $|W|_{t_k,t_{k+1}}$ for the time interval $[t_k, t_{k+1}]$), the swivel angle of the human arm for a specified wrist position is optimized by the following cost function:

$$C = |W|_{t_k,t_{k+1}} = \sum_{i=1}^{4} |W_i|_{t_k,t_{k+1}} \tag{103}$$

where $|W_i|_{t_k,t_{k+1}}$ denotes the work done by the i-th joint.

4 Experiments

4.1 Experiment Setup

3D Spherical Workspace: A 3D spherical workspace is set up in order to examine the swivel angle estimation performance of the above two criteria in more general conditions. This spherical workspace, with its center denoted by the green point in Figure 6, is calculated as a part of the surface of a virtual sphere. The calculated target locations, denoted by blue marks, fall within a red circle, whose size is limited by the width of the back frame of the experiment table. By varying the radius of the virtual sphere, the spherical workspace can be resized, resulting in a new distribution

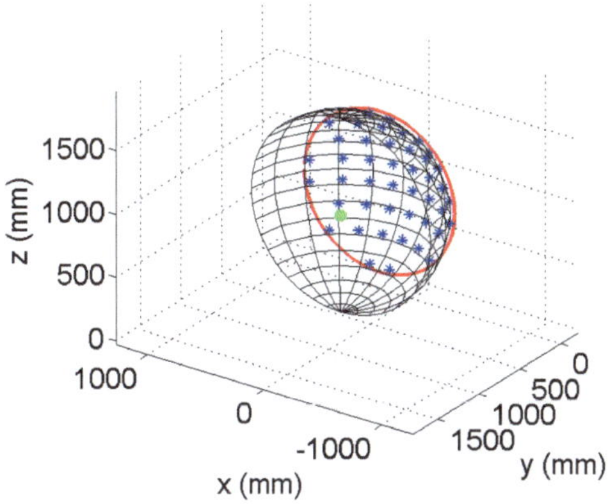

(a) Simulation of a 3D spherical workspace.

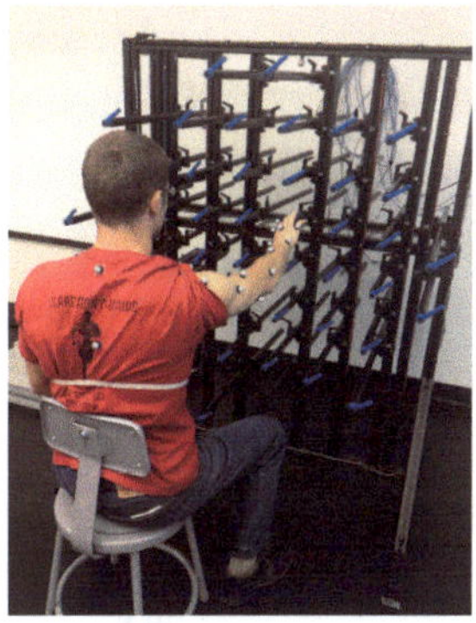

(b) Experimental setup.

Fig. 6 3D spherical workspace: (a) targets (shown as blues points) are arranged on the surface of a virtual sphere, the center of which is shown as the green point; (b) the spherical workspace is installed on the back frame of a table customized for reaching and grasping experiments

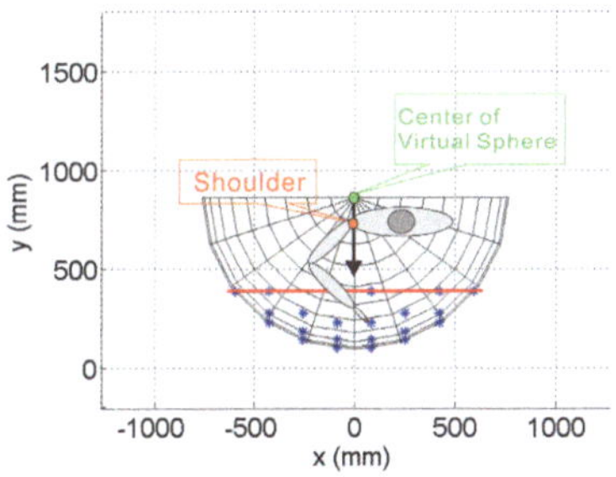

(a) Top view

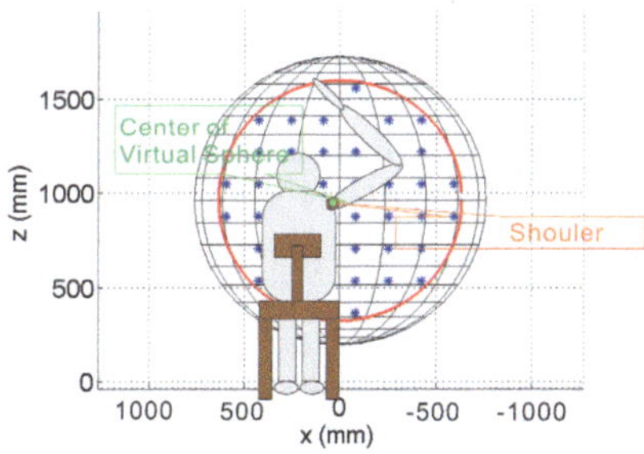

(b) Front view

Fig. 7 Experiment setup for reaching movements in a 3D spherical workspace

of target locations and a different target density. The spherical workspace is installed on the back frame of an experiment table customized for reaching and grasping experiments.

To maximize the use of the back frame of the experiment table, this experiment allocates the radius of the virtual sphere to be slightly larger than the width of the back frame. The origin of the virtual sphere (as well as the shoulder of the subject) is equidistant between the left and right boundaries of the frame (Figure 7(a)). With an adjustable chair, the right shoulder of a subject can be aligned at the height of the center of the virtual sphere. As shown in Figure 7(b), the center of the virtual sphere and the shoulder of the subject overlap in the front view.

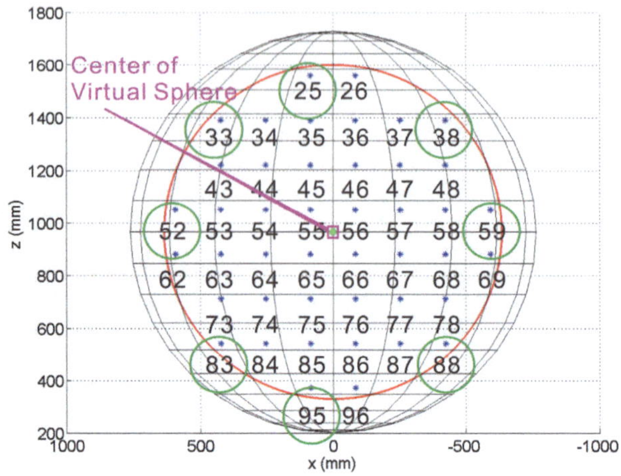

Fig. 8 Targets involved in reaching experiments, marked by green circles

The spherical workspace is evenly discritized and 44 targets are allocated with the same interval in both the vertical direction (i.e. along the z-axis) and the horizontal direction (i.e. along the x-axis). Each target is numbered by its row and column.

Subjects: This experiment involves five healthy volunteer right-handed subjects (three males and two females; age range 22-38 years old; average age 28 years old) without any clinical symptoms or any history of motor, sensory or neurological disorders. The subjects are naive as to the purpose the experiment, and are only instructed to perform point to point reaching movements in a naturally self-paced way.

Experiment Protocol: In this experiment, subjects are expected to conduct eight sessions of reaching movements with their right arms. The targets involved in this experiment are 25, 33, 38, 52, 59, 83, 88, and 96, as shown in Figure 8. Each of the eight sessions chooses an involved target to be the destination (i.e. the end target) of the reaching movements for the whole session. In each session, a subject iteratively starts from one of all the other involved targets and reaches to the end target of that session, according to the trial sequence denoted in Equation (104). Each session consists of five repetitions of seven different reaching movements (35 trials in total).

$$5\ repeats \times \begin{cases} Session\ 1: 33 \to 25,\ 38 \to 25,\ 52 \to 25,\ 59 \to 25,\ 83 \to 25,\ 88 \to 25,\ 96 \to 25 \\ Session\ 2: 25 \to 33,\ 38 \to 33,\ 52 \to 33,\ 59 \to 33,\ 83 \to 33,\ 88 \to 33,\ 96 \to 33 \\ Session\ 3: 25 \to 38,\ 33 \to 38,\ 52 \to 38,\ 59 \to 38,\ 83 \to 38,\ 88 \to 38,\ 96 \to 38 \\ Session\ 4: 25 \to 52,\ 33 \to 52,\ 38 \to 52,\ 59 \to 52,\ 83 \to 52,\ 88 \to 52,\ 96 \to 52 \\ Session\ 5: 25 \to 59,\ 33 \to 59,\ 38 \to 59,\ 52 \to 59,\ 83 \to 59,\ 88 \to 59,\ 96 \to 59 \\ Session\ 6: 25 \to 83,\ 33 \to 83,\ 38 \to 83,\ 52 \to 83,\ 59 \to 83,\ 88 \to 83,\ 96 \to 83 \\ Session\ 7: 25 \to 88,\ 33 \to 88,\ 38 \to 88,\ 52 \to 88,\ 59 \to 88,\ 83 \to 88,\ 96 \to 88 \\ Session\ 8: 25 \to 96,\ 33 \to 96,\ 38 \to 96,\ 52 \to 96,\ 59 \to 96,\ 83 \to 96,\ 88 \to 96 \end{cases} \tag{104}$$

During the experiment, a subject sits in a chair with a straight back support. The placement of the chair enables the subject to point at the targets with comfort and with his/her elbow flexed to roughly 90° (as shown in Figure 6(b)). The height of the chair is adjustable such that the right shoulder of the subject is aligned with the height of the center of the spherical workspace. The right arm is free for pointing movements, but the body of the subject is bounded to the chair back, in order to minimize shoulder displacement. During the pointing movements, the subject is asked to keep the pointing finger in line with the forearm to minimize wrist flexion.

The subjects are instructed to point with the tip of the index finger at their own paces. At the beginning of each trial, a subject is informed of the targets that the trajectory starts with and ends at, i.e., the start target and end target. After receiving a "start" command, the subject points his/her index finger from the start target to the end target.

A motion capture system records an individual file for each trial, starting from the time when the subject puts his/her index finger on the start target and ending after the index finger becomes steady at the end target. To avoid the effect of fatigue, the subject can take a rest after completing each session and can take a rest during a session if he/she feels like it.

5 Results and Discussion

This section presents the results of the swivel angle estimation based on different criteria, and compares them with the measurements of swivel angles from the reaching experiment. The performance of different swivel angle estimation algorithms is evaluated and compared using their estimation errors at targets and during movements.

Figure 9 provides an example of the swivel angle profiles of a subject reaching between two targets: moving forward from Target 1 to Target 7, and moving backward from Target 7 to Target 1. Note that when the measured swivel angle (the blue lines in Figure 9), which denotes the arm postures of the subject in the experiment, is approximately symmetric for moving forward and backward between the two targets. This symmetry of the arm postures is better addressed in the swivel estimation by Criterion I (the green lines), than that by Criterion II.

Section 5.1 examines the swivel angle profiles of all the trials for each subject. The experimentally measured swivel angle demonstrates posture consistency for each individual subject, i.e. the same subject tends to use the same arm posture (measured by the swivel angle) to reach for a target, regardless of whether the subject is moving toward or moving away from the target. In addition, reaching movements are symmetric when comparing the profiles of measured swivel angles for the reaching movements between the same two targets and in opposite directions. The posture consistency and trajectory symmetry are related. Particularly, the posture consistency is an important characteristic of human arm movements, such that the swivel angle estimation of a good control strategy for human arm movements should be able to address it.

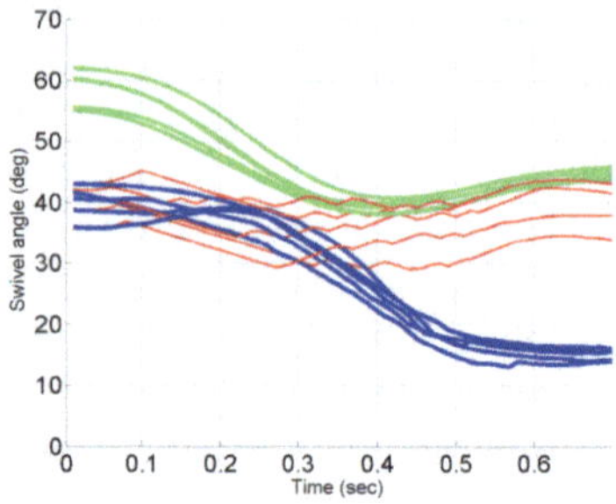

(a) From Target 1 to Target 7

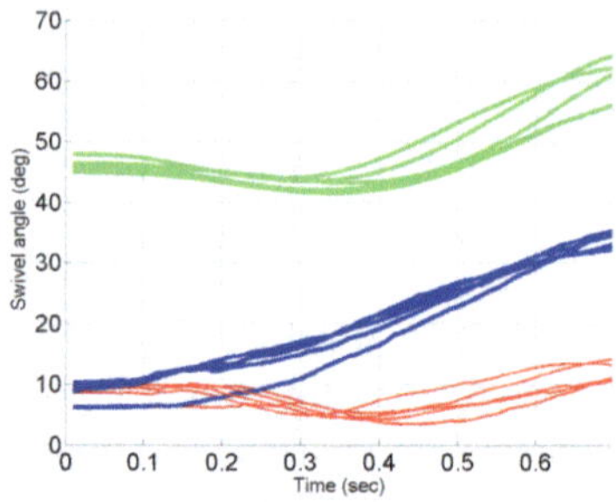

(b) From Target 7 to Target 1

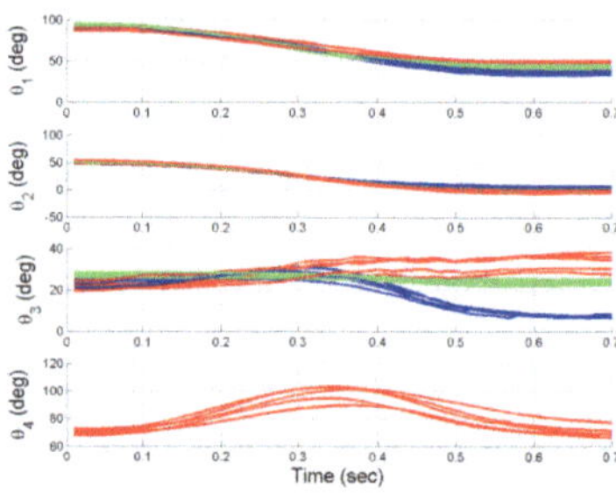

(c) From Target 1 to Target 7

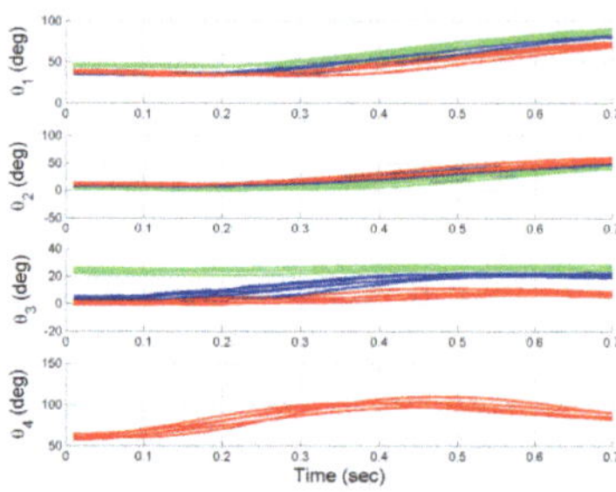

(d) From Target 7 to Target 1

Fig. 9 Figure 9(a) and Figure 9(b) show the experimentally measured swivel angle profiles and the estimated swivel angle profiles by different criteria; Figure 9(c) and Figure 9(d) are the profiles of joint angles correspondingly. In all the figures, the blue lines are the measured swivel angle profiles, the green lines are the profiles of the swivel angle estimated based on criterion I (the efficiency of arm movement), and the red lines are the profiles of swivel angle estimated based criterion II (minimizing work in the joint space).

5.1 Posture Consistency in Human Arm Movements

Posture Consistency: The Regularity in Human Arm Movements: Posture consistency is an interesting characteristic of human arm movements. Without any specific instructions or manipulation intentions, a subject moving his/her arm in free space tends to use the same arm posture to reach the same position, regardless of whether the subject is moving toward or away from the target.

Posture consistency is important because it implies that for reaching movements in a free space, a unique redundancy resolution corresponding to a unique arm posture is associated with each wrist position. Among all the hypotheses of control strategies for human arm movements, a control strategy that addresses posture consistency can be systematically adjusted to achieve improved estimation accuracy, while control strategies that do not address posture consistency may not represent the characteristics of human arm movements.

To study the arm postures at each target, we define the swivel angle that a subject takes when he/she moves away from the target as the "start posture", and the swivel angle that a subject takes when he/she moves toward the target as the "end posture". Figure 10 presents the statistics of target postures for each subject, showing that at each target the start and the end postures are very close to each other.

Figure 11 shows the statistics of averaged posture differences measured during the experiment for each subject. The averaged posture difference $||\phi_{start} - \phi_{end}||$ is computed for each involved target, as the difference between the averaged start

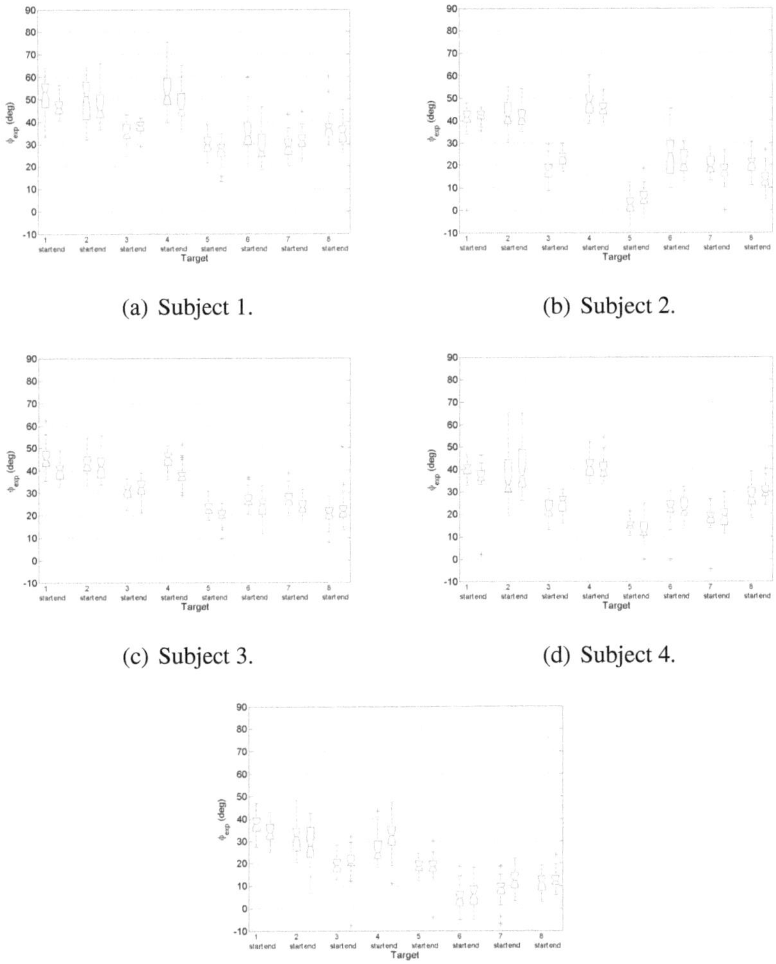

(a) Subject 1.

(b) Subject 2.

(c) Subject 3.

(d) Subject 4.

(e) Subject 5.

Fig. 10 Swivel angle difference between the start and the end postures

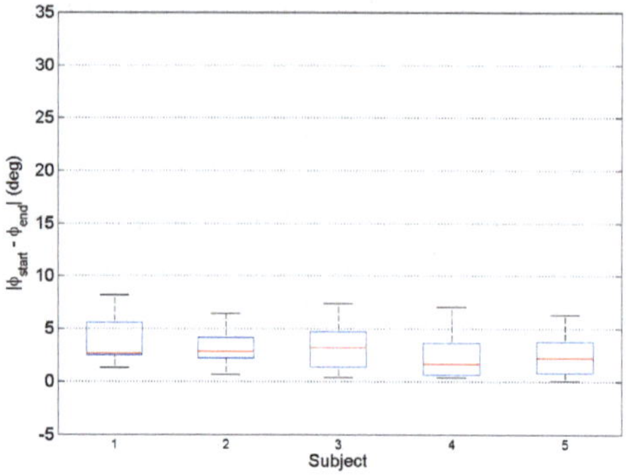

Fig. 11 Posture consistency measured in the experiments. Each subject had a median average posture difference of less than 5° across all targets, and the maximum average posture difference was less than 10° for any target.

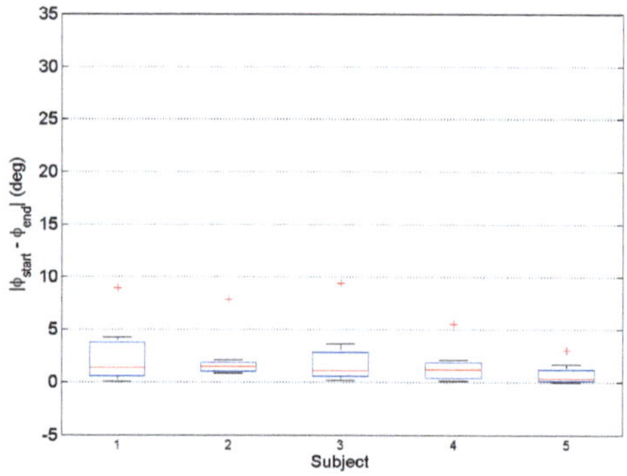

Fig. 12 Posture consistency by Criterion I. Each subject had a median average posture difference of less than 3° across all targets, and the maximum average posture difference was less than 5° for any target.

posture at a target and the averaged end posture at the same target. As shown in Figure 11, all subjects have a median of the averaged posture difference (across all targets) of less than 5°, and the maximum of the averaged posture difference for any target was less than 10°.

The average posture difference can also be calculated for the estimated target postures according to the different estimation criteria. The average predicted posture differences for each subject are generally smaller in Figure 12 than in Figure 13. It can be concluded that posture consistency is addressed better by Criterion I than by Criterion II. Under Criterion I all subjects have a median of the averaged posture difference less than 3°, and the maximum of the averaged posture difference across all subjects is less than 5°, while using Criterion II all subjects have a median of the

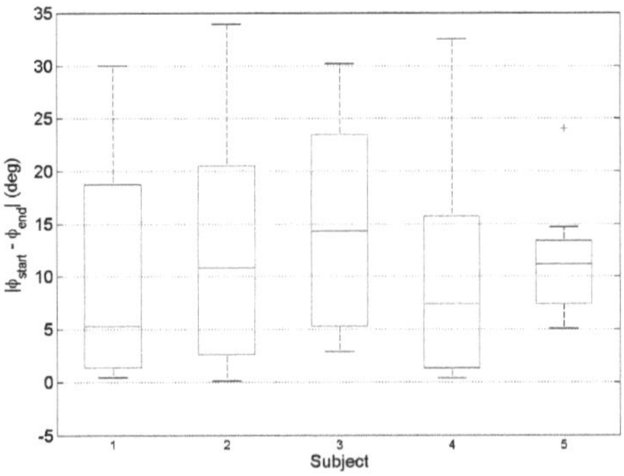

Fig. 13 Posture consistency by Criterion II. Each subject had a median average posture difference of less than 15° across all targets, and the maximum average posture difference was less than 35° for any target.

averaged posture difference less than 15°, and the maximum of the averaged posture difference is less than 35°. Criterion I has a better performance than Criterion II on posture consistency.

Analysis of Systematic Errors in Swivel Angle Estimation: Figure 14 demonstrates the systematic error in swivel angle estimation by different criteria, and for start and the end posture, respectively. ϕ_{exp} is the average swivel angle measured in the experiment, while ϕ_{est} and ϕ_{dyna} are the average swivel angle estimated by Criterion I and Criterion II, respectively. Note that the perfect posture consistency demonstrated in Figure 14(b) is because Criterion II intentionally uses the measured swivel angle as the initial value to start its estimation.

Both Figure 14(a) and Figure 14(c) show that the systematic error of swivel angle estimation based on Criterion I can be related to the horizontal position of targets with respect to shoulder location. In general, Criterion I tends to overestimate the swivel angle and therefore expects higher elbow position than the experimental measurements; however, given targets of the same height, the overestimation is more significant for targets to the left of the shoulder than for the targets to the right of the shoulder. Note that in the workspace, target 3, 5 and 7 (called "right targets") are to the right of shoulder, target 2, 4, 6 (called "left targets") are to the left of shoulder, and target 1 and 8 (called "middle targets") are aligned with the shoulder. The systematic overestimation related to the horizontal position of the targets can be found by comparing target postures at the left target and at right target of the same height in pairs (e.g., target 2 VS target 3, target 4 VS target 5; and target 6 VS target 7). Note that the arm postures at target 1 are more overestimated than at target 8.

The systematic estimation error of Criterion I may be explained by the following fact: when reaching to the targets to the right of shoulder, the right arm moves in free space and therefore the effect of gravity is more significant, while reaching to the targets to the left of the shoulder, the movements of right arm will be blocked

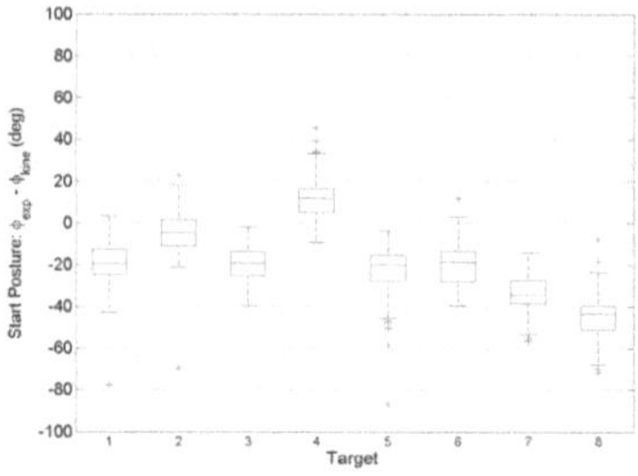

(a) Criterion I: start posture.

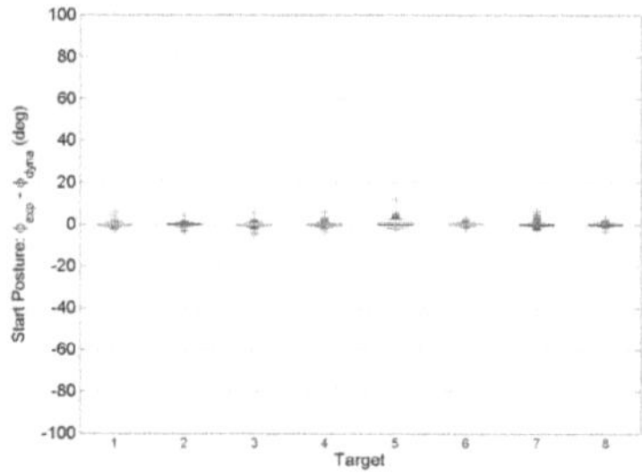

(b) Criterion II: start posture.

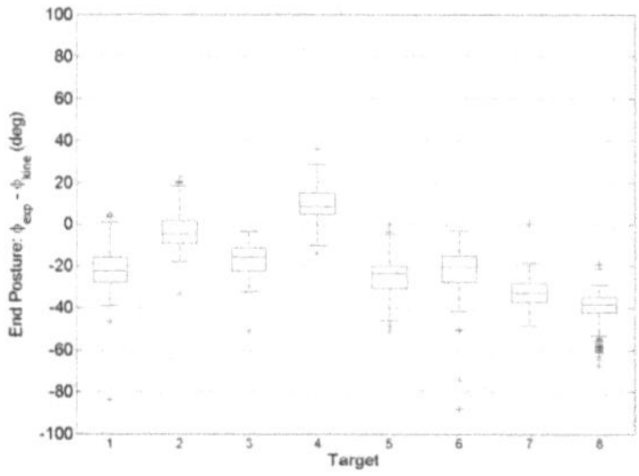

(c) Criterion I: end posture.

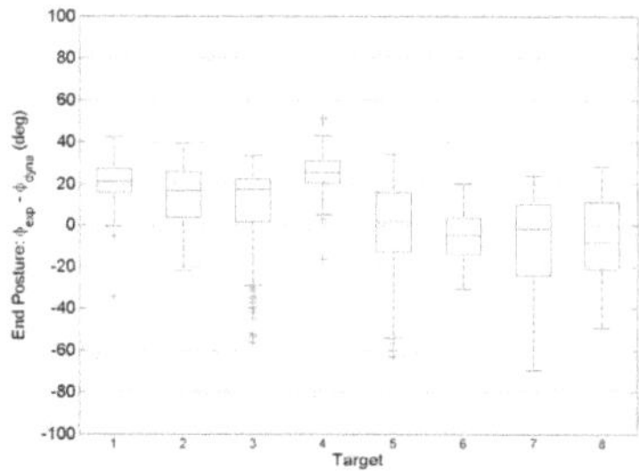

(d) Criterion II: end posture.

Fig. 14 Systematic error in swivel angle estimation for different criteria

by the torso and the lap. The effect of gravity is partially countered by the force produced by joint limits and workspace constraints.

In general, Criterion II tends to underestimate the swivel angle and expect lower elbow positions compared to those measured in the experiments. Figure 14(d) shows that the underestimation of swivel angles is related to the heights of targets with respect to the height of the shoulder, targets 1, 2, and 3 (called "higher targets") are above the shoulder, targets 6, 7, and 8 (called "lower targets") are below the shoulder, and targets 4 and 5 (called "middle targets") are aligned at the same height as the shoulder. In Figure 14(d), it is shown that the swivel angles at the higher targets and middle target 4 are more underestimated, while the lower targets and middle target 5 may be slightly underestimated, with the estimated swivel angles close to measured swivel angles.

5.2 Estimation Error during Reaching Movements

As noted in the experimental protocol, each subject conducted five repeats of each trajectory (i.e. reaching movements starting from the same target and ending at the same target). The estimation performance of different criteria are further evaluated by the standard deviation of estimation errors during the movements.

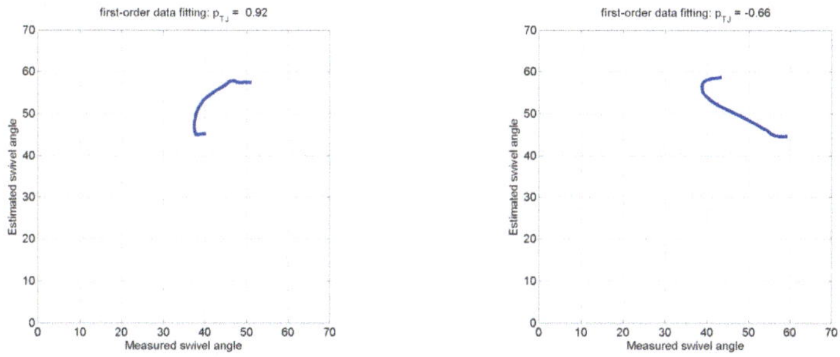

(a) Good swivel angle estimation. (b) Bad swivel angle estimation.

Fig. 15 Example of the different performance of swivel angle estimation. p_{TJ} is calculated by first-order regression to represent how well the estimated swivel angle fits the measured swivel angle along a trajectory. In (a) $P_{TJ} = 0.92$, indicating the estimated swivel angle fits the measured swivel angle well; in (b) $P_{TJ} = -0.66$, indicating the estimated swivel angle does not fit the measured swivel angle.

Figure 15 shows examples of evaluating the performance of a criterion on the swivel angle estimation along a trajectory. For different trajectories created by the same subject, the same criterion may produce swivel angle estimations that are better for some trajectories than for others. The evaluation of estimation performance is conducted by linear regression of the estimated swivel angle versus the measured swivel angle along a trajectory. The slope of linear regression (p_{TJ}) shows how much the estimated swivel angle fits the measured swivel angle. In the best case, the slope of linear regression is supposed to be 1, indicating that the estimated swivel angle fits the measured swivel angle all along the trajectory.

As shown in Figure 16, all the trajectories of reaching movements are further categorized by the slope p_{TJ} of the first order regression of the measured swivel angles with respect to the estimated swivel angles along a trajectory. In Figure 16, the targets are numbered in sequence and trajectories are denoted by colored vectors: trajectories with $p_{TJ} < -0.6$ are in red; trajectories with $p_{TJ} > 0.6$ are in green; and trajectories with $0.6 < p_{TJ} < 0.6$ are in yellow. The estimation performance at each target, measured by $\overline{MEAN}_{Target}$ and calculated by averaging the absolute values of the estimation errors at a target, are also categorized and marked by different colors.

The targets with $\overline{MEAN}_{Target} < 10°$ are green, the targets with $\overline{MEAN}_{Target} > 20°$ are red, and the targets with $10° < \overline{MEAN}_{Target} < 20°$ are yellow. By comparing the estimation performance of the two criteria, it can be found that Criterion II outperforms Criterion I, indicating that the control strategy for human arm movements has is linked to energy, which is mostly related to the effect of gravity. Therefore, a pure kinematic model such as Criterion I can not estimate the swivel angle very well by itself. On the other hand, the estimation performance of Criterion I shows the blocking effect of the human body on human arm movements, particularly for the trajectories with poor swivel angle estimation (the cluster on the left side of the workspace, mostly related to Target 4). (Note that when reaching to Target 4, the

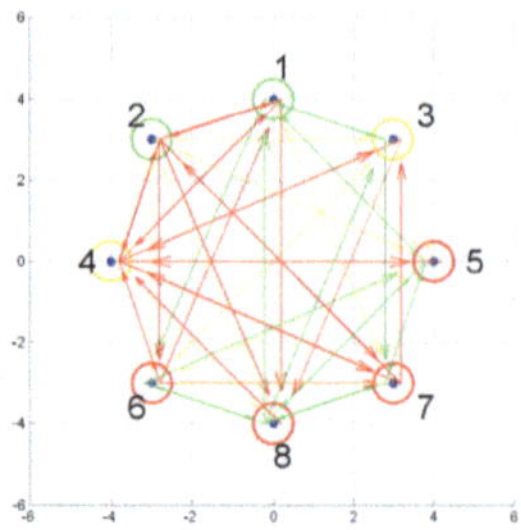

(a) Subject 1.

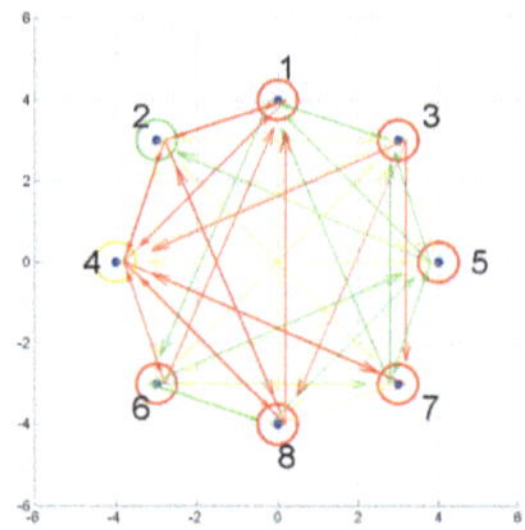

(b) Subject 1.

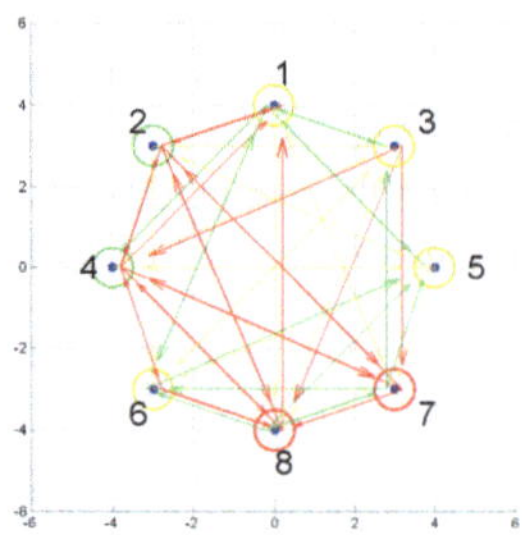

(c) Subject 3.

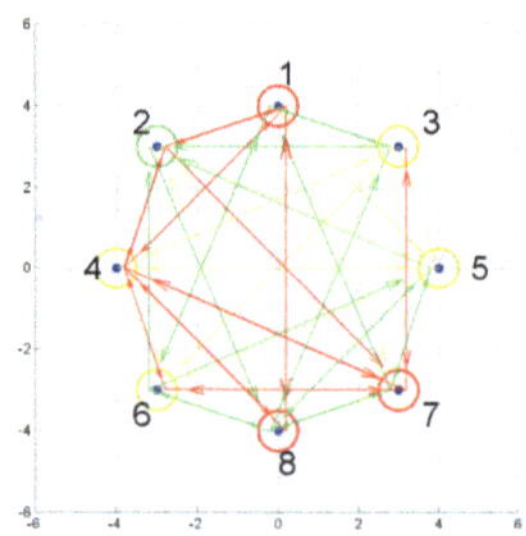

(d) Subject 4.

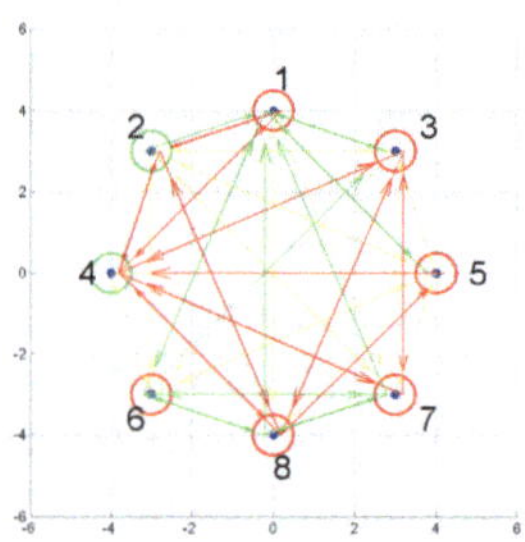

(e) Subject 5.

Fig. 16 Criterion I: categorization of reaching movement trajectories by estimation performance. The performance of estimation is evaluated by the slope p_{TJ} of the first order regression between measured swivel angles and estimated swivel angles along a trajectory. Colored vectors connecting targets denote the corresponding trajectories of movements: trajectories with $p_{TJ} < -0.6$ are in red; trajectories with $p_{TJ} > 0.6$ are in green; and trajectories with $0.6 < p_{TJ} < 0.6$ are in yellow. The estimation performance at each target, measured by $\overline{MEAN}_{Target}$ and calculated by averaging the absolute values of estimation errors at a target, are also categorized and marked by different colors: the targets with $\overline{MEAN}_{Target} < 10°$ are green, the targets with $\overline{MEAN}_{Target} > 20°$ are red, and the targets with $10° < \overline{MEAN}_{Target} < 20°$ are yellow.

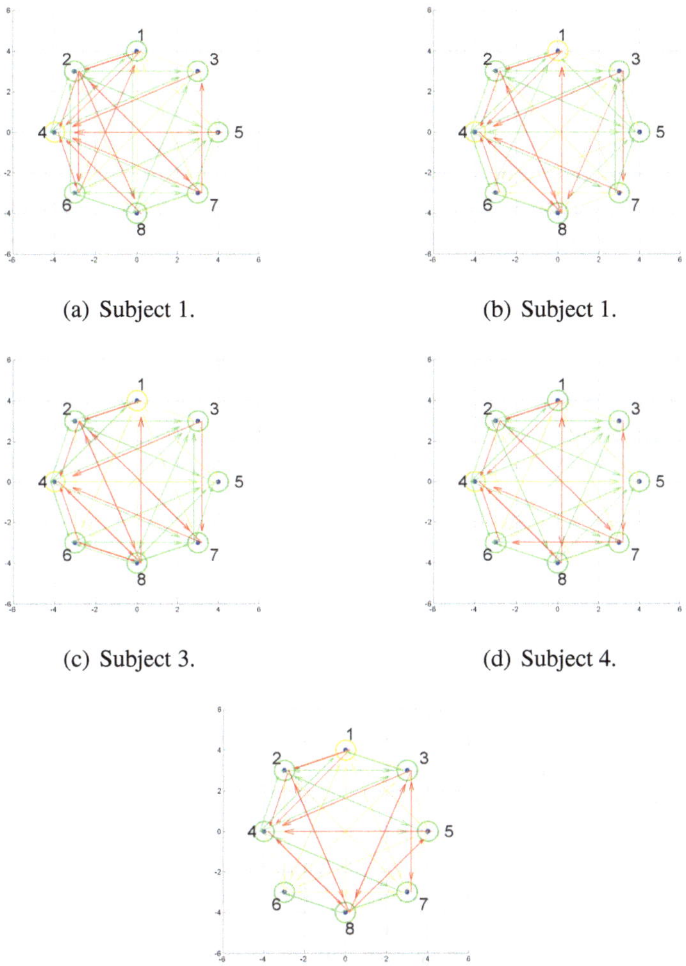

(a) Subject 1.

(b) Subject 1.

(c) Subject 3.

(d) Subject 4.

(e) Subject 5.

Fig. 17 (Continue Figure 16) **Criterion II**: categorization of reaching movement trajectories by estimation performance

right human arm is blocked by torso to the largest extent.) The estimation performance of Criterion I also demonstrates the effect of gravity, since the poor estimation of target posture happens for the lower targets on the right of the workspace (i.e. Target 5 and Target 7). There, the human arm moves in free space, and the measured swivel angles are much lower than the estimated swivel angles.

Comparing Figure 16 and Figure 17, Criterion II demonstrates a better performance for the estimation of arm postures at targets and along trajectories. This experimental result may seem to contradict the results of [23], in which Criterion I provided a better swivel angle estimation than Criterion II. However, this apparent contradiction can be explained by the difference in the experimental protocol. In [23], the targets are on the surface of the table and on frames, and the subjects are asked to touch the targets slightly with their hands. During the experiments, subjects touched the targets with the finger pads of the three middle fingers. The orientation of the hand affected the elbow position and resulted in a larger swivel angle. In addition to the hand orientation, the subjects avoided collision with the table surface and therefore chose higher elbow positions unconsciously. In addition, note that under Criterion II, the swivel angle estimation algorithm takes the measured swivel angle at the start target as an initial condition, which contributes to a higher performance of swivel angle estimation for the whole trajectory. At the same time, Criterion I estimates the initial swivel angle and does not require measuring the initial swivel angle.

6 Conclusion

Posture consistency is an important characteristic of human arm movements. Without intent to manipulate, a subject tends to use the same arm posture to reach the same target in a 3D free space, regardless of whether the subject is moving toward or moving away from the target. This regularity in human arm movements, previously revealed by Donders' law, is confirmed by the experimental data presented in this chapter. The subjects involved in the reaching experiments have their posture difference median at the targets smaller than 5°, and posture difference maximum across all the targets less than 10°. This posture consistency implies that given the kinematic redundancy in the human arm, human motor control chooses a unique arm posture associated with a 3D hand position.

Previously proposed control strategies for controlling the redundant degree of freedom provide different predictions of the arm postures corresponding to the same wrist position in a 3D space. This chapter examined the arm posture predictions of two control strategies: one that maximizes the movement efficiency towards the head, particularly toward the mouth (Criterion I); and the other that minimizes the energy consumption in joint space at each time step (Criterion II). The predictions of arm postures by the two control strategies are compared and the prediction errors for each control strategy are analyzed: posture consistency is a better addressed by Criterion I, while Criterion II has smaller estimation error along the arm movement trajectories and therefore demonstrates better performance when predicting the "movement trend". Further work will evaluate arm posture predictions based on other criteria or based on their combinations to achicvc improved arm posture predictions.

References

1. Angeles, J., Lopez-Cajun, C.S.: Kinematic isotropy and the conditioning index of serial robotic manipulators. Int. J. Robotics Research 11(6), 560–571 (1992)
2. Angeles, J., Park, F.C.: Performance evaluation and design criteria. In: Siciliano, B., Khatib, O. (eds.) A Handbook of Robotics, 1st edn. Springer Handbook of Robotics, pp. 229–244 (2008)
3. Asada, H., Granito, J.: Kinematic and static characterization of wrist joints and their optimal design. In: Proc. 1985 Int. Conf. Robot. Automat., St. Louis, Missouri, pp. 244–250 (March 1985)
4. Badler, N.I., Tolani, D.: Real-time inverse kinematics of the human arm. Presence 5(4), 393–401 (1996)
5. Bernstein, N.A.: The coordination and regulation of movements. Pergamon Press, Oxford (1967)
6. Bicchi, A.: On the mobility and manipulability of general multiple limb robots. IEEE Journal of Robotics and Automation 11(2), 215–228 (1995)
7. Boullion, T.L., Odell, P.L.: Generalized inverse matrices. Wiley-Interscience (1971)
8. Craig, J.J.: Introduction to Robotics: Mechanics and Control, 3rd edn., ch.1. Prentice Hall (2003)
9. Desmurget, M., Grea, H., Prablanc, C.: Final posture of the upper limb depends on the initial position of the hand during prehension movements. Current Biology 119(4), 511–516 (1998)
10. Dum, R., Strick, P.: Motor Areas in the frontal Lobe:The Anatomical Substrate for the Central Control of Movement. In: Motor Cortex in Voluntary Movements. CRC Press (2005)
11. Flash, T., Hogan, N.: The coordination of arm movements: an experimentally confirmed mathematical model. Journal of Neurophysiology 5, 1688–1703 (1985)
12. Motion Genesis. Autolev
13. Gordon, J., Ghilardi, M.F., Ghez, C.: Accuracy of planar reaching movements. Experimental Brain Research 99(1), 97–111 (1994)
14. Gosselin, C.M., Angeles, J.: A new performance index for the kinematic optimization of robotic manipulator. In: Proc. ASME Mec. Conf., Kissimmee, FL, pp. 441–447 (1988)
15. Harris, C.M., Wolpert, D.M.: Signal-dependent noise determines motor planning. Nature 194, 780–784 (1998)
16. Harwood, M.R., Mezey, L.E., Harris, C.M.: The spectral main sequence of human saccades. Journal of Neurophysiology, 19
17. Haustein, W.: Considerations on listing's law and the primary position by means of a matrix description of eye position control. Biological Cybernetics 60(6), 411–420 (1989)
18. Hermens, F., Gielen, S.: Posture-based or trajectory-based movement planning: a comparison of direct and indirect pointing movements. Experimental Brain Research 159(3), 340–348 (2004)
19. Hogan, N.: An organizing principle for a class of voluntary movements. Journal of Neuroscience 4(2), 2745–2754 (1984)
20. Kane, T.R., Levinson, D.A.: Dynamics: Theory and Applications. McGraw-Hill, New York (1985)
21. Kang, T., He, J., Helms Tillery, S.I.: Determining natural arm configuration along a reaching trajectory. Exp. Brain Res. 167, 352–361 (2005)

22. Kawato, M.: Bidirectional theory approach to integration. In: Inui, T., McClelland, J.L. (eds.) Attention and Performance XVI: Information Integration in Perception and Communication, pp. 335–367. The MIT Press, Cambridge (1996)
23. Kim, H., Li, Z., Milutinovic, D., Rosen, J.: Resolving the redundancy of a seven dof wearable robotic system based on kinematic and dynamic constraint. In: ICRA, St. Paul, Minnesota, USA (2012)
24. Kim, H., Miller, L., Rosen, J.: Redundancy resolution of a human arm for controlling a seven dof wearable robotic system. In: IEEE International Conference on EMBC, Boston, USA, August 30-September 3 (2011)
25. Krebs, H.I., Ferraro, M., Buerger, S. P., Newbery, M.J., Makiyama, A., Sandmann, M., Lynch, D., Volpe, B. T., Hogan, N.: Rehabilitation robotics: pilot trial of a spatial extension for mit-manus. J. Neuro Engineering & Rehabilitation (October 2004)
26. Li, Z., Latash, M.L., Zatsiorsky, V.M.: Force sharing among fingers as a model of the redundancy problem. Experimental Brain Research, 119
27. Li, Z., Glozman, D., Milutinovic, D., Rosen, J.: Maximizing dexterous workspace and optimal port placement of a multi-arm surgical robot. In: Proc. IEEE Int. Conf. on Robotics and Automation, Shanghai, China, pp. 3394–3399 (May 2011)
28. Maciejewski, A.A.: Dealing wit the ill-conditioned equations of motion for articulated figures. In: IEEE Computer Graphics and Applications, pp. 63–71 (1990)
29. Medendorp, W.P., Crawford, J.D., Henriques, D.Y.P., Van Gisbergen, J.A.M., Gielen, C.C.A.M.: Kinematic strategies for upper armčforearm coordination in three dimensions. J. Neurophysiology 84(5), 2302–2316 (2000)
30. Aerospace medical research laboratory. Investigation of inertial properties of human body. Technical report (March 1975)
31. Messier, J., Kalaska, J.F.: Comparison of variability of initial kinematics and endpoints of reaching movements. Experimental Brain Research 125(2), 139–152 (1999)
32. Miller, L.M., Rosen, J.: Comparison of multi-sensor admittance control in joint space and task space for a seven degree of freedom upper limb exoskeleton. In: Proceedings of the 3rd IEEE RAS & EMBS International Conference on Biomedical Robotics and Biomechatronics, Tokyo, Japan, pp. 26–29 (September 2010)
33. Graziano, M.S., Tylor, C.S., Moore, T.: Complex movements evoked by micro stimulation of precentral cortex. Neuron 30, 34(5), 841–851 (2002)
34. Muller, H., Sternad, D.: Motor learning: changes in the structure of variability in a redundant task. Advances in Experimental Medicine and Biology, Progress in Motor Control 629, 439–456 (2009)
35. Nakano, E., Imamizu, H., Osu, R., Uno, Y., Gomi, H., Yoshioka, T., Kawato, M.: Quantitative examinations of internal representations for arm trajectory planning: Minimum commanded torque change model. The Journal of Neurophysiology 81(5), 2140–2155 (1999)
36. Nelson, W.L.: Physical principles for economies of skilled movements. Biological Cybernetics, 46
37. Pandy, M.G., Garner, B.A., Anderson, F.C.: Optimal control of non-ballistic muscular movements: a constraint-based performance criterion for rising from a chair. Journal of Neurophysiology 117
38. Perry, J.C., Rosen, J.: Design of a 7 degree-of-freedom upper-limb powered exoskeleton. In: IEEE/RAS-EMBS International Conference on Biomedical Robotics and Biomechatronics, Pisa, Italy (February 2006)

39. Perry, J.C., Rosen, J., Burns, S.: Upper-limb powered exoskeleton design. Mechatronics 12(4), 408–417 (2007)
40. Perry, J.C., Rosen, J., Burns, S.: Upper-limb powered exoskeleton design. Mechatronics 12(4), 408–417 (2007)
41. Roberts, S.L., Falkenberg, S.K.: Biomechanics: Problem Solving for Functional Activity. St. Louis, Mosby (1992)
42. Rosenbaum, D.A.: Human Motor Control. Academic Press (1990)
43. Rosenbaum, D.A., Meulenbroek, R.J., Vaughan, J., Jansen, C.: Posture-based motion planning: applications to grasping. Psychological Review 108(4), 709–734 (2001)
44. Sabes, P.N., Jordan, M.I., Wolpert, D.M.: The role of inertial sensitivity in motor planning. Journal of Neuroscience, 18
45. Saha, S.K.: Introduction to Robotics. Tata McGraw-Hill Publishing Company Limited (2008)
46. Schaal, S., Sternad, D.: Origins and violations of the 2/3 power law in rhythmic 3d movements. Experimental Brain Research 136, 60–72 (2001)
47. Schmidt, R.A., Zelaznik, H., Hawkins, B., Frank, J.S., Quinn, J.T.: Motor-output variability: a theory for the accuracy of rapid motor acts. Psychology Review, 47
48. Scholz, J.P., Schoner, G.: The uncontrolled manifold concept: identifying control variables for a functional task. Experimental Brain Research 126(3), 289–306 (1999)
49. Scholz, J.P., Schoner, G., Latash, M.L.: Identifying the control structure of multijoint coordination during pistol shooting. Experimental Brain Research 135(3), 382–404 (2000)
50. Sciavicco, L.: A dynamic solution to the inverse kinematic problem for redundant manipulators. In: ICRA, vol. 4, pp. 1081–1087 (March 1987)
51. Sciavicco, L.: A solution algorithm to the inverse kinematic problem for redundant manipulators. IEEE Journal of Robotics and Automation 4(4), 403–410 (1988)
52. Smeets, J.B., Brenner, E.: A new view on grasping. Motor Control 3, 237–271 (1999)
53. Soechting, J.F., Buneo, C.A., Herrmann, U., Flanders, M.: Moving effortlessly in three dimensions: does donders' law apply to arm movement? J. Neuroscience 15(9), 6271–6280 (1995)
54. Soechting, J.F., Flanders, M.: Parallel, interdependent channels for location and orientation in sensorimotor transformations for reaching and grasping. Journal of Neurophysiology 70(3), 1137–1150 (1993)
55. Sutton, G.G., Sykes, K.: The variation of hand tremor with force in healthy subjects. Journal of Physiology, 191
56. Terzuoloa, C.A., Viviania, P.: Determinants and characteristics of motor patterns used for typing. Journal of Neuroscience 5, 431–437 (1982)
57. Tillery, S.I., Ebner, T.J., Soechting, J.F.: Task dependence of primate arm postures. Experimental. Brain Researchy 104(1), 1–11 (1995)
58. Todorov, E.: Cosine tuning minimizes motor errors. Neural Computation 14
59. Todorov, E., Jordan, M.I.: Optimal feedback control as a theory of motor coordination. Nature Neuroscience 5, 1226–1235 (2002)
60. Todorov, E., Jordan, M.I.: Smoothness maximization along a predefined path accurately predicts the speed profiles of complex arm movements. Journal of Neuroscience 880, 696–714 (1998)
61. Tolani, D., Goswami, A., Badler, N.I.: Real-time inverse kinematics techniques for anthropomorphic limbs. Graphical Models and Image Processing 62(5), 353–388 (2000)

62. Uno, Y., Kawato, M., Suzuki, R.: Formation and control of optimal trajectory in human multijoint arm movement - minimum torque-change model. Biology Cybernetics 61, 89–101 (1989)
63. Vaillancourt, D.E., Newell, K.M.: movement variability and theories of motor control. In: Latash, M., Zatsiorsky, V. (eds.) Classics in Movement Sceience, pp. 409–435. Human Kinetics (2001)
64. Viviani, P., Terzuolo, C.: Trajectory determines movement dynamics. Journal of Neuroscience 7, 431–437 (1982)
65. Wada, Y., Kaneko, Y., Nakano, E., Osu, R., Kawato, M.: Quantitative examinations for multi joint arm trajectory planningusing a robust calculation algorithm of the minimum commanded torque change trajectory. Journal of Neural Networks 14, 381–393 (2001)
66. Woodworth, R.S.: The Accuracy of Voluntary Movement. The Macmillan Company (1899)
67. Yoshikawa, T.: Dynamic manipulability of robot manipulators. In: Proc. Int. Conf. Robot. Automat., St. Louis, Missouri, pp. 1033–1038 (March 1985)
68. Yoshikawa, T.: Foundations of Robotics: Analysis and Control. The MIT Press (1990)

Author Index

MIX
Papier aus verantwortungsvollen Quellen
Paper from responsible sources
FSC® C105338

If you have any concerns about our products,
you can contact us on
ProductSafety@springernature.com

In case Publisher is established outside the EU,
the EU authorized representative is:
Springer Nature Customer Service Center GmbH
Europaplatz 3, 69115 Heidelberg, Germany

Printed by Libri Plureos GmbH
in Hamburg, Germany